AVIATION MENTAL HEALTH

Aviation Mental Health
Psychological Implications for Air Transportation

Edited by

ROBERT BOR
Royal Free Hospital, London, UK

and

TODD HUBBARD
Oklahoma State University, USA

ASHGATE

© Robert Bor and Todd Hubbard 2006

All rights reserved. No part of this publication may be reproduced, stored in a retrieval system or transmitted in any form or by any means, electronic, mechanical, photocopying, recording or otherwise without the prior permission of the publisher.

Robert Bor and Todd Hubbard have asserted their right under the Copyright, Designs and Patents Act, 1988, to be identified as the editors of this work.

Published by
Ashgate Publishing Limited
Wey Court East
Union Road
Farnham
Surrey GU9 7PT
England

Ashgate Publishing Company
Suite 420
101 Cherry Street
Burlington, VT 05401-4405
USA

Ashgate website: http://www.ashgate.com

British Library Cataloguing in Publication Data
Aviation mental health : psychological implications for air
   transportation
   1.Aviation psychology 2.Flight crews - Psychology
   I.Bor, Robert II.Hubbard, Todd P.
   155.9'65

Library of Congress Cataloging-in-Publication Data
Aviation mental health : psychological implications for air transportation / edited by Robert Bor and Todd Hubbard.
      p.cm.
   Includes index.
   ISBN 978-0-7546-4371-5
   1. Aviation psychology. I. Bor, Robert. II. Hubbard, Todd.

   RC1085.A92 2006
   155.9'65--dc22
                                                                    2006005838

ISBN 978 0 7546 4371 5

Reprinted 2009, 2010

Printed and bound in Great Britain by
MPG Books Ltd, Bodmin, Cornwall.

Contents

List of Figures	*vii*
List of Tables	*viii*
List of Contributors	*ix*
Foreword	*xix*

1   Aviation Mental Health: An Introduction   1
       Robert Bor and Todd Hubbard

PART 1   **Psychological Issues of Flight and Cabin Crew**

2   Psychological Stress and Air Travel: An Overview of Psychological
       Stress Affecting Airline Passengers   13
       Vivien Swanson and Iain B. McIntosh

3   Psychological Factors Relating to Physical Health Issues: How Physical
       Factors in Aviation and Travel Affect Psychological Functioning   27
       Paul Richards, Jennifer Cleland, and Jane Zuckerman

4   Psychological Problems Among Passengers and On-Board Psychiatric
       Emergencies   39
       Graham Lucas and Tony Goodwin

5   The Nature, Characteristics, Impact, and Personal Implications
       of Fear of Flying   53
       Elaine Iljon Foreman, Robert Bor, and Lucas van Gerwen

6   Flight or Fright? Psychological Approaches to the Treatment
       of Fear of Flying   69
       Elaine Iljon Foreman, Robert Bor, and Lucas van Gerwen

7   Posttraumatic Stress Reactions Following Aircraft Disasters   83
       Man Cheung Chung

PART 2   **Psychological Processes Among Passengers and Crew**

8   Psychiatric Disorders and Syndromes Among Pilots   107
       Jennifer S. Morse and Robert Bor

9	The Psychiatric Evaluation of Air Crew *Gordon J. Turnbull*	127
10	Psychological Assessment and Reporting of Crew Mental Health *Robert Bor*	145
11	Psychological Factors in Cockpit Crew Selection *Robert A. Roe and Pieter H. Hermans*	161
12	Psychological Aspects of Selection of Flight Attendants *Ferenc Albert*	195
13	How Cabin Crew Cope with Work Stress *Carina Eriksen*	209
14	Psychological Problems Among Cabin Crew *Chris Partridge and Tracy Goodman*	227

PART 3   Related Themes in Aviation

15	Psychological Aspects of Astronaut Selection *David M. Musson*	243
16	Occupational Factors in Pilot Mental Health: Sleep Loss, Jet Lag, and Shift Work *Jim Waterhouse, Ben Edwards, Greg Atkinson, Thomas Reilly, Mick Spencer, and Adrian Elsey*	255
17	Legal Aspects of Aviation Health: The Changing Landscape *D Anthony Frances*	285
18	Aviation Psychology in a Developing Country: South Africa *Johann Coetzee*	303
19	How We Explain Misfortune: Psychological Implications for Mental Health in Aviation Professionals *Todd Hubbard*	315

Epilogue	335
Index	*337*

List of Figures

Figure 2.1	Transactional Framework for Psychological Effects of Air Travel	15
Figure 11.1	Competence Architecture Model	164
Figure 11.2	Basic Design Model	173
Figure 12.1	Effective Competence Model	198
Figure 12.2	Valid Assessment based on Different Data Sources	201
Figure 13.1	The Interactions between Lower Level Processes and their Connection to the Higher Level Central Executive	222
Figure 16.1	Core Temperature v. Time	256
Figure 16.2	Shifts of Body Clock caused by Light	258
Figure 16.3	Causal Links between Routine Deteriorations	262
Figure 16.4	Day-time Rhythms	263
Figure 16.5	Alertness v. Time of Day	265
Figure 16.6	Predicted Alertness v. Time	266
Figure 16.7	Times of Sleep and Naps, 1 Hour Advance	269
Figure 16.8	Times of Sleep and Naps, 6 Hour Delay	270
Figure 16.9	Times of Sleep and Naps, 7 Hour Advance	270

List of Tables

Table 2.1	Reported Anxiety for Aspects of Air Travel	17
Table 2.2	Frequency of Anxiety Reduction Methods for Air Travel Anxiety	22
Table 5.1	Danger of Flying, Reported by Mode	58
Table 11.1	Extract from a Civil Pilot Competence Profile	166
Table 11.2	Meta-analysis from Hunter and Burke (1994) and Martinussen (1996)	170
Table 11.3	Selection System for KLM Royal Dutch Airlines (1994–1999)	183
Table 11.4	Some Typical Aircrew Applicant Expectations and Recommended Actions	185
Table 11.5	Do's and Don'ts in the Design of Aircrew Selection Systems	188
Table 16.1	Body Clock	279
Table 19.1	Comparison of Industries	326

List of Contributors

Dr Ferenc Albert PhD has a background in both clinical and academic psychology. He has held his position as assistant professor at the University of Stockholm for several years, first at the Department of Education and later the Department of Social Work. He was trained by and has worked under Professor Arne Trankell – one of the European pioneers in the field of Aviation Psychology. He is Visiting Professor at Kasetsart University, Bangkok, where he teaches courses in Organizations and Leadership at an International MBA. He is also an authorized senior psychologist on the Swedish National Board of Health and Welfare and has practiced as a clinical psychotherapist for almost 20 years at the Institute of Psychotherapy, Stockholm including consultancy for air crew members. He is an authorized Aviation Psychologist for the Swedish Board of Civil Aviation and member of the European Association for Aviation Psychology. As an aviation psychologist he has worked as consultant for major Scandinavian and Asian airlines. He has been involved in the selection of pilots, in designing and implementing cabin selection models and has trained various airline staff in selection and assessment techniques. Presently, he runs Albert Consulting Ltd, a consulting firm with assignments both in selection and development of executive leaders worldwide and consultancy in aviation psychology specializing in selection of air crew.

Greg Atkinson co-ordinates the Chronobiology Research Group within the Research Institute of Sport and Exercise Sciences (RISES) at Liverpool John Moores University and has researched human circadian rhythms for 15 years. He is a BASES accredited scientist (research), a Fellow of the Royal Statistical Society and a member of The American College of Sports Medicine, The Ergonomics Society, The Physiological Society and The American Physiological Society. He is also interested in research methods, being especially interested in how measurement error impacts on performance tests and research, as well as how sports scientists should analyze data that are collected over time.

Professor Robert Bor DPhil CPsychol FBPsS FRAeS has worked as both a clinical and academic psychologist, and also holds a pilots' licence. He is presently a consultant clinical psychologist at the Royal Free Hospital, London. He is Visiting Professor at City University, London where he teaches on the MSc in Air Transport Management and Emeritus Professor of Psychology at London Metropolitan University. He also contributes to the MSc in Travel Health and Medicine at the Royal Free and University College Medical School in the Academic Department of Travel Medicine and Vaccines. He is a Chartered Clinical, Counselling and Health Psychologist and a Fellow of the British Psychological Society. He is a UKCP Registered Family Therapist, having completed his specialist training at the Tavistock Clinic, London. He provides a specialist consultation service for air crew and their families for several leading international airlines and is also involved in the selection of airline pilots, www.crewcare.org. He offers a treatment service at the Royal Free Travel Health Clinic in London for passengers who have a fear of flying. He has also served as an expert witness in some aviation legal cases, relating to air rage, post incident traumatic stress reactions and crew behavior. He serves on the editorial board of several leading international academic journals and has authored or co-authored numerous textbooks, the most recent being: Passenger Behaviour (2003); Psychological Perspectives on Fear of Flying (2003); and Doing Therapy Briefly (2003). He holds the Freedom of the City of London, is a Fellow of the Royal Aeronautical Society, a Member of the European Association for Aviation Psychology, British Travel Health Association and is a Liveryman of the Guild of Air Pilots and Air Navigators. He is also a member of the Association for Family Therapy, The Institute of Family Therapy, London, American Association for Marital & Family Therapy, American Psychological Association and American Family Therapy Academy. Robert Bor is a Churchill Fellow.

Dr Man Cheung Chung PhD earned his BA in Psychology and Sociology at the University of Guelph, Canada, and PhD at the University of Sheffield, United Kingdom. He worked as a Research Psychologist at University College London. He then took on a Research Fellowship at the University of Birmingham and subsequently held lectureships at the universities of Wolverhampton and Sheffield. He is now a Principal Lecturer in the School of Psychology at the University of Plymouth. His main research interests focus on posttraumatic stress disorder and the history and philosophy of psychology. He has published numerous articles and chapters and has delivered many conference papers related to the above topics, as well as topics such as challenging behavior, schizophrenia, court diversion scheme and stress and burnout.

Dr Jennifer Cleland is a Senior Clinical Lecturer, School of Medicine, University of Aberdeen. As a practicing Clinical Psychologist and keen high-altitude trekker/climber, her specific areas of interest are cognitive functioning (e.g., decision making) at altitude and the psychological impact of extreme environments. Her academic research is the management of chronic respiratory disease in primary care, with particular emphasis on training healthcare staff in novel techniques for improving communication within the consultation.

Professor Johann Coetzee is a practicing aviation psychologist in South Africa and also a professor in both Industrial Psychology and Organization Behaviour at the universities of North West and the Free State respectively. He is also director of the Institute for Aviation Safety and the Institute for Aviation Psychology in South Africa.

He is also an active helicopter pilot and director/owner of Henley Air in South Africa, a prominent helicopter organization, operating throughout Southern Africa.

Dr Ben Edwards PhD is a lecturer in chronobiology and environmental physiology at John Moores University (Liverpool), having previously qualified with a BSc (Hon), MSc and PhD in Sports Science at this university. Ben teaches on the first, second and third year respectively, of the Sport and Exercise Science degree as well as the school Masters of Research programme. Ben has full or part authorship in over 24 international journals, two reports to the British Olympic Committee and several book chapters. His research interests include circadian rhythms and performance, jet-lag and effects of exercise in extreme environments.

Captain Adrian Elsey is a practicing airline Captain with 35 years of experience and 19,000 flying hours, gained mostly from long-haul flying. He has been a member of the Medical Study Group of the British Airline Pilots Association since 1987. His main area of expertise lies in sleep and fatigue. He has also acted as an Aviation Consultant on sleep and fatigue, participating in crew vigilance research for Airbus Industry.

Carina Eriksen, originally from the North of Norway, is currently completing a Practitioner Doctorate in Counselling Psychology at London Metropolitan University (UK). Her clinical area is mainly adult mental health, utilizing cognitive-behavioral therapy with reference to both humanistic and psychodynamic practice. She also works as long haul cabin crew for a major UK airline. It was this combination of clinical practice and experience in the field that inspired her current research interest in cabin crew mental health. Robert Bor has been supervising Carina's work and was the first to see potential in her contribution.

Elaine Iljon Foreman is a Chartered Clinical Psychologist who specializes in the treatment of fear of flying and other anxiety related problems. Elaine is a regular expert on television and radio and speaks on the subject of anxiety and phobias at conferences around the world. She has researched the treatment of anxiety at Middlesex Hospital Medical School and invitations to present her research in this field have taken her to Europe, the Americas, Australia, the Middle and the Far East. The Freedom to Fly™ programme has been developed by her and is based on Cognitive Behavior Therapy techniques developed from over 20 years of clinical research into anxiety and phobias.

D Anthony Frances LLM. AMRAeS, is a Senior Associate in the Aviation and Shipping Department of Clyde & Co. He specializes in aviation insurance, aviation lease, regulatory and finance matters. He has worked in Australia, the United Kingdom and the Middle East both in the aviation and shipping industries, as well as in private practice. He acts on behalf of a number of airlines, operators, insurers and shipping companies and has recently been involved in the Kuwait Airways

Corporation v Iraqi Airways Company litigation, regarding the loss of aircraft and spares during the Gulf War in 1990.

Tracy Goodman is the clinical supervisor, trainer and manager of the Crewcare Counseling Service at British Airways. She is a BACP accredited counselor and supervisor who has managed her own practice in West London for ten years where her specialist arena is eating disorders. She is currently completing a Msc in Integrative Psychotherapy at the Metanoia Institute. Previously, Tracy was employed as a senior crew member on the long haul division for British Airways.

Dr Tony Goodwin qualified at Guy's Hospital in 1962, then trained at the Royal Air Force Institute of Aviation Medicine at Farnborough before taking up a three-year posting as Station Medical Officer at Changi. Following a brief spell as a professional racing driver he entered general practice, becoming senior partner in a large London practice in 1979, while still acting as relief Medical Officer for British Airways and Caledonian Airways. He left general practice in 1993 to become senior partner at Airport Medical Services at Gatwick, specializing in aircrew licensing on behalf of the UK Civil Aviation Authority, subsequently the Joint Aviation Authorities, and those in the US, Australia, Hong Kong, UAE and Japan. He has recently retired as Medical Adviser to Virgin Atlantic, which included devising and supervising guidelines for carriage of invalid passengers. He continues as Medical Adviser to the British Airline Pilots' Association (BALPA).

Pieter H. Hermans (49) developed selection methods for pilots, cabin crew, traffic-controllers, divers and astronauts. As an experimental psychologist with the Netherlands Airforce and Navy, he was both scientist and practitioner in the field of assessment. In the 1990s, Pieter joined the Saville & Holdsworth Group. He developed videogame simulations to cover crew resource management. With Hemmo Mulder he published a step-by-step guide to interviewing candidates for aircrew positions. Pieter is now a senior consultant with the Zeelenberg Advisory Group. He helps companies to develop simulation based selection methods and he continues to work internationally as a trainer for interviewers, and as a director of large group assessment centres. Pieter lives and works in Amsterdam, Paris and Frankfurt.

Dr Todd Hubbard EdD, Lt. Col. USAF (retired) is an associate professor and principal investigator for aviation-oriented research projects in educational psychology, cognitive psychology, ergonomics and sociology at Oklahoma State University, where he holds the Clarence E. Page Endowed Chair. During his more than 3400 hours of military flight time, he has held ratings in the KC-135, T-37, T-38, U-2R and TR-1A aircraft. A near death experience, in which he was forced to eject from a U-2 aircraft, moments after takeoff, profoundly changed him. The resultant journey through episodes of posttraumatic stress, fear of flying, bundled with bipolar disorder, stirred within him an interest in the psychology of aviation and the treatment of military pilots who silently suffer. In 2001, he created the Federal Aviation Administration Academy's peer-reviewed journal, International Journal of Applied Aviation Studies and became its first editor-in-chief. He is listed among the Jaycee's Outstanding Young Men in America and is a member of the Association for Aviation Psychology, the University Aviation Association, Order of Daedalians, Phi Kappa Phi honors fraternity and the Aircraft Owners and Pilots Association.

E. Graham Lucas MB, FRCP, FRCPsych, FFOM, RCP, D(Obst), RCOG is Visiting Professor Postgraduate Medical School, University of Surrey; Consultant in Occupational Mental Health, Foreign & Commonwealth Office; Priory Healthcare Services; Health Supervisor to the General Medical Council; Mental Health Act Commission Second Opinion Appointed Doctor; Visiting Lecturer Department of Aviation Medicine, King's College London; and a Member of the International Association of Physicians for the Overseas Services. He was formerly Secretary to the Interdepartmental (Department of Health and Home Office) Advisory Committee on Drug Dependence; Consultant Psychiatrist at King's College and Maudsley Hospitals, and to the Civil Aviation Authority; Adviser in Mental Health to the Health and Safety Executive (HSE); Chair of HSE Working Parties on Mental Health and Drug Abuse at Work respectively; a member of the Faculty of Occupational Medicine Working Party on Testing for Drugs of Abuse in the Workplace; Chief Consultant Psychiatrist to the Ex-Services Mental Welfare Society/Combat Stress; Medical Member Appeals Service, and Mental Health Review Tribunal; Major, Royal Army Medical Corps. His special interests include aviation psychiatry, work related stress, post traumatic stress disorder, alcohol and substance abuse, anxiety and depression. His publications are on these subjects.

Dr Iain B. McIntosh BA (Hons) MBChB DGMRCP DRCOG is a part-time medical practitioner in St Ninians, Stirling. He is a sessional lecturer for post graduate students in medical education, working on behalf of NHS Education, Scotland. A principal area of interest at post graduate level is travel related medicine and health studies. Iain is a tutor and examiner for the Diploma of Travel Medicine, Royal College of Physicians. He is also a GP assessor, auditing and appraising post graduate projects in Medical Education, working on behalf of NHS, Scotland. He is the editor of the Journal of British Travel Health Association and associate editor with Vertex, Journal of Psychology of Argentina. He also serves on the editorial boards of Journal of Travel Medicine, Geriatric Medicine, and the Travelwise newsletter and is a member of the International Society of Travel Medicine, the Society of Authors, the Society of Medical Writers and the Association of Broadcasting Doctors. He is chairman of the British Travel Health Association and an In-Flight Physician with the Global Ambulance Association. He has authored five books on travel related medicine, four books on clinical management, six chapters in medical books in the United Kingdom, one Argentine book, thirty-five scientific articles, many on travel medicine in peer-reviewed journals, and over 150 travel health articles in the medical press.

Dr Jennifer S. Morse MD is a consultant in Aerospace Psychiatry. Her medical career has included clinical experience in Primary Care medicine, Aerospace Medicine, and more recently Psychiatry. She formerly served as the head of Aviation Psychiatry, Naval Aerospace Medical Institute. She was the recipient of the 2000 Aerospace Medical Association William F. Longacre Award for outstanding accomplishments in Aerospace Psychiatry. A former US Naval Flight Surgeon, she recently retired from the US Navy after 24 years of service. A diplomat of the American Board of Psychiatry and Neurology, she is an Assistant Clinical Professor of Psychiatry, non-salaried, University of California San Diego Medical School.

Dr David M. Musson PhD is a research scientist with the Human Factors Research Project at the University of Texas at Austin. He received his MD from the University of Western Ontario in 1988 and served as a flight surgeon with the Canadian Air Force from 1989 until 1994. In 2003, he received his PhD in psychology from the University of Texas. His current research looks at human behavior in high performance work groups, including astronauts, Antarctic personnel and surgical teams. His areas of interest include the relationship between personality and performance, human error, crew resource training, and cultural influences on group behavior.

Professor Anthony N. Nicholson is Professor of Aviation Medicine, School of Biomedical and Health Sciences, King's College London, and Chairman of Trustees for the Confidential Human Factors Incident Reporting Programme (CHIRP) concerned with the reporting of air and maritime incidents in the United Kingdom. He was the Commandant (Air Commodore) and Director of Research, Royal Air Force Institute of Aviation Medicine from 1993–99. His research interests are concerned with the physiology and pharmacology of the sleep-wakefulness continuum. He has published widely on sleep disturbance with respect to civil and military air operations and on the safe use drugs by aircrew for both therapeutic and operational purposes. He graduated in medicine from the University of Birmingham and holds the degrees of Doctor of Philosophy and Doctor of Science. He was awarded a Doctorate (Honoris Causa) by the Russian Academy of Sciences through the Russian Institute of Aviation and Space Medicine. He is a fellow of the Royal Colleges of Physicians of London and Edinburgh, the Royal Aeronautical Society and the Aerospace Medical Association. He holds the Freedom of the City of London and is a Liveryman of the Society of Apothecaries and of the Guild of Air Pilots and Air Navigators. He is an Officer of the Most Excellent Order of the British Empire and was an Honorary Physician to HM The Queen from 1994–99.

Chris Partridge commenced his counseling career in 1986 on a telephone helpline and became a member of the British Airways Crewcare service in 1994. Since then Chris has completed a Diploma and Degree in Person Centred Counselling at the Metanoia Institute in West London. He subsequently became interested in Jungian psychology and obtained a post-graduate diploma in Transpersonal Psychotherapy from the Whittington School of Psychotherapy in London. Chris is qualified in hypnotherapy and EMDR for trauma aftercare. He also has an interest in Hellinger's family systems therapy. Chris has a private counseling practice in Richmond, West London and is a workshop facilitator for stress awareness and counseling skills training at British Airways. Chris wrote and produced a video on basic counseling skills used for training at various British Airways seminars. Chris enjoys writing, particularly on the spiritual aspects of therapy and the emotional literacy of organizations. Chris combines his counseling career with flying as a senior crew member for British Airways.

Dr Thomas Reilly PhD is Director of the Research Institute for Sport and Exercise Sciences at Liverpool John Moores University. He holds an MSc in Ergonomics, a PhD in Exercise Physiology and a DSc in chronobiology related to exercise. He is a Fellow of the Institute of Biology, a Fellow of the Ergonomics Society and a Fellow of the European College of Sport Science. From 1992 to 2005 he acted as Chair of the British Olympic Association's Exercise Physiology Steering Committee and is currently Chair of the Scientific Committee of the European College of Sport Science.

Dr Paul Richards MBChB MRCGP DFFP MSc is a General Medical Practitioner and MSc Clinical Tutor in Travel Medicine at the Academic Centre for Travel Medicine and Vaccines, WHO Collaborating Centre for Travel Medicine at the Royal Free & University College Medical School, University College London. A Fellow of the Royal Geographical Society, he is a mountaineer and a director of Medical Expeditions, a charity with the remit to promote research and education into high altitude physiology and medicine. He is an organizing faculty member of the British Diploma in Mountain Medicine and has participated and contributed to numerous research expeditions to high altitude. Research interests include the effect of altitude and hypoxia on diabetes and sleep.

Professor Robert A. Roe is Professor of Organizational Theory and Organizational Behavior at the University of Maastricht, the Netherlands. He has been Professor of Work & Organizational Psychology at the Dutch universities of Delft, Tilburg and Nijmegen. He has also been director of the Work and Organization Research Center in Tilburg and director of the Netherlands Aeromedical Institute in Soesterberg. His publications cover books, book chapters and journal articles in areas such as personnel selection and appraisal, performance, motivation, competence, organizational assessment and other issues. He holds a masters degree in Psychology and a PhD in Social Sciences from the University of Amsterdam.

Mick Spencer is a Principal Consultant in the QinetiQ Centre for Human Sciences. His particular interest is the development of models to predict the alertness and performance of shift workers and of those exposed to irregular schedules of work and rest. He has undertaken, on behalf of the Health and Safety Executive, the construction and the validation of a Fatigue Index for the assessment of the risks associated with fatigue in different patterns of shift work, and has developed, for the UK Civil Aviation Authority, a computer program to predict levels of alertness of the civil airline pilot. He is currently advising international airlines and aircraft manufacturers on the likely development of aircrew fatigue in the new range of super jumbos.

Dr Vivien Swanson PhD is a Chartered Health Psychologist currently working as a Senior Lecturer in the Department of Psychology, University of Stirling. She is also Course Director for the MSc in Health Psychology, which means spending much of her time on teaching, training of health psychologists and supervision of research. As a health psychologist, she has become involved in promoting the application of psychological theories and methods to improving health and well-being, and raising awareness of what psychology can offer to improving health. This approach is manifested several related areas of research. The first research area of interest was in occupational stress and health, and this was the topic of her PhD thesis in 1997. She has investigated stress and carried out stress management interventions in several occupational groups. Early research in this area was carried out with medical professionals, and more recent work is looking at ways that students manage to combine employment and studies to maximize their adjustment and satisfaction with university life. Other studies of occupational stress have investigated stressors and satisfactions related to organizational change, stress between work and home, and the role of social support in work contexts. It was this interest which first led to collaborations in travel health, specifically in relation to stress and air travel. She worked for several years with a General Practice Research Group, carrying out several research projects in primary care settings, and is a founder member of the Anxiety and Stress Research Centre at Stirling University. She has over 50 publications in a range of areas. Other main areas of research interests include attitudinal and social influences on health behavior, and decision making about medical treatment. She is currently involved in research on Breastfeeding, Obesity, Diet in Disadvantaged Children, and Diabetes. Originally from Lancashire, Scotland has been her adopted country for over 20 years.

Professor Gordon J. Turnbull BSc MB ChB FRCP FRCPsych FRGS FRSA is a graduate of Edinburgh University in 1973, Professor Gordon Turnbull started his psychiatry at the Neuropsychiatric Centre, Princess Alexandra Royal Air Force Hospital, Wroughton in Wiltshire, in 1980 after previous post-graduate experience in General Medicine with Membership of the Royal College of Physicians (MRCP) in 1979. Fellowship of the Royal Geographical Society (FRGS) in 1977 followed a Joint Services Expedition to Antarctica. Membership of the Royal College of Psychiatrists (MRCPsych) was added in 1982. He was appointed Consultant in 1986 and Head of Psychiatric Division at RAF Wroughton, and ran the main in-patient and out-patient psychiatric facility in the RAF as a Wing Commander from 1988 to 1993 when he retired from the RAF. FRCP and FRCPsych followed in 1995 and Fellowship of the Royal Society of Arts in 1999. His experience of managing psychological trauma began in the aftermath of the Lockerbie Air Disaster in 1988 with the RAF Mountain Rescue teams and then active service in the Gulf War of 1991 as RAF psychiatric adviser in the field led to psychological debriefing of British prisoners-of-war following their release and then released British hostages from the Lebanon later that same year. He developed treatment strategies for traumatic stress reactions within the RAF and pioneered a novel group treatment strategy. He left the RAF in 1993 and developed trauma services for civilians and military veterans at Ticehurst House Hospital in East Sussex. Currently, he has four roles: Consultant Psychiatrist developing psychological trauma services at Capio Nightingale Hospitals in London and Consultant Psychiatrist at the Ridgeway Hospital in Wroughton, Wiltshire; Consultant Psychiatrist to the Civil Aviation Authority (CAA) and Visiting Professor to the MSc in Psychological Trauma at the University of Chester. His main interests are the assessment and treatment of psychological trauma, aviation psychiatry, psychosomatic medicine and the medico-legal aspects of psychotrauma. He lectures on these subjects in the UK and overseas to a wide range of audiences including the emergency services, occupational health physicians and lawyers. He is currently helping to develop an innovatory MSc in Psychological Trauma at the University of Chester. He has published extensively in the fields of interest in mainstream psychiatric, medical and legal journals and contributed chapters to several textbooks. He received the Whittingham Memorial Prize for Aviation Medicine in the RAF in 1990, joined a 'People to People' delegation in 1992 to China and in 1993 received the 'People of the Year' Award from the Royal Association for Disability and Rehabilitation in recognition of his work in the field of psychological trauma.

Dr Lucas van Gerwen PhD is a clinical psychologist, psychotherapist and a professional pilot. He is the managing director of the VALK Foundation, a collaborative venture between the University of Leiden, KLM Royal Dutch Airlines and Amsterdam Airport Schiphol with the aim of helping people to overcome their fear of flying. The Foundation also has the goal to help prevent fear of flying and helps other organizations to develop their own programmes. As an EAAP registered aviation psychologist, he provides stress management training for ab initio student pilots with the KLM flying school. Together with the University of Leiden, he conducts research into the fear of flying, passenger behavior, psychological problems among aircrew and the psychological aftermath of aircraft accidents. He is the organizer and founding father of the International World Conferences on Fear of Flying. Dr van Gerwen is the author of several books on the topic of fear of flying and has published numerous papers in scientific journals on assessment and treatment of the problem.

Professor Jim Waterhouse is Professor of Biological Rhythms at the Research Institute for Sport and Exercise Sciences, Liverpool John Moores University. He gained his doctorate from the University of Oxford (1969), and since then he has worked in the field of biological rhythms, first at Manchester University and then at Liverpool. His main interests have been how circadian rhythms can be used as markers of the body clock, how they are altered following time-zone transitions and during shift work, and how they differ in neonates and in old age. He has been President of the European and International Societies for Chronobiology.

Dr Jane Zuckerman is Director of the Academic Centre for Travel Medicine and Vaccines, a WHO Collaborating Centre for Reference, Research and Training in Travel Medicine. She holds the positions of Sub-Dean and Senior Lecturer and Honorary Consultant at the Royal Free and University College Medical School and the Royal Free Hampstead NHS Trust, London. She is also the Medical Director of the Royal Free Travel Health Centre. Her major fields of interest include the evaluation of new travel related and more general vaccines including those of hepatitis A and B. Dr Zuckerman's interests include research into different aspects of travel health and medicine including issues of staying well when travelling, and occupational risks of exposure to health hazards, particularly blood borne viruses. Dr Zuckerman received the award of UK Hospital Doctor of the Year 2001 and the UK Hospital Doctor Innovation Award for 2001. She was also listed in 2001 by the British Medical Association as one of the 82 "Pioneers in Patient Care: NHS Consultants Leading Change".

Foreword

> Sometimes we are devils to ourselves
> When we will tempt the frailty of our powers,
> Presuming on their changeful potency.
>
> Troilus and Cressida

I was most pleased to be asked by the Editors to write the foreword to this book as many of the authors are long-standing colleagues, and some are involved in my postgraduate courses in Aviation Medicine at King's College, London. These courses cover the medical and psychological requirements for the instruction of medical practitioners who will have the responsibility for the medical surveillance of air personnel. Ensuring the health of aircrew is central to the safety of air operations, and in this endeavour the disciplines of clinical psychology and psychological medicine are crucial. The Editors are to be congratulated in bringing together the main issues in mental health that impinge upon the well being of both flight and cabin crews, as well as passengers. They have provided a focus that will stimulate discussion within the disciplines of clinical psychology and psychological medicine, and the means by which others in the world of aviation can become familiar with the psychological dimension.

The behavioral sciences have been involved in aviation since the early days, and the early days were much concerned with the selection of aircrew. The chapters in the present book on selection, nearly a century later, prove to be particularly revealing. It is evident that much uncertainty exists with the appropriateness of current selection techniques for aircrew. Competency in technical procedures is no longer the overriding skill demanded of flight crews: management, leadership, interpersonal and representational skills are of increasing value. Similarly, in the case of cabin crew the selection process attempts to ensure, not only the necessary interpersonal skills to reassure and to interact sympathetically with the passengers, but, at the same time, the likelihood to act effectively in an emergency.

Indeed, ensuring safety is a recurrent theme throughout this book, and the role of the selection procedure in predicting the qualities required of flight crews is an intriguing one. Uncertainty exists concerning the relative value in selection processes of the determination of competencies that presumably provide objective information and subjective information that can be obtained from the opinions of experienced interviewers. Further, it is emphasized that selection processes have tended to be concerned with technical competencies and that the psychological well being of the candidate has received little attention. Those concerned with selection express doubt whether some procedures are sufficiently rigorous or even relevant to the rapidly changing world of aviation, and whether personality assessments can be a reliable indicant of behavior, present or future.

Nevertheless, consideration of adverse personality traits and the possibility of developing psychopathology are relevant to the selection of aircrew. They may proceed to a managerial role perhaps at a time in their life that is likely to be complicated by adverse personal events, and may even one day be in command of an aircraft coping with an incident. Unfortunately, assessments that would predict appropriate behavior have not yet been developed, and it may only be the emergence of the stress, either managerial or operational, that will reveal the adequacy or inadequacy of the individual. Even so, it is argued that, despite the present unsatisfactory situation, some attempt should be made to assess mental health. Clearly this area of uncertainty demands the concerted attention of the disciplines of clinical psychology and psychological medicine, and must surely be an important area for behavioral research.

There is little doubt that psychological and possibly medical problems may arise in cabin crew, and the authors point out that these may appear against a background of fatigue and sleep disturbance. Nevertheless, it is important to separate problems that are incidental to the lifestyle from those that may arise specifically from the stress of the work itself. As far as cabin crew are concerned the former would appear to be amenable to counseling, but those that may arise from fatigue and sleep disturbance need more consideration. Hard data is needed to establish whether personal relationships suffer more in aviation than in employments with greater regularity of duty hours, and whether the nature of the work itself impairs health. We can no longer rely on impressions gained from interviews and subjective reports. An interdisciplinary approach is needed to establish the facts and this must involve the disciplines of psychology and medicine working more closely together. Nevertheless, one cannot but ponder whether the personal problems experienced by female cabin crew need special attention. The demands placed on cabin crews may be much less acceptable to the partners of female attendants than to the partners of male attendants. Further, it may be much more difficult for females to accept the possibility that their home based personal relationships may be continuously disturbed and that the crews at work will be forever changing.

It must also be appreciated that managerial style and initiatives can influence, for better or worse, the wellbeing of crews. It is evident that counseling services for cabin crew and programmes concerned with drug and alcohol abuse have been success stories and are much appreciated. Unfortunately, the interplay between management and air crew does not always favour easy communication, and it is in this context that confidential reporting systems have an important role to play in air safety. Confidential reporting of air incidents has provided a means whereby crews bring issues to the attention of independent expertise that has access to management and the regulatory authority. It encourages the dissemination of sensitive information related to air safety (that can be described as 'but for the Grace of God go I') to other aircrew. In the general context of management some companies have introduced programmes concerned with absence from work. The reasons behind such programmes are understandable – though the initiative has to be handled carefully. It can easily exert pressure on staff who are finding it difficult to cope if the potential behavioral and medical issues are not taken into consideration. It is appreciated that in all management initiatives the details of individual cases must remain confidential, but the information so gathered would provide useful insights into the ongoing mental health of the crews.

As far as passengers are concerned the prevalence of psychopathology is uncertain, but, beyond what may be considered as 'reasonable concerns', 'Fear of Flying' would appear to be a relatively common problem. However, with 'Fear of Flying' there are issues of definition, and there are also uncertainties whether such a condition is specific, whether it is one manifestation of a phobic disorder or whether it is related to other psychopathology such as depression and anxiety. It is evident that much more needs to be known about pre-existing and co-morbidity in those presenting with the complaint. There would appear to be little consensus on its aetiology or nature, but if the incidence of this problem in passengers is that which is claimed – possibly up to 40 percent, then this is certainly an area that needs further attention.

Cognitive behavioral therapy is used extensively in the treatment of 'Fear of Flying' and is now central to the work of many clinical psychologists. It has changed the approach to the treatment of phobias. It attempts to cope with symptomatology whereas the psychodynamic approach attempted to link the fear to unconscious processes. However, it would appear to come in many guises and whether there is a single effective component has yet to be established, though it would appear that the skills of the therapist may decide largely whether the treatment will be successful. Behavioral therapy is also used in the treatment of the posttraumatic stress disorder where earlier intervention, psychological debriefing, has been advocated – but has gained little enthusiasm. Much remains to be understood concerning cognitive behavioral and associated therapies. The usefulness of the various methodologies need to be studied further and the value of programmes run by airlines need to be critically assessed.

It is evident from the editing of this book that the successful practice of mental health in the world of aviation is dependent on the input of many disciplines beyond those of psychological medicine and clinical psychology. These encompass aviation physiology and clinical pharmacology, and as far as air operations are concerned they include airmanship, human factors and accident investigation. An excellent example of the need for a multi-disciplinary approach is the problem of sleep disturbance – a recurrent issue of the authors. The nature of sleep disturbance in aviation was initially investigated about 40 years ago by aviation physiologists, but it became clear that relating the complexity of work patterns to acceptable sleep – essential to operational safety – needed the input of the mathematicians with their skills in dealing with highly variable data and modeling.

A further example of the need for a multi-disciplinary approach to disturbed sleep is the understanding of its adverse effects on behavior. This has, by and large, been concerned with psychomotor and cognitive impairment. Less easily measured effects have been ignored, even though sleep disturbance may adversely affect coping strategies in lifestyles that involve irregularity of the sleep–wakefulness continuum. Management has the initial responsibility to ensure that the time-lines are broadly acceptable, but individuals who have difficulties, not experienced by other aircrew, in coping with work schedules may need help. Specialist sleep investigations may be needed, the individual may have a psychological or a medical problem, and medication may be needed for a limited period of time. As far as the latter is concerned it is important to stress that the pharmacological profile of any drug and its potentially adverse effects on behavior must be well understood, and that applies to any medication used in aviation. An understanding of the input of many disciplines is essential to those involved in advising aircrew with problems in coping with their work–rest schedules.

The authors within this volume have provided a refreshingly critical approach to the practice of clinical psychology and psychological medicine in aviation. In this way they have identified areas of uncertainty and, importantly, areas of mental health that need more attention. The inclusion of authors outside the disciplines of clinical psychology and psychological medicine indicates that those involved in aviation mental health appreciate the value of inter-disciplinary research and the need for a multidisciplinary approach to the problems of both flight and cabin crews. It is hoped that this book will be read, not only by those involved in the well being of aircrew, but also by the flight and cabin crews they seek to serve. The book will also be of much help to those involved in management where an appreciation of the frailty of man (and woman) is vital. We must not presume on our changeful potency.

Professor Anthony N. Nicholson OBE
Professor of Aviation Medicine, School of Biomedical and Health Sciences, King's College London and Chairman of Trustees, United Kingdom Confidential Human Factors Incident Reporting Programme, Farnborough, Hampshire.

Chapter 1
Aviation Mental Health: An Introduction

Robert Bor and Todd Hubbard

A book combining *aviation* with *mental health* places one in triple jeopardy. Firstly, in such a highly regulated, safe and successful industry (including both commercial air transportation and military air operations), it would appear from the outside, at least, that mental health issues have little or no relevance. Secondly, many of those employed within commercial and military aviation, and especially pilots, have a deep distrust of psychologists and psychiatrists and are dismissive of anything that hints at "psychobabble." These negative views may have been formed as a consequence of bad experiences during selection, training, crew licensing or in the course of trying to manage day-to-day personal problems. Thirdly, mental health is not a static or precise science. Much like aviation itself, mental health as a field or specialty keeps evolving. It has long been accepted that definitions of mental illness are culturally relative and have also shifted over the course of time. While this will come as no surprise to most trained physicians, psychologists and psychiatrists, such apparent vagaries do not sit comfortably with those who are used to precision and unequivocal clarity, such as pilots. This book seeks to present a modern, informed, balanced and useful application of mental health issues in aviation and to challenge outdated and negative impressions held by some about what mental health insights can offer to aviation. It is about the mental health of the millions of professionals worldwide responsible for flight. It is not, however, a book about aviation human factors.

As authors we appreciate that some of the negative reactions to mental health issues as well as mental health professionals lies with those responsible for employee mental health, either because the issues have been ignored or because the presence of problems has been used to disadvantage or even terminate the career of an individual. It would seem that mental health issues only appear to have relevance when "things go wrong" with a flight, pilot, air traffic controller, aircraft maintenance engineer, and so on. This perception is regrettable as most crew members accept and appreciate from their human factors courses and crew resource management training that psychological factors have an important role to play in safe and efficient flight. Mental health is also a key area for assessment when crew undergo routine medical checks for licensing and is therefore as much a concern for authorized medical examiners as it should be for crew. Indeed, one could argue that mental health issues are at the heart of challenges to modern air travel and possibly the primary obstacle to coping with the increasing number and length of commercial flights, as well as coping with the unique demands of space tourism and exploration.

There have been numerous psychological and physical challenges to flight for both crew and passengers from the earliest days of controlled and powered flight, just over a century ago. Man has not evolved naturally to fly, as the psychologist, James Reason reminds us (1974). Even though as a species we have evolved over millions of years, our bodies are largely still designed to hunt and gather in small groups in the open plains. We remain a species that is best designed and equipped to be self-propelling at a few miles an hour under the conditions of terrestrial gravity (Reason, 1974). There are several obstacles and "physical evolution barriers" to our position or motion senses, as well as our capacity for processing information, that is apparent to both the novice air traveller and the seasoned pilot. While there have been remarkable achievements in engineering over the past century that have made air travel both possible and highly accessible within the span of a single lifetime, this has not been without its challenges. When evolutionary barriers to motion are exceeded, numerous penalties are exacted, the most common of which are motion sickness, jet lag, fatigue, as well as increased arousal and stress. For flight crew, there may be additional problems relating to judgment, decision-making, perception and concentration, among others. Air travel brings us into close contact with strangers: it also forces us to depend upon and fully trust the input of groups of unseen professionals, and an understanding of the social psychology of behavior within groups and teams is therefore relevant. Air travel also disrupts human relationships: shift-work, short or prolonged absences from home, as well as stress can all exact a toll and demand resilience and unique coping behaviors.

These insights help us to understand that there are five main sources of mental health problems among aviation employees. They include (a) stresses associated with coping, safety and survival, (b) stress that emanates from workload, how work is organized and the organizational climate (e.g. rostering, frequency of flights, jet lag, pensions and financial challenges), (c) personal problems that stem from disruption to personal relationships, which clinical research suggests should act as a buffer to work stress, (d) ever-present concerns about loss of license as a consequence of the onset of a disqualifying medical condition, and (e) *normal* psychological problems that occur naturally in the everyday life of the population at large. As Jones and his colleagues have pointed out (1997), not only do pilots have to deal with the unique pressures of flying aircraft, but they also have to contend with the normal pressures of daily life as well as job insecurity. At present, there is no published data from longitudinal or cross sectional studies which help us to understand the relative weighting of each of the different causes of mental health problems among aviation personnel and how these may have changed over time. Irrespective of the source of the mental health problems, however, the outcome can be just as serious, if not devastating, for the individual concerned.

The standard source of mental health diagnoses is the American Psychiatric Association's Diagnostic and Statistical Manual IV (APA, 1994). However, those familiar with this taxonomy with appreciate the complexity of defining what is and what is not a psychiatric problem. It was devised with the general population in mind and therefore it does not specifically address those problems that are most likely to affect aviation personnel. It does not specify the standards of mental health required among different occupational groups to work in aviation. It is also of limited value to the authorized medical examiner as it does not list approved treatments for disorders or aftercare required and the likely contra indications for work either as a consequence of the diagnosis or whilst treatment is being provided (Jones & Marsh, 2001).

The mental health of employees in all organizations and industries is directly related to standards of safety and productivity. The airline industry is no exception. In the post 9/11 era, it might be argued that in the light of increased pressure in the workplace, greater security risks and threats, the need for tighter and more robust selection and ongoing appraisal methods of staff, more demanding and challenging passengers and the profound and rapid changes within airlines companies and the concomitant impact on employees as well as passengers, mental health issues assume an even higher priority.

As in many work settings, mental health issues are also a matter of some sensitivity due to social stigma as well as practical and legal consequences for both the affected individual and the organization where problems have been detected. This is certainly true in the airline industry and psychological fitness to work is embedded within the practices and rules that operate in all work places and across occupational groups. A key difference among airline and military aircraft crews is the regularity and stringency of health checks and the ever-present threat of loss of license. The exclusions for medical certification are broadly similar across aircrew licensing authorities the world over. However, some psychological problems are transient or reactive and therefore either amenable to treatment or disappear with the passing of time. There have been significant advances recently in the treatment of certain commonly presenting psychological problems, such as depression and anxiety, while psychological counseling for those suffering from stress and relationship difficulties is more effective than was the case several years ago (Roth & Fonagy, 1996). These and related advances in psychological assessment and treatment should be noted for several reasons. Firstly, it may be possible for some crew members to return to work either during the course of or following treatment for certain psychological disorders. Secondly, the stigma associated with mental health problems and the fear of its consequences among air crew clearly deters some who are affected from seeking professional help from the relevant sources. They either conceal their problems or take the problem to a professional outside of the aviation medical context for treatment which may lead to incorrect advice being given, or inappropriate treatment or monitoring being provided, which may threaten safety. Thirdly, emerging problems such as stress (due to extended periods of duty, uncertainty in the aviation economic climate and threats to jobs, the pensions crisis, etc.) and their impact on psychological health, air crew functioning and safety can be more closely studied.

We hope that readers of this book will gain further understanding of a wide range of contemporary mental health issues that affect airline pilots, as well as others who work in the industry. The contents of the book also highlight for those who work in the industry some of the important mental health issues that may affect passengers. We envisage the primary audience of the book to come from those who work professionally with employees in the airline industry. This includes authorized medical examiners, psychologists, psychiatrists, employee assistance counselors, human resources specialists, crew rostering and operations personnel, air accident investigators, managers, trainers, aviation lawyers as well as those personnel worldwide employed by their respective aviation authorities or air forces. A secondary audience includes those employees within the airline industry about whom this book is concerned, including pilots (commercial, military and private), cabin crew, air traffic controllers and aircraft maintenance engineers, among many others, who may have an interest in the issues raised within this book. We envision that researchers as well as behavioral scientists in related fields, including those concerned with human factors, and others, will also have some interest in the contents of this book. It is possible that some airline passengers might also wish to learn more about mental health issues in aviation.

Those with an interest in mental health issues in aviation have, hitherto, had to consult a wide range of specialist books, journals and professional magazines, often in quite dispirit fields and specialties, to learn about many of the issues raised in this book. Academic papers concerning a wide range of the issues raised have been published in specialist medical, aviation, psychology, psychiatry and human factors journals, as will be seen in the list of references at the end of each chapter. A further aim of this book was to bring some of these insights together into a single authoritative text to assist those interested in learning about current knowledge and trends pertaining to these issues.

It is perhaps surprising that, to the best of our knowledge, there are few – if any other – contemporary specialist books available that address this topic. The first comprehensive collection of papers on the topic, entitled *Psychiatry in Aerospace Medicine*, and edited by Carlos Perry was published more than 35 years ago in 1967. The topic of psychiatry features as a specialist chapter in most of the core texts in aerospace medicine (see for example, *Fundamentals of Aerospace Medicine* (De Hart, 1996)), whilst the related problem of psychiatric and behavioral problems among airline passengers are dealt with in more recent related texts, including *Aviation Medicine and the Airline Passenger* (Nicholson & Cummin, 2002), *Passenger Behavior* (Bor, 2003) and *Psychological Perspectives on Fear of Flying* (Bor & van Gerwen, 2003). Readers who have an interest in an historical perspective in aviation mental health should consult the fascinating book by H. Graeme Anderson (1919) entitled *The Medical and Surgical Aspects of Aviation* which offers what are arguably the first authoritative professional insights into crew selection, the psychology of aviation and "aeroneurosis" (akin to a fear of flying among pilots), and was published at around the time of the birth of the aviation industry. It is remarkable how many of the insights offered by the author nearly a century ago, still apply. There is one notable exception: there was slightly greater tolerance (if not encouragement) of alcohol use among pilots in those early days to subjugate anxiety from the great risk of injury or death from accidents, though there was also recognition that excess alcohol use would ultimately ruin the career of the pilot.

Pilots and others employed within commercial and military aviation are in many ways a unique occupational group. Their training is both intense and vigorous, and the tasks performed by many demand good physical health and psychological stability. As individuals, crew have to be proficient in handling complex systems on board aircraft, as well as have an ability to work as part of a small team or crew and usually within an organization. As shift workers, they do not usually follow the same routine, and a pilot's *office* is normally a cramped flight deck on board an aircraft at 35,000 feet in the air. Increasing automation on the flight deck over the past decade has altered the role of the modern pilot. He or she must be an efficient and well-organized manager, proficient at communicating with customers and fellow crew and adept at computer programming as much as demonstrating traditional "stick and rudder" skills. When operating as part of a crew, a pilot's actions are subject to the close monitoring of other crew members, similar to an incessant driving test. Regular simulator and line tests, as well as medical assessments for physical and psychological fitness, a comparatively low retirement age, increased uncertainty in the job market as well disruption to one's domestic life, all add to the stress of the job. In an era where litigation is increasingly an option where a pilot's actions have been brought into question, appropriate psychological assessment of pilots within their six monthly medical examination has to be considered.

Psychological factors that are relevant to the selection of pilots have been fairly well documented. The assessment and management of psychological problems among pilots presenting for medical licensing or who have been referred with specific problems is less clearly understood. While severe psychological disturbance among air crew is generally considered rare, air crew may, nonetheless, suffer from the full spectrum of psychiatric disorders ranging from a phobic fear of flying (Medialdea & Tejada, 2005) to acting on suicidal thoughts, even when flying (Bills, Grabowski & Li, 2005). Psychological problems such as mood disorder (Jones & Ireland, 2004), anxiety as a symptom of another psychological problem or from occupational stress (Cooper & Sloane, 1985; Girodo, 1988), relationship problems (Raschmann, Patterson, & Schofield, 1990), sexual dysfunction (Grossman et al., 2004) alcohol misuse (Harris, 2002), and sleep disturbance (Waterhouse, Reilly, & Atkinson, 1997) present with greater frequency and require careful assessment and treatment. After physical disorders, psychiatric disturbance has been reported to be the most common source of attrition and loss of license among pilots (Bennet, 1983; Smith, 1983). In a cohort of 136 members of cabin crew deemed to be medically unfit to fly, 12.5 percent presented with psychiatric disorders while 30.1 percent and 21.3 percent respectively were grounded for otorhinolaryngological and musculoskeletal problems (Pombal et al., 2005).

Many airline and military pilots regard the potential value of psychology as quite risible. This deep skepticism of psychiatry and psychology stems, perhaps, in large measure from the long-standing historical use of psychometric testing in both the initial selection of pilots for training, and later, in job promotion. Some airline workers hold a rather narrow notion of the breadth and scope of the mental health profession, and it is hoped that this book will help to better explain the relevance of mental health issues among pilots. Especially in an industry which privileges physical safety and robustness of pilots, airline workers may be loathe to seek professional help for mental health difficulties or anxieties, and understandably, regard mental health workers with suspicion. In a profit-driven industry, mental health welfare tends to be marginalized and regarded as a luxury, unless a pilot becomes overtly depressed or alcoholic. Airlines have not been particularly knowledgeable about or sensitive to brewing stresses in their employees. For this reason, accessible, proactive and preventive mental health services need to be developed within the industry.

Aviation mental health is concerned with six main tasks: (a) selecting out those found to be psychologically unfit to fly or work within the industry; (b) monitoring the psychological health of those who enter into training and employment; (c) assessing and treating those who develop psychological problems in the course of their work; (d) determining whether and for how long an individual is unfit to fly or work within aviation; (e) emotionally supporting those deemed to be unfit to fly, whether temporarily or in the longer term; and, (f) preventing mental health problems through proactive intervention, health promotion and research.

The field of aviation mental health should not be seen to be limited to the diagnosis and treatment of psychopathology and psychiatric problems. A book of this scope is also concerned with the prevention of psychological problems, especially among crew. For this reason, the book includes chapters on the need for and approaches to crew (pilots, cabin crew and astronaut) selection. It also presents an overview of those aspects of passenger behavior that may adversely affect crewmembers in the course of their duties. As editors, we have sought to include as many areas that are relevant to the topic of aviation mental health as possible. However, this has not been without its limitations. We recognize that there is a paucity of published literature in several key areas. Readers will note, for example, that there are no chapters which specifically concern or address air traffic controllers or aircraft maintenance engineers. Almost all published aviation mental health research pertains to pilots and to a lesser extent cabin crew, but regrettably there is a paucity of literature on air traffic controllers and maintenance engineers, and this unintended bias is reflected in this book. We had also hoped to include specialist chapters on alcohol misuse among airline employees, the history of aviation mental health and how family relationships are affected by shift work and absences within the airline industry. These omissions are because, in spite of our best intentions, we could find no experienced researchers within these specialisms to undertake the task of writing a chapter. We hope that this first edition will encourage further research among our colleagues within the field of aviation mental health and we very much hope to include chapters on these topics and professional groups in future editions.

Introduction to Chapters

In order to help you to navigate through this text, we present a short description of the scope and direction of each chapter.

This text is divided into three parts: psychological issues of flight and cabin crew, psychological processes among passengers and crew, and related themes in aviation. Swanson and McIntosh lead off in Part 1 with a transactional theoretical model of stress and coping as a means to explain how air travel influences the psychological wellbeing of each airline passenger. Richards, Cleland, and Zuckerman continue the theme of stress in air travel with their insights into the psychological factors relating to physical health. They conclude that the flight environment produces physical strains upon flight crew which may result in undesirable psychological effects. Lucas and Goodwin tell us in their chapter that the psychological problems that plague passengers while on the ground become more acute and have lingering effects when they present in flight. Iljon Foreman, Bor, and van Gerwen contributed two chapters, both on the theme of fear of flying. In their first chapter on the nature of fear of flying, the authors make some interesting observations about how aircrew and passengers share common elements, events, and personality factors that contribute to the onset of fear of flying. In their chapter on the treatment of fear of flying, the authors present an overview of the literature on treatment programs that are available to those who suffer from fear of flying. They note that fear of flying treatment generalizes to other fears and difficulties as well. Chung completes Part 1 with his review of the existing research on posttraumatic stress resulting from aircraft disasters. In particular, he examines the differentiated symptoms of PTSD, based on proximity to the event.

Whereas Part 1 focuses on psychological issues among aircrew and passengers, Part 2 directs our focus to the mental health issues among crew. Morse and Bor begin with an explanation of psychiatric disorders and syndromes among pilots that can be threatening to individual performance and the safe conduct of a flight. Their contribution provides the backdrop for further discussion on pilot assessment and crew selection. For example, in Turnbull's chapter on mental state assessment of pilots, he stresses the importance of working collaboratively with pilots who are referred to him for assessment and the need to put them at ease in what is, for most, a difficult clinical encounter. Continuing on the theme of assessment, Bor points out the differentiation among assessment reports for pilots. This is a sensitive topic and he points out that some are written from a clinical psychologist's point of view, while others are written from an occupational or counseling psychologist's point of view. In any report, the author should keep to medico-legal standards of writing. Roe and Hermans outline the principles and methods of cockpit crew selection and discuss how effective selection systems can be developed. Albert continues on the theme of selection, but moves away from cockpit crew to flight attendants. This unique insight into the nature of the cabin crew environment provides clinical specialists interesting perspectives as they contemplate how aircraft design and crew composition influence crew preparedness for psychological problems in flight. Eriksen extends the discussion about cabin crew by describing four threats to emotional wellbeing. She points out that shift work, irregular routine, interruptions to one's personal life, and the constant demands of unreasonable passengers can have more dramatic effects on individuals who have no means to combat these occupational intimidations to healthiness. Partridge and Goodman complete Part 2 with an interesting first-person view of what it is like to be a flight attendant. They use their real life experience to examine how the British Airways Crewcare counseling service has supported cabin crew for the past 20 years.

Part 3 is for readers who may already have some knowledge of the literature regarding psychological issues for pilots, cabin crew, and passengers. The chapters in this section are wide-ranging and serve to introduce a range of specialist topics. Musson starts with an interesting view of the psychological aspects of astronaut selection, this being an area of potential future growth. He suggests that astronauts and cosmonauts on long missions might need to bring along a crew counselor or therapist. Waterhouse, Edwards, Atkinson, Reilly, Spencer, and Elsey concentrate on the factors of sleep loss, jet lag, and shift work as a means to describe some of the problems confronted by crew and other aviation employees. In keeping with the diverse views considered in Part 3, Frances presents some unique perspectives into the legal aspects of air travel. He makes the insightful deduction that physical and psychological problems are directly related to cheaper tickets and the diminishing space for economy class passengers. He also suggests that these problems translate directly to misdemeanors and felonies by disquieted passengers. Coetzee takes us in another direction altogether. He explains how psychology and aviation have joined together to form a unique blend in South Africa. Although many industrial nations already embrace aviation psychology as a research interest, not everyone is as equally disposed or resourced. Hubbard completes Part 3 with his insights into how aviation professionals explain misfortune and react to the inevitable nature of error making by pilots. He suggests that biological, sociological, and psychological issues influence how we compensate for error proneness.

We have addressed, in part, the psychological implications of humankind's exposure to air transportation: how it shapes and influences those who fly and how those who fly have influenced the very nature of flight. It is therefore fitting that we reexamine these implications periodically. We look forward to collaborating with our readers in the future.

References

American Psychiatric Association. (1994). *Diagnostic and statistical manual of mental disorders*, 4th edition (DSM-IV). Washington DC: American Psychiatric Press.

Anderson, H. (1919). *The medical and surgical aspects of aviation.* London: Oxford University Press and Hodder & Stoughton.

Bennett, G. (1983). Psychiatric disorders in civilian pilots. *Aviation, Space and Environmental Medicine, 54*, 588-589.

Bills, C., Grabowski, G. & Li, G. (2005). Suicide by aircraft: a comparative analysis. *Aviation, Space and Environmental Medicine*, 76, 8, 715–719.

Bor, R. (ed.) (2003). *Passenger behaviour*. Aldershot: Ashgate.

Bor, R., & van Gerwen, L. (eds.) (2003). *Psychological perspectives on fear of flying*. Aldershot: Ashgate.

Cooper, C. & Sloan, S. (1985). Occupational and psychosocial stress among commercial aviation pilots. *Journal of Occupational Medicine*, 27, 8, 570–576.

DeHart, R. (ed.) (1997). *Fundamentals of aerospace medicine*. Baltimore MD: Williams & Wilkins.

Girodo, M. (1988). The psychological health and stress of pilots in a labor dispute. *Aviation, Space and Environmental Medicine*, 59, 6, 505–510.

Grossman, A., Barenboim, E., Azaria, B., Sherer, Y., & Goldstein, L. (2004). Oral drug therapy for erectile dysfunction: an overview and aeromedical implications. *Aviation, Space and Environmental Medicine*, 75, 11, 997–1000.

Harris, D. (2002). Drinking and flying: Causes, effects and the development of effective countermeasures. *Human Factors in Aerospace Safety*, 2, 4, 297–317.

Jones, D. & Ireland, R. (2004). Aeromedical regulation of aviators using selective serotonin reuptake inhibitors for depressive disorders. *Aviation, Space and Environmental Medicine*, 75, 5, 461–470.

Jones, D., Katchen, M., Patterson, J., & Rea, M. (1997). In R. DeHart (ed.) *Fundamentals of aerospace medicine* (pp. 593–642). Baltimore MD: Williams & Wilkins.

Jones, D. & Marsh, R. (2001). Psychiatric considerations in military aerospace. *Aviation, Space and Environmental Medicine*, 72, 2, 129–135.

Medialdea, J. & Tejada, F. (2005). Phobic fear of flying in aircrews: Epidemiological aspects and comorbidity. *Aviation, Space and Environmental Medicine*, 76, 6, 566–568.

Nicholson, A. N. & Cummin, A. R. C. (Eds.) (2002). *Aviation medicine and the airline passenger*. London: Arnold Publishing.

Perry, C. (ed.) (1967). Psychiatry in aerospace medicine. *International Psychiatry Clinics*, 4, 1. Whole issue.

Pombal, R., Peixoto, H., Lima, M., & Jorge, A. (2005). Permanent medical disqualification in airline cabin crew: causes in 136 cases, 1993–2002. *Aviation, Space and Environmental Medicine*, 76, 10, 981–984.

Rachsmann, J., Patterson, J., & Schofield, G. (1990). A retrospective study of marital discord in pilots: The USAFSAM experience. *Aviation, Space and Environmental Medicine*, 61, 1145–1148.

Reason, J. (1974). *Man in motion: The psychology of air travel*. New York: Walker & Co.

Roth, A. & Fonagy, P. (1996). *What works for whom? A critical review of psychotherapy research*. New York: Guilford Press.

Smith, R. (1983). Psychiatric disorders as they relate to aviation: the problem in perspective. *Aviation, Space and Environmental Medicine*, 54, 586–587.

Waterhouse, J., Reilly, T., & Atkinson, G. (1997). Jet lag. *Lancet*, 350, 1611–1615.

PART 1
Psychological Issues of Flight and Cabin Crew

Chapter 2

Psychological Stress and Air Travel: An Overview of Psychological Stress Affecting Airline Passengers

Vivien Swanson and Iain B. McIntosh

Introduction

A recent review of health issues in relation to air travel suggested that "for many, air travel is a way of life" (Dehart, 2003, p. 134). It also cataloged an impressive list of minor and serious potential health risks associated with passenger air travel. Although air travel is commonplace, it can be a source of worry, stress or anxiety for many. Air travel is the preferred method of travel for the majority of foreign travelers (Steptoe, 1998). Since stress is related to ill health via its effects on the body, including the cardiovascular and immune systems, it is important to recognize potential sources of psychological stress in relation to air travel. These include fear of the physical sensations of being airborne, take-off and landing, and anxieties related to relatively minor hassles on the ground, due to airport delays, airport congestion, and security procedures. More recently, the unpredictable phenomenon of air terrorism, including hijack and bomb threats, coupled with widespread media reporting of events, has made air travel appear potentially more risky, with a marked public reaction.

Immediately after the September 11, 2001 events in the USA, the number of air passengers fell dramatically. Airlines registered 16 percent fewer international and domestic flights in the immediate aftermath of the disaster. A year later these had not reached pre-disaster figures, with a 15 percent shortfall in traffic between the US and Europe. Three years later, these rates appear to be returning to previous levels. In a general population survey carried out in the UK one year post-September 11 (Gauld, Hirst, McIntosh & Swanson, 2003) the vast majority of participants (85 percent) said the September 11 events would not affect their future air travel. In contrast, fewer Americans are traveling on transatlantic routes than previously.

In addition to increasing perceptions of risk of air terrorism, more media coverage has been given in recent years to the potential health risks of flying, including the potential for deep vein thrombosis (DVT), cardiac problems, infection risk (e.g. SARS), and passenger disruption (air rage). For the majority of travelers, the previously glamorous image of air travel has been replaced by perceptions of threatened disaster, airport delays and restrictions, poor on-flight conditions, and potential ill-health as a consequence. All of these factors can influence passenger anxiety.

Individual perceptions of personal risk reflect more than an objective rational calculation of the probability of such events occurring, but also involve a more subjective emotional or evaluative process (Illjon Foreman, 2003; Slovic, 2001; Slovic, Finucane, Peters & Macgregor, 2004). Where risks are outwith personal control to a very high degree, such as with air travel, individuals may be less likely to tolerate risk (Campbell, 2004). This is evidenced by increased road travel in the USA immediately after September 11, leading to a corresponding increase in road traffic accidents, presenting a much greater statistical risk than air travel at that time (Gray & Ropeik, 2002; Gigerenzer, 2004) A key factor in people's judgment is that responses are formulated on the basis of the *perceived* and not the *true* risk of an activity (Gewertz, 1996). Contrary to common perceptions, one has a greater statistical chance of dying if one avoids flying and stays at home than of being killed in a plane crash (Greco, 1989).

Despite this, the number of people undertaking air travel has increased exponentially in recent years, and for a decade there has been a year on year increase in global air traffic (Bor, Parker, Papadopoulos, 2001; Bor, Russell, Parker, & Papadopoulos, 2001). In addition to the economic benefits of this expansion for airlines and businesses (including the global leisure market), this has meant that more and more people have had access to fast, cheap and relatively safe transport around the globe. Although all modes of transport have associated risks, flying is perceived by the general public as the most dangerous of current common methods of travel. Apart from potential anxieties associated with flying itself, i.e. fear of heights or being in enclosed spaces, travelers are also affected by in-transit worries and fears. Being away from home and in what may be an unfamiliar and uncontrollable environment means that some travelers may be exposed to considerable stress at a time of maximal vulnerability (Pollit, 1986).

Air travel is generally perceived as a stressful experience and travel associated anxieties and fears are common (McIntosh, Power & Reed, 1996). However, there is a continuum from mild to intense fear; and some people experience more severe travel related anxieties or phobias, leading to anticipatory dread of flying, avoidance, or a reliance on prescribed anxiolytic medication. In an age where air travel is increasingly common, especially in developed countries, an inability to fly because of anxiety can have a serious negative impact on quality of life, working life, and personal relationships. Additionally, airline and airport managements may be unwilling or unprepared to advise their customers of potential health risks and strategies to reduce risk, for example in relation to development of DVT.

It is therefore important for the individual traveler, and for the economic future of the air travel industry that increased consideration is given to the wellbeing and satisfaction of the passenger with the travel process.

The term *stress* is often used ambiguously by lay people, the media, and in the research literature, to define both causes (i.e. sources of demand, hassle, or

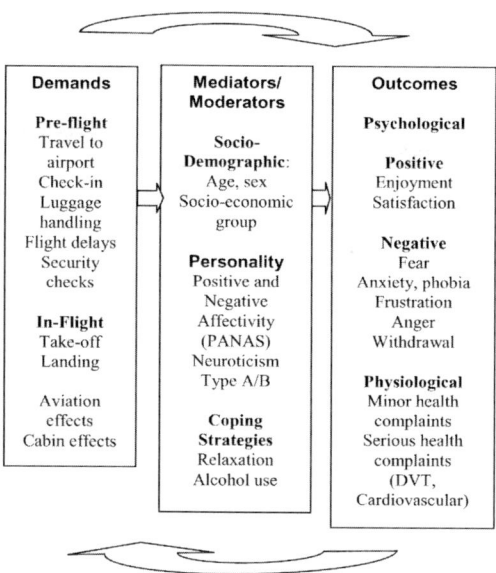

Figure 2.1 Transactional Framework for Psychological Effects of Air Travel

pressure) and outcomes (i.e. impact of these sources on the individual). The ability to distinguish between cause and outcome is important for those who treat stress and it is suggested that the term *stressor* be used to describe causal factors, and *stress* to describe the psychological outcome. It should be noted that stressors can be both physical and environmental (e.g. noise, temperature) or psychosocial (e.g. conflicting time demands, or relationships with others) – and objective or subjective.

An individual's enjoyment of travel depends upon a predisposition to cope well with a variety of physical and psychological stresses (Locke & Feinsod, 1982). This suggests that under the same stressor conditions, individuals will cope differently according to their own characteristics and resources. These can include coping resources such as skills and experience, support from others, or demographic characteristics such as age, gender, socio-economic circumstances or health, which might mediate or moderate travel outcomes.

This chapter considers these aspects; firstly, sources of stress (stressors) as possible influences on traveler wellbeing, and secondly, the impact of individual differences on people's experience of the travel process. This theoretical approach is based on the *transactional* theory of stress and coping which emphasizes the importance of inter-relationships between sources of stress and individual factors (Folkman, Lazarus, Gruen & DeLongis, 1986). Transactional theory suggests that the impact of different stressors on stress outcomes will vary according to the individual's appraisal of the seriousness of the threat, and their own ability to cope with it. Once an individual has made a primary appraisal of the problem, a secondary appraisal will assess whether current coping resources are sufficient. If not, the individual will experience psychological distress, or stress in some form. A diagrammatic representation, adapted to represent psychological stress in air travelers, is shown in Figure 2.1. The focus is on three components of the transactional model:

- potential objectively measurable sources of stress in the air travel process,
- possible outcomes for passengers in terms of physical and mental wellbeing,
- possible mediators or moderators of this relationship, including socio-demographic factors, individual psychological differences, and coping strategies.

Sources of Air Passenger Stress

Sources of air travel stress can be divided into two broad categories; those related to the travel process and those related to being airborne. Within the first category, stressors are mainly psychological (although they may have physiological outcomes, such as increase in blood pressure or raised heart rate) and relate to the whole air travel process. They may be associated with anticipatory anxieties related to fear of flying, pre-flight stressors resulting from travel to the airport, handling luggage, check-in, flight delays, customs and security checks – and post-flight stressors related to landing and baggage reclaim. Additionally, there is considerable potential for frustration and anger to develop at several points during the travel process, leading to inter-personal conflicts with staff and other travelers. Pre-flight and post-flight stressors may be equally or more hazardous to health than being in-flight (Neumann, 1996).

The physiological stress of the in-flight environment should also be acknowledged. Health risks were catalogued and reviewed in a recent article by DeHart (2003), which categorizes the flight environment in relation to aviation effects (noise/vibration; reduced oxygen supply; reduced atmospheric pressure; low temperature) and also to cabin environmental factors (air quality; seating conditions; motion sickness; exposure to disease vectors). It is likely that these factors will interact with psychological stressors, such as stress, anxiety, frustration or anger, and traveler characteristics such as amount of flying experience, personality and availability of coping strategies, to impact adversely on health. However, there is currently little research that investigates these relationships in more detail. Additionally, there have been few studies which have evaluated travelers' perceptions of different aspects of the air travel process in relation to psychosocial stress.

One large scale UK study of intended travelers in the general public (McIntosh, Swanson, Power, Raeside & Dempster, 1998) revealed the extent to which different aspects of the air travel process caused respondents to feel anxious. Items are sequentially ordered from travel to airport, to baggage reclaim as shown in Table 2.1.

All of the aspects of air travel in Table 2.1 were rated as a source of anxiety *sometimes* or *always* by some respondents, suggesting a high level of perceived

Table 2.1   Reported Anxiety for Aspects of Air Travel

Item	Never (percentage)	Sometimes, Often, Always (percentage)
Travel to airport	166 (70.3)	70 (29.6)
Check-in	165 (69.9)	71 (30.1)
Flight delays	115 (49.6)	117 (50.5)
Transfer between terminals	156 (67.2)	76 (32.8)
Waiting in lounge	172 (72.9)	64 (27.1)
Boarding flight	182 (77.1)	54 (12.9)
Take-off	137 (57.8)	100 (42.2)
During flight	149 (63.4)	86 (37.6)
Landing	133 (56.6)	102 (43.4)
Baggage reclaim	141 (60.0)	94 (40.0)
Customs	154 (65.3)	82 (34.8)

stress for most of these components. Overall, flight delays were most frequently rated as a source of anxiety, with just over 50 percent of travelers reporting anxiety for this item. Boarding the flight was the least anxiety-provoking aspect of air travel. As one might expect, take off and landing were rated as being *often* or *always* a source of anxiety by as many as 42 percent and 43 percent of travelers respectively, but baggage reclaim was not far behind at 40 percent. This suggests that it is important to take account of all aspects of the air travel process as potential sources of psychosocial stress, and not to focus only on in-flight situations. Although useful, this study was not able to establish which particular aspects or components of situations were stressful. The stress of flight delay may have been due to lack of information, time loss, personal inconvenience and a loss of control over events. For example, delays may have been attributed to aircraft faults, or related to late arrival at destination. For claustrophobic individuals, delays may mean being enclosed for longer. The perceived meaning of the delay is likely to differ between individuals and an understanding and evaluation of such meanings would be beneficial for airlines looking for ways of reducing airport passenger stress.

This study rated perceived flight related anxiety in the general public, including take-off and landing, but did not ask about *fear of flying* per se, which would categorize those with diagnosable fears or phobias. Additionally, only anxiety was measured as a possible psychological outcome, whereas other factors such as helplessness, frustration or anger may be equally important. A similar approach considering aspects of the travel process was adopted in a recent study with people seeking treatment for fear of flying (Kraaij, Garnefski & van Gerwen, 2003), who used a self-report anxiety scale with three subscales; Anticipatory Flight Anxiety, In-Flight Anxiety, and Generalized Flight Anxiety. It was not possible to rank these aspects, since mean values were not presented, but each factor was found to correlate significantly with negative cognitive coping strategies, such as self-blame and catastrophizing.

Outcomes of Passenger Stress

Psychological Outcomes

For some, the stress of air travel may be a positive and stimulating experience, appraised as a challenge rather than a threat, and seen as part of the excitement of being in a new environment, whether for business or leisure purposes. As such, it may lead to positive emotions, including enjoyment and satisfaction with the flying process. However, potentially positive outcomes of stress have been little researched in the psychological literature, and a search revealed no references to beneficial effects in relation to air travel. Negative psychological outcomes related to flying have been described to a greater extent, but these have generally focussed on severe and clinically diagnosable outcomes of anxiety related to fear of flying or flying phobias.

As noted, estimates of the prevalence of flying phobias vary from 10-40 percent of the general population (Agras, 1969; van Gerwen & Diekstra, 2000). The upper end of this estimate appears unreasonably high, but the wide range in these figures is probably due to differences in definitions, assessments of anxiety, use of different categorization systems (e.g. DSM IV, ICD 10), and estimates based on different populations. Additionally, many people with fear of flying choose to avoid flying altogether, and a large proportion probably remains undiagnosed. Population estimates based only on self-report may also be unreliable. For those individuals who do travel by air, levels of anxiety can vary from very mild to very severe, and may be associated with physical or psychosomatic symptoms. One problem with research in this area is that diagnosis of flying phobia or anxiety is often not clearly defined, or measured using valid and standardized instruments. The symptoms of fear of flying can span several diagnostic categories (van Gerwen, Spinhoven, Diekstra & van Dyck, 1997), making treatment potentially complex. A history of previous psychological disorder (i.e. anxiety or depression) also appears to be related to travel anxiety, but with some exceptions (notably, van Gerwen et al. 1997), there is little evidence of studies taking account of this in their assessments or evaluations of outcome. In a review of treatment programs for air travel passengers, van Gerwen and Diekstra (2000) noted these and other methodological difficulties in this area, and question the evidence for the efficacy of many existing programs on the basis of their review. Their own study also suffered from difficulties of low response rates. Methodological problems identified include the issues of patient self-selection to programs, lack of controls, variability in the quality of programs offered, lack of suitably trained therapists, and failure to base treatments on a psychologically derived theoretical approach. Although generally behaviorally based, the programs evaluated differed substantially in content and efficacy. There is a need for more high quality research in this area, to provide reliable evidence and pointers for standardized intervention programs.

Although fear and anxiety are notable psychological outcomes of air travel stress, other emotional outcomes should also be considered. These include frustration, anger and withdrawal (depression). Systematic research on these topics is very limited

(Bor, Parker et. al., 2001), although anecdotal reports exist regarding the increasing prevalence of in-flight passenger disturbances or *air rage*, disruptive behaviors and aggression towards airline staff. Anger and frustration are also linked with cardiovascular illness, and individuals with certain trait characteristics (such as *Type A* personality) and anger expression may exhibit greater stress reactivity. Although not a direct cause of cardiovascular illness, *Type A* behavior may also precipitate or exacerbate symptoms in individuals with cardiovascular disease. Triggers for air rage are thought to include environmental stressors, alcohol consumption and mental instability (Anglin, Neves, Giesbrecht & Kobus-Matthews, 2002). Although the third factor is difficult to change, arguably, the first two could be managed by airport and airline authorities to reduce the occurrence of air rage incidents (See also chapter by Lucas and Goodwin, in this text).

Physical Outcomes

Air travel related environmental stressors can have a negative impact on physical health, particularly where travelers have pre-existing physical health problems. Environmental factors such as levels of noise, temperature, air quality, dehydration and immobility can be objectively measured and related to illness outcomes. Certain categories of passenger have been identified as being at risk during air travel, such as those with cardiovascular problems and respiratory disorders who might respond negatively to these environmental stressors. Although clear links have been reported between chronic psychosocial stress, anxiety and physical health outcomes in longitudinal studies, it is more difficult to establish causal links between short-term psychosocial stresses associated with air travel and morbidity, since reporting of such stressors by the individual is subjective, and depends on appraisal. For example, although links have been shown between prolonged sitting and increased DVT, when pre-existing risk factors such as smoking and family history are controlled, research evidence has not definitively linked this increase with increased air travel (DeHart, 2003). Although the impact of stress and arousal on the cardiovascular system is well documented, it is difficult to partial out the impact of psychosocial stress on DVT in relation to air travel, and more research is needed in this area. Similarly, perceived stress may affect the impact of the environmental stressors detailed above.

A recent focus in the stress literature has been on the relationship between stressors and immune system function (psychoneuroimmunology – PNI). Studies have unequivocally shown links between psychosocial stress and poorer immune responses, for example in relation to wound healing, and the common cold (Kiecolt-Glaser et. al., 1995; Kiecolt-Glaser et. al., 2002). Although airborne and other pathogens are controlled to some extent in the aircraft environment, there is a potential for transmission of viruses and bacteria in this closed environment, particularly on longer haul flights. A literature search revealed no research studies which have investigated the potential for psychosocial stressors to influence individual immune responses in an air-travel environment, although such work may be of interest in promoting better passenger health and wellbeing.

There is also a paucity of studies investigating the frequency of minor health complaints in air travelers. In our general population study of intended travelers (McIntosh et al., 1998) a large proportion of people reported a range of mild symptoms related to air travel, many of which may have had a somatic component, and there was a strong correlation (Pearson $r = 0.51$) between overall anxiety and frequency of health problems. The most common health problems reported were ear problems, reported by 55 percent of respondents, for both long-haul and short-haul flights. Headache (41 percent), stuffy nose (31 percent), and swollen ankles (31 percent) were also very common (all were more predominant in female than male travelers). However, other items which may have had a psycho-somatic component related to stress or arousal were also endorsed – including nausea (23 percent), palpitations (19 percent), muscular pain (16 percent), dizziness (11 percent), and breathlessness (8 percent), were also reported to have been experienced *sometimes* or *often*. All of these problems were more common on long-haul than short-haul flights.

It is not possible to establish causal links between these minor health complaints and stress and anxiety, since the somatic component of such complaints is not distinguishable. Some reported symptoms were those usually associated with stress or anxiety, such as headache, palpitations, muscular tension, or breathlessness, but may equally have been a result of environmental factors such as poor air quality, low humidity, or in relation to pre-existing physical illness. Nevertheless, the overall number of health complaints reported is cause for some concern, suggesting attempts to reduce psychosocial stress in relation to air travel would be beneficial in reducing symptoms.

Mediators and Moderators: Passenger Differences

As noted, not all passengers react in the same way to stressors in the air travel process. Many factors will affect the overall experience of stress, some of which are discussed below. Coping adaptations may also be beneficial or deleterious. Many anxious travelers resort to alcohol consumption to combat stress and help them through the journey, but over indulgence by some may result in aggressive behavior and air rage. People do continue to fly despite marked fears and 20 percent of these fearful travelers use alcohol or sedatives to cope with severe anxiety (Greist, 1981). These coping strategies may ameliorate fear but exaggerate other comorbidities in the aircraft environment. Alcohol misuse is also considered to be implicated in 25 percent of cases of air rage incidents (Bor, Russell, et al., 2001).

Demographic Differences

Although some studies have evaluated differences in stress outcomes between male and female travelers, many fail to report gender differences. It is generally found that women tend to experience more stress, worry, anxiety and fear of flying than men. This is in line with the greater prevalence of anxiety disorders in women in general.

A large (1,650 participants) general practice based study in the UK found that women were significantly more worried about flying than men (p<0.001) (McIntosh et al., 1996) and also reported more general travel worries. In our study of anxiety and health problems related to air travel, women also reported more *in-flight anxiety* and more anxiety about the aircraft landing than men (McIntosh et al., 1998). They also reported significantly more flight related health problems overall (p<0.001) and specifically experienced more swollen ankles, sore eyes, stuffy noses and headaches than men. Again, this is in line with research suggesting women in general report more minor health symptoms than men (Walters, 1993).

Anxieties related to air travel have been researched in adults but data on younger people is very limited. Fears and phobias are more common in children than older people (Locke & Feinsod, 1982), and air phobia affects more children than adults. In addition, the potential for air accidents or terrorist events may have more impact on younger travelers (Gauld et al., 2003). They may also have different perceptions of personal risk, due to a lack of knowledge or inability to put risks into a rational context. People who travel by air frequently, such as businessmen, may have a reduced risk perception due to their personal experience of continued safe air travel.

Earlier retirement, increased wealth and a potentially longer life-span mean that many more elderly passengers now undertake air travel. A recent survey comparing older people, (over 65 years), people of working age and schoolchildrens' attitudes to air travel post-September 11, suggested that older people were more anxious about air travel than both the children and working-age population groups (Gauld et al., 2003). In addition, older travelers may justifiably exhibit greater anxiety about possible in-flight health problems, particularly in relation to cardiovascular illness and DVT, musculo-skeletal problems, and susceptibility to infection. However, this whole area is very under-researched in relation to stress, anxiety and air travel.

Psychological Factors

Dispositional characteristics and current psychological state are likely to affect perceptions of the stressfulness of air travel, and individual outcomes. Underlying characteristics such as neuroticism or negative affectivity have been shown to have a large effect on stress and mental health disorders such as anxiety and depression, and no doubt have a similar influence on air travel anxieties and phobias. There is also a degree of comorbidity for flying phobia and other anxiety disorders. Other dispositional or behavioral characteristics may also have an important influence on air travel stress. For example, *Type A* behavior has been associated with occupational stress and risk factors for coronary heart disease (Hayes & Feinleib, 1980), although links are not directly causal and there are some methodological problems with this approach (Evans, 1990). One component of this is internal/externalized anger, suggesting that *Type A* personality may be a factor in instances of passenger disruption or *air rage*. However, there is little published research evidence in this area.

Certain personality traits may be related to involvement in air rage incidents. Common traits and features include a difficulty in managing appropriate boundaries

Table 2.2 Frequency of Anxiety Reduction Methods for Air Travel Anxiety

Coping method	Use sometimes (percentage)	Use often/ always (percentage)
Distraction	90 (38.6)	19 (8.2)
Alcohol	67 (28.9)	19 (8.2)
Relaxation	38 (16.5)	6 (2.6)
Cigarettes	16 (6.9)	20 (8.6)
Non-prescribed medication	11 (4.7)	4 (1.7)
Prescribed medication	9 (3.9)	1 (0.4)

and impulsivity, and a tendency to act quickly without due consideration of possible consequences (Heller, 2003). They are also likely to drink alcohol in excess and thus present a challenge to air and ground staff. Greater awareness of the potential behavioral risks posed by such individuals, particularly when associated with alcohol misuse, is required. Refining the criteria for boarding fitness and training in conflict management to de-escalate threat would help to diminish the occurrence of dangerous air rage incidents.

Coping Strategies

Passengers use many different strategies in attempts to mediate or moderate the impact of air travel stressors on anxiety. At the extreme end of the continuum, programs to tackle flying phobias are run by many major airlines, although the quality and long-term effectiveness of some of these has been questioned (van Gerwen & Diekstra, 2000). Individuals can use cognitive coping strategies aimed at rationalization, management or distraction from their fears which may be either adaptive or maladaptive in managing anxiety. Identification of adaptive beliefs could help to inform treatment interventions to reduce flight anxiety (Kraaij et al., 2003). Passengers also use a range of other strategies to reduce travel stress and anxiety. These were investigated in our study of intended travelers (McIntosh et al., 1998). Reported methods were classified as shown in Table 2.2.

The most common strategy was distraction, used by almost half (47 percent) of participants. Use of relaxation techniques was also quite common, but worryingly, alcohol use and cigarette smoking were also utilized by a high number of travelers, with 8 percent of participants reporting that these methods were used almost always. Individuals may also rely on prescribed psychotropic medication or over the counter medications to treat their anxiety. Relaxation techniques are easily learned, and appear effective. Information about use of such techniques, and the problems of using more maladaptive strategies could easily be made available to prospective travelers via general practice based travel clinics, in combination with pre-travel health advice (MacDougall and Gyorkos, 2001) or from travel agents or airlines at booking.

Conclusion

For many people, air travel is a way of life, yet this mode of travel has potential psychological health risks and can be a source of stress, worry or anxiety, as well as enjoyment. This chapter has considered possible influences on air travelers' psychological wellbeing, using an established transactional theoretical model of stress and coping as a framework. Using this model allows sources of stress to be identified and measured, and differentiated from their effects. It focuses on individual coping to explain why different individuals can have very different reactions to potential sources of stress within the air travel process. Nevertheless, there are many unanswered questions, and the chapter has highlighted a need for more high quality studies in this generally under-researched area.

Whether air travel is experienced as enjoyable or stressful will depend on an individual's predisposition to cope with the more negative aspects of modern air travel, which include perceptions of threatened disaster and poor in-flight conditions. The increase in pre-flight stressors created by additional security, air traffic control restrictions, delay and recourse to alcohol consumption may fuel incidents of air rage in predisposed individuals. Good customer management by the industry should encompass the training of staff in identifying potential situations where control may be lost, recognizing passengers likely to become a threat and management of conflict between staff and customers to ensure a controlled safe outcome.

The transactional model provides a useful framework for interventions to alleviate such stress and anxiety by identifying stressors, and raising awareness of possible outcomes and potential mediators/moderators of stress. With appropriate warning, advice and precaution the traveler can adopt personal strategies to anticipate and minimise personal health risk. Good pre-travel health advice from health professionals may help travelers to anticipate problems and develop effective coping strategies. Additionally, the air travel industry could assist passengers to more accurately and realistically identify risk, and take appropriate precautions (for example to avoid DVT and reduce alcohol intake). Similarly pre-flight, in-flight and post-flight stressors could be identified and reduced by improved organization and communication by airlines. Paying attention to the some of the points identified in this chapter is in the best interests of passengers and the air travel industry, leading to reductions in passenger stress, and improved enjoyment of the air travel experience.

References

Agras, S., Sylvester, D., & Oliveau, D. (1969). Epidemiology of common fears and phobias. *Comprehensive Psychiatry*, *10*, 2, 151.

Anglin, L., Neves, P., Giesbrecht, N., & Kobus-Matthews, M. (2002). Alcohol-related air rage: From damage control to primary prevention. *Journal of Primary Prevention*, *23*, 3 283–297.

Bor, R., Russell, M., Parker, J., & Papadopoulos, L. (2001) Survey of the world's airlines about managing disruptive passengers. *International Civil Aviation Organisation Journal*, *56*, 2, 21–30.

Bor, R., Parker, J., & Papadopoulos, L. (2001). Aviation psychology. In J. Zuckerman (ed.) *Principles and practice of travel medicine* (pp. 237–246). West Sussex, UK: John Wiley & Sons.

Campbell, M. (2004). In communicating risk. *Pulse Apr. 5*, pp 51–52.

DeHart, R. (2003). Health issues of air travel. *Annual Review of Public Health*, *24*, 133–151.

Evans, P.D . (1990) Type A behaviour and coronary heart disease: When will the jury return? *British Journal of Psychology*, *81* (2), 147–157.

Folkman, S., Lazarus, R.S., Gruen, R.,J., DeLongis, A. (1986). Appraisal, coping, health status, and psychological symptoms. *Journal of Personality and Social Psychology*, *50*, 3, 571–579.

Gauld, J., Hirst, M., McIntosh, I.B., & Swanson, V. (2003). Attitudes to air travel after terrorist events. *British Travel Health Association Journal*, *3*, 62–67.

Gewertz, B. (1996). Vascular surgery in the next decade. *Journal of Vascular Surgery*, *23*, 5, 745–748.

Gigerenzer, G. (2004). Dread risk, Sept 11 and fatal traffic accidents. *Psychological Science*, *15*, 4, 286.

Gray, G.M. & Ropeik, D.P. (2002). Dealing with the dangers of fear: The role of risk communication. *Health Affair,. 21*, 6, 106–116.

Greco, T. (1989). A cognitive behavioural approach to the fear of flying. *Phobia Practice and Research Journal*, *2*, 3–15.

Greist, J. & Greist, G. (1981). *Fearless flying. A passenger guide to modern travel.* Chicago. Nelson Hall.

Hayes, S., G. & Feinleib, M. (1980). Women, work and coronary heart disease: Prospective findings from the Framingham Heart Study. *American Journal of Public Health*, *70*, 2, 133–141.

Heller J. (2003). Psychological and psychiatric difficulties among airline passengers. In R. Bor (ed.) *Passenger behaviour*, (p.60). Aldershot: Ashgate Publishing.

Illjon Foreman, E. (2003). Just plane scared. In R. Bor (ed.) *Passenger behaviour*, (pp. 46–47). Aldershot: Ashgate Publishing.

Kiecolt-Glaser, J.K., Marucha, P.T., Malarkey, W.B., Mercado, A.M., & Glaser, R. (1995). Slowing of wound healing by psychological stress. *Lancet*, *346*, 1194–1196.

Kiecolt-Glaser, J.K., McGuire, L., Robles, T.F., & Glaser, R. (2002). Emotions, morbidity, and mortality: new perspectives from psychoneuroimmunology. *Annual Review of Psychology*, *53*, 83–107.

Kraaij, V., Garnefski, N., & van Gerwen, L. (2003). Cognitive coping and anxiety symptoms among people who seek help with fear of flying. *Aviation, Space and Environmental Medicine*, *74*, 3, 273–277.

Locke, S.A. & Feinsod, F. M. (1982). Psychological preparation for young travelers travelling abroad. *Adolescence*, *17*, 815–819.

Lucas, G.E. (1987). Psychological aspects of travel. *Travel Medicine International, 99*, 104.

MacDougall, L. & Gyorkos, T. (2001). Promoting travel clinic referrals: Exploring partnerships for healthier travel. *Social Science and Medicine, 53*, 1461–1468.

McIntosh, I.B., Power, K.G., & Reed, J. M. (1996). Prevalence, intensity and sex differences in travel related stressors. *Journal of Travel Medicine, 3*, 96–102.

McIntosh, I.B., Swanson, V., Power K.G., Raeside, F., & Dempster, C. (1998). Anxiety and health problems related to air travel. *Journal of Travel Medicine, 5*, 198–204.

Neumann, K. (1996). Turbulence in the air-flight related health problems. *Travel Medicine International, 14*, 3, 113.

Pollitt, J. (1986). The mind and travel. *Travel Medicine International, 72*, 4, 72.

Slovic, P. (2001). The risk game. *Journal of Hazardous Materials, 86*, 17–24.

Slovic, P., Finucane, M.L., Peters, E., & MacGregor, D.G. (2004, April 24). Risk as analysis and risk as feelings. *Risk Analysis*, 311–322.

Steffen, R. & Du Pont, H. (1994). Travel medicine, What's That? *Journal of Travel Medicine, 1*, 1–3.

Steptoe A. (1988). Managing flying phobia. *British Medical Journal, 296*, 25.

van Gerwen, L.J. & Diekstra, R.W.F. (2000). Fear of flying treatment programs for passengers: An international review. *Aviation, Space and Environmental Medicine, 71*, 4, 430–437.

van Gerwen, L.J., Spinhoven, P., Diekstra, R.F.W., & van Dyck, R. (1997). People who seek help for fear of flying: Typology of flying phobics. *Behaviour Therapy, 28*, 237–251.

Walters, V. (1993). Stress, anxiety and depression: Women's accounts of their health problems. *Social Science and Medicine, 36*, 4, 393–402.

Chapter 3

Psychological Factors Relating to Physical Health Issues: How Physical Factors in Aviation and Travel Affect Psychological Functioning

Paul Richards,[1] Jennifer Cleland,[2] and Jane Zuckerman[3]

Introduction

The "office" for flight crew comprises a unique working environment which places physiological strains upon the occupants. In addition to physical effects these can result in psychological changes which themselves become more significant than they might be for other occupations due to the flight requirement of good psychomotor and cognitive skills. This chapter gives examples of how hypoxia and jet lag, two inevitable effects of aircraft flight, can affect psychological functioning. Furthermore, foreign destinations can expose any traveler to additional hazards such as malaria, but medication commonly used may have undesirable psychological consequences for flight crew. Examples are given from malaria prophylaxis, antibiotics, and antihistamines.

Hypoxia

Although commercial aircraft typically fly at 35,000–40,000 feet (10,500–12,000m), regulations stipulate that cabin pressure must not fall below 8000ft (2,438m) equivalent. This cabin altitude was chosen to maintain oxygen saturation in the

---

1 Clinical Tutor, Academic Centre for Travel Medicine and Vaccines, WHO Collaborating Centre for Travel Medicine, Royal Free & University College Medical School, University College London.

2 Senior Clinical Lecturer/ Department of General Practice and Primary Care, University of Aberdeen.

3 Senior Lecturer and Director, Academic Centre for Travel Medicine and Vaccines, WHO Collaborating Centre for Travel Medicine, Royal Free & University College Medical School, University College London.

majority of healthy individuals at 85–91 percent. (The British Medical Association [BMA]. 2004). Even at this altitude symptoms can be experienced on long flights. Evidence of ventilatory acclimatization has been found after an eight-hour flight (Fatemian, Kim, Poulin, & Robbins, 2001).

The impact of altitude on human performance is a critical concern in the field of aviation. At higher altitudes, humans suffer from a variety of symptoms that result from oxygen deficiency (Dean, Yip, & Hoffman, 1989; Dean, Yip, & Hoffmann, 1990; Honigman et al., 1993). These symptoms of hypoxia can include impaired psychological functioning.

The effect of hypoxia on cognitive performance has been examined during mountain climbs and simulations, mostly at high altitude. Lasting cognitive deficits found in climbers after a high-altitude expedition include decreased memory performance (Cavaletti & Tredici, 1993), mild impairment in concentration, verbal learning and memory and cognitive flexibility (Bonnon, Noel-Jorand, & Therme, 1995; Regard, Oelz, Brugger, & Landis, 1989) and decline in visual and verbal learning and memory (Hornbein, 1992). Whether or not any of these deficits become permanent after repeated exposure to extreme altitude is uncertain (Clark, Heaton, & Wiens, 1983; Jason, Pajurkova, & Lee, 1989), given that findings from different studies have reached opposing conclusions.

Temporary impairments in cognitive functioning found at high altitude include deterioration of the ability to learn, remember and express information verbally (Townes, Hornbein, Schoene, Sarquist, & Grant, 1984), impaired concentration and cognitive flexibility (Regard et al., 1989), and mild impairment in either short-term memory or conceptual tasks (Regard et al., 1991). Impairments in grammatical reasoning and in pattern comparison have also been reported (Kennedy, Dunlap, Banderet, Smith, & Houston, 1989). Cognitive impairments have often led to accidents due to improper evaluation of danger or other poor judgments (Nelson et al., 1990), suggesting some frontal lobe involvement. It is of note that, while Lieberman, Protopapas, and Kanki (1995) found that performance on a verbal fluency task was not significantly affected by testing location, they did not take a baseline measure of functioning at sea level or compare results with the normative data available.

The findings from extreme altitudes may be very different to those experiences in 6000–8000ft or so cabin pressure. Additionally, the adverse environmental conditions experienced during a mountain climb could also contribute to the cognitive deficits observed. Thus, other researchers have used breathing mixtures to induce hypoxia at sea level. Fowler, Prlic, and Brabant (1994) gave participants breath mixtures of oxygen and nitrogen to reduce SaO_2 saturation in the blood to 64–66 percent (13,900–14,400ft). In a separate testing condition participants were allowed to breathe normal air. Participants where tested under both conditions in one session with a 20-minute interval when switching from one condition to another. The results indicated that participants responded significantly slower in the hypoxic compared with normoxic condition on a memory task (asked if a probe item was a member of a learned memory set), but the rate of scanning short-term memory was similar in both groups. The brief time, however, between normoxic and hypoxic conditions may have resulted in carryover effects. The potential for practice effects also in this situation was clearly documented by Bonnon, Noel-Jorand, and Therme (1995). This has led to a probable underestimation of the impact on aviators of hypoxia at lower altitudes.

This issue was addressed by Bartholomew et al. (1999). This study looked at the effects of moderate altitudes 12,500 ft and 15,000 ft on short-term memory in 72 student pilots and instructors, in comparison to 2,000 ft. Participants performed a 30-minute vigilance task while listening to an audiotape with instructions to recall radio calls prefaced by their assigned call sign. Half the radio calls were high memory (at least four pieces of information), and half were low memory loads (no more than two pieces of information). Altitude did not affect the vigilance task. However, for readbacks of high memory load, significant deficits in recall were observed at 12,500 ft and 15,000 ft, whereas no effect of altitude was observed on recall of readbacks with low memory load. These results indicate that, at moderate altitudes, short-term memory was exceeded for the readbacks requiring a larger amount of information to be recalled. As there were no significant altitude effects for low memory loads, it is unlikely that the difference for high memory loads was due to some physical factor such as diminished auditory sensitivity. This study supports the conclusion that cognitive deficits, specifically the amount of cognitive resources available to process information, are found at lower altitudes than previously observed. This could lead to dangerous situations, such as missed indications of engine problems, incorrect reading of instruments and added difficulty in handling unusual situations such as emergencies. The risk of accident may increase in situations with high workload, such as descent and approach. Participants in this study had a relatively low number of flight hours, an average of fewer than 550, so the data may not apply to those with more experience, such as airline and military pilots. Future research is required to examine the generalizability of Bartholomew's results.

While certain decreases in cognitive performance may take place below 3,000m (Bartholomew et al., 1999; Brierley, 1976), and indeed there is evidence that performance on cognitive tests of mental arithmetic decreases steadily as altitude increases (Wu, Li, Han, Wang, & Wei, 1998), many studies in natural environments (Nelson, 1982) and simulated high altitudes (Cahoon, 1972) indicate that 3,000–3,500m and above is the zone where most adverse alterations appear.

Acute Hypoxia and Cognition

Cognitive decrements appear to follow a specific time course after exposure to altitude. Performance can be degraded as early as 0–6 hours after exposure (acute hypoxic state), before the onset of other medical effects (chronic hypoxic state). Initial impairments in performance are then followed by a progressive return to baseline. For example, Bonnon et al. (1995) found an improvement in the performance of the experimental group as their high-altitude stay lengthened. This supports the hypothesis that there are different stages of adaptation to altitude, which, in turn, depend on factors associated with respiratory function (Schoene, 2001). Thus, a decline in cognitive functioning may be particularly apparent during the acute stages of hypoxia (Leifflen et al., 1997).

One study found decrements on seven cognitive tasks employed 1–6 hours after simulated ascent to 4,600m. At 14h to 19 hours, four tasks were still impaired and by 38 or 43 hours two were still below normal. Thus, even after individuals are acclimated, cognitive decrements, especially in psychomotor tasks, may not be totally eliminated (Cahoon, 1972). However, our own data suggests that functioning in the areas of visual and verbal memory, planning and mental flexibility, motor speed and verbal expression returns to normal: at Day 1 of arrival at 5,100m after a 20-day walk-in from 410m, scores on a variety of cognitive tests were significantly impaired from baseline (sea level) (Harris, Cleland, Collie, Bennell, & McCrory, 2004). Scores had significantly improved, back to baseline levels on re-test at Day 7. We used parallel forms of test materials wherever possible to control for learning effects (Harris et al., 2004) as tests that have a large speed component, require an unfamiliar mode of response, or have a single solution are most likely to show significant practice effects (Matarazzo & Herman, 1984), as do tests which involve learning (Lezak, 1995).

Interestingly, Schlaepfer, Bartsch, and Fisch (1992) found mild improvements in cognitive function in acute hypoxic states. They studied the effect of an acute altitude challenge (rapid helicopter transport to the Jungfraujoch, 3,450m) and an acute exposure to mild hypoxia (fractional inspiratory oxygen concentration 14.5 percent) on a simple test of cognitive performance (the time needed to read briefly displayed letters) in ten healthy subjects. Under both hypoxic conditions the time needed to read briefly presented letters decreased, from 12.1 +/- SD 3.8 ms to 8.3 +/- 1.5 ms ($P < 0.01$) in the first experiment, and from 11.9 +/- 1.9 ms to 8.1 +/- 1.1 ms ($P < 0.01$) in the second. They concluded that a rapid and mild hypoxic challenge seems to improve a simple measure of cognitive performance above normal values. However, data is not available on the time frame of this improvement and there is no research currently available on the effects of acute hypoxic state on more complex cognitive tasks, such as the high memory load task examined by Bartholomew et al. (1999).

Thus, perhaps the safest conclusion is that aviators at risk of exposure to acute hypoxia should have available supplementary oxygen: many studies on respiratory sensations indicate that alleviation of hypoxia may result in an improvement in neuro-cognition (Krop, Block, & Cohen, 1973; Newsom-Davis, Lyall, Leigh, Moxham, & Goldstein, 2001).

There is the question of who is fit to administer cognitive tests at altitude or simulated altitude: it is likely that the individual administering the tests at altitude is also likely to have temporary cognitive deficits unless they have acclimatized. Given this, it is worthwhile considering the use of computerized testing, which has been found to be more sensitive to mild cognitive impairment (Harris et al., 2004).

Jet Lag

Desynchronization of normal circadian rhythms due to transmeridian flight produces a syndrome of physiological and psychological symptoms. The latter include daytime sleepiness, difficulty in sleeping in the local night, irritability, reduced attention span and general malaise (Herxheimer & Waterhouse, 2003). Susceptibility increases with the number of time zones crossed and exhibits inter-individual variability, but appears to increase with age.

One review by Herxheimer, of ten randomized placebo controlled trials (nine in passengers, one in cabin aircrew) found that destination bedtime Melatonin reduced jet lag scores in eight of the studies (Herxheimer & Petrie, 2005). The aircrew study of 52 cabin crew found reduced sleep disturbance and jet lag scores compared with placebo, but only when melatonin was taken after destination arrival and not when it was taken both before and after arrival (Petrie, Dawson, Thompson, & Brook, 1993). The review authors however, felt it was difficult to generalize this result as the aircrew had such complicated circadian disruption due to serial flights. Due to sedative effects, flight crew are not permitted to use Melatonin or hypnotics such as Zopiclone, Zolpidem, and Temazepam (Joint Aviation Authorities [JAA] Europe, 2005).

Medication

Whilst commonly occurring effects, such as sedation with some antihistamines, are well recognized, any individual's idiosyncratic side-effect response to medication is difficult to predict. Most pharmacological research is conducted at sea level and the additional effect of the aviation environment of hypoxia, noise, vibration, dehydration (controversial), and fatigue is largely unknown. Those drugs which may produce central nervous system depression or stimulation, disturbance of the autonomic nervous system, opthalmological or labyrinthine effects are of particular concern. Many drugs are thus restricted or prohibited for aircrew use. It is also noted that the underlying disorder for which the medication is being taken may itself preclude flying.

US Federal Aviation Administration (FAA) regulations permit medication for some chronic conditions where aviation safety is not compromised. For short term conditions some pharmaceuticals may be permitted, usually with the stipulation that the pilot does not fly during the time when symptomatic or treatment is required. It is also to be noted that many medications have long half lives and will remain active for many hours after the last dose which may produce impairment after the patient believes themselves to be clear of the drug. Herbal medicines cannot be considered safe as the active ingredients may not be documented or even known (Civil Aviation Authority, United Kingdom [CAA, UK], 2005). Whilst issuing guidance, the FAA cautions against use of medication information lists as sole determinants of fitness to fly or which might encourage self-assessment of risks posed by medical conditions. Air crew should obtain an authoritative sound medical assessment which balances the various factors based

Antimalarials

Mefloquine Mefloquine shows an increase in neuropsychiatric side effects compared with other antimalarials. A systematic review by Croft in 2000 of ten Randomized Controlled Trials (RCT) (n=2750) found commonly reported side effects included headache (16 percent), insomnia (15 percent), and fatigue (8 percent) (Croft & Garner, 2000). (It is to be noted that five of these trials were field trials in mostly male soldiers and findings may not be transferable to the general population). A retrospective postal and telephone questionnaire survey of tourist and business travelers found sleep disturbance and neuropsychiatric symptoms occurred in 0.7 percent (333/1214) of those taking mefloquine compared with 0.09 percent (189/1181) of those taking choroquine with proguanil (p<0.001) (Barrett, Emmins, Clarke, & Bradley, 1996).

A randomized controlled trial of 976 non-immune travelers showed significantly more adverse effects attributed to mefloquine than to atovaquone plus proguanil (204/483, 42 percent vs 149/493, 30 percent. p=0.001) (Overbosch et al., 2001). Specifically, mefloquine increased the incidence of "strange or vivid dreams" (66/483, 13 percent) compared with atovaquone with proguanil (33/493, 7 percent), insomnia (65/483, 13 percent vs 15/493, 3 percent), dizziness or vertigo (43/483, 9 percent vs 11/493, 2 percent), visual difficulties (16/483, 3 percent vs 8/493, 2 percent), anxiety (18/483, 4 percent vs 3/493, <1 percent), and depression (17/483, 4 percent vs 3/493, <1 percent) (Overbosch et al., 2001). Psychosis has also been reported (Stuiver, Ligthelm, & Goud, 1989).

Data specific to flight crews is sparse. One small double blind placebo controlled cross-over study of 23 trainee commercial air pilots found one withdrawal during the mefloquine loading dose phase due to dizziness, diarrhea and flu-like symptoms. Three further volunteers reported non-serious sleep-related adverse effects. There was no significant difference in flying performance after three weeks of treatment (mean total number of errors 12.6 with mefloquine versus 11.7 with placebo), psychomotor functions or mean sway for any of three tested positions. Reductions in mean total nocturnal sleep were non significant (mefloquine 450mins versus placebo 484mins). Mood scores indicated a predominance of positive states particularly of vigour (Schlagenhauf et al., 1997).

Mefloquine is currently not recommended for use where situations involve fine motor coordination or spatial discrimination, or in persons with a history of fits or psychiatric disorder (Roche Pharmaceuticals, 2005). It is currently not allowed for use by European civilian pilots (Joint Aviation Authorities Europe, 2005)

Atovaquone and Proquanil Atovaquone and Proquanil appears to have no significant neuropsychological effects, impairment of psychomotor performance, mood changes, sleeplessness or fatigue (see above). It is already used by several airlines and at the time of writing is approved for pilots by the FAA.

Doxycycline Doxycycline is used for prophylaxis of malaria (Bradley & Bannister, 2003) and travelers' diarrhea (Farthing, 1991; Rendi-Wagner & Kollaritsch, 2002). One RCT of malaria prophylaxis in Indonesian soldiers reported side effects of headache (16 percent), cough (31 percent) and dermatological problems (33 percent) (Ohrt et al., 1997). A retrospective questionnaire survey of 28 aviators and 15 non-aviator aircrew comparing mefloquine to doxycycline found compliance was better with mefloquine (100 percent vs 75 percent) and side effects, mostly gastrointestinal, occurred more frequently with doxycycline (39 percent cf 13 percent). Of the pilots, 7 of 28 reported abdominal pain and 5 of 28 reported fatigue (Shamiss, Atar, Zohar, & Cain, 1996).

Currently doxycycline malaria prophylaxis is not officially approved for European civilian pilots, but is used by military pilots in high risk areas due to lack of effective alternatives (Joint Aviation Authorities Europe, 2005).

Considering the difficulties in choice of antimalarial for aircrew, it is reassuring that behavioral and non-chemotherapy prophylaxis, such as short layovers, remaining in air-conditioned hotels, use of air-conditioned ground transport and good compliance with personal protection methods, results in an attack rate in one study of crews flying to sub-Saharan Africa, much less than that of general travelers at 1.6 cases per 100,000 nights (Byrne & Behrens, 2004).

Antihistamines

Antihistamines are commonly used by the general population for allergic conditions, such as hayfever (seasonal allergic rhinitis), and by travelers, for example, for insect bites. Traditional antihistamines such as Diphenhydramine are sedative and their use would preclude flying, particularly as they have been implicated in driving and air accidents (Kay & Quig, 2001). Some second generation antihistamines do not so readily cross the blood-brain barrier and are less sedative and produce less anticholinergic side effects such as dry mouth. There are differences, however, between the side effect profiles of preparations within the group. Concerns have been expressed regarding Astemizole and Terfenadine which appear to increase electrical conduction times within the heart (QT interval) predisposing to rare, but fatal, arrhythmias particularly if taken in combination with certain other drugs. Cetirazine has been shown to cause sedation and psychomotor impairment, albeit to a lesser degree than first generation antihistamines (Gonzalez & Estes, 1998), whereas Loratidine does not appear to impair cognitive or psychomotor performance at standard doses (Kay & Harris, 1999).

Desloratidine, the biologically active primary metabolite of loratidine, has shown no clinically significant effects on wakefulness or psychomotor performance (Geha & Meltzer, 2001) and no impairment of driving performance (Vuurman, Rikken, Muntjewerff, de Halleux, & Ramaekers, 2004). Indeed, in a "real-world" randomized, double-blinded cross-over study involving information processing capacity and complex decision-making tasks, Desloratidine actually restored performance impaired by the symptoms of hayfever in sufferers to that of non sufferer (placebo) levels in six of the nine parameters measured, and improved performance in the remaining three parameters. (Satish, Streufert, Dewan, & Voort, 2004).

Antibiotics

The use of antibiotics for the prophylaxis of travelers' diarrhea, whilst effective, is usually only recommended for short term use for those with time critical itineraries, such as sports persons competing at events or businessmen with important meetings where even a short illness could be extremely disruptive (Al Abri, Beeching, & Nye, 2005). Commonly used drugs are doxycycline, which is discussed above, and ciprofloxacin. Data on the effect of ciprofloxacin on psychomotor performance is sparse, but one placebo controlled trial of three days treatment showed no impairment and no significant effect on concentration or vigilance compared with placebo (Kamali, 1994).

Conclusion

The flight environment produces physical strains upon flight crew which may result in undesirable psychological effects. Similarly, medication which might be commonly used by other travelers, or available over the counter without prescription, can have deleterious effects on safe crew functioning. Aircrew need to be mindful of these effects and seek appropriate qualified medical advice before medicating.

References

Al Abri, S. S., Beeching, N. J., & Nye, F. J. (2005). Traveler's diarrhoea. *Lancet Infectious Diseases*, 5, 349–360.

Barrett, P. J., Emmins, P. D., Clarke, P. D., & Bradley, D. J. (1996). Comparison of adverse events associated with use of mefloquine and combination of chloroquine and proguanil as antimalarial prophylaxis: postal and telephone survey of travelers. *British Medical Journal*, 313, 525–528.

Bartholomew, C. J., Jensen, W., Petros, T. V., Ferraro, F. R., Fire, K. M., Biberdorf, D. et al. (1999). The effect of moderate levels of simulated altitude on sustained cognitive performance. *International Journal of Aviation Psychology*, 9, 351–359.

Bonnon, M., Noel-Jorand, M. C., & Therme, P. (1995). Psychological changes during altitude hypoxia. *Aviation Space and Environmental Medicine*, 66, 330–335.

Bradley, D. J. & Bannister, B. (2003). Guidelines for malaria prevention in travelers from the United Kingdom for 2003. *Communicable Disease and Public Health*, 6, 180–199.

Brierley, J. (1976). Cerebral Hypoxia. In *Greenfield's Neruopathology* (pp. 43–85). Chicago: Yearbook Medical Publishers.

Byrne, N. J. & Behrens, R. H. (2004). Airline crews' risk for malaria on layovers in urban sub-Saharan Africa: risk assessment and appropriate prevention policy. *Journal of Travel Medicine*, *11*, 359–363.

Cahoon, R. L. (1972). Simple decision making at high altitude. *Ergonomics*, *15*, 157–163.

Cavaletti, G. & Tredici, G. (1993). Long-lasting neuropsychological changes after a single high altitude climb. *Acta Neurologica Scandinavica*, *87*, 103–105.

Civil Aviation Authority (UK) (2005). Medical Safety Regulation. *Civil Aviation Authority* (UK) [On-line]. Available: http://www.caa.co.uk/

Clark, C. F., Heaton, R. K., & Wiens, A. N. (1983). Neuropsychological functioning after prolonged high altitude exposure in mountaineering. *Aviation, Space & Environmental Medicine*, *54*, 202–207.

Croft, A. M. & Garner, P. (2000). Mefloquine for preventing malaria in non-immune adult travelers. *Cochrane Database Systematic Review*, CD000138.

Dean, A., Yip, R., & Hoffman, R. (1989). An Epidemic of Acute Mountain Sickness Among Epidemiologists: High Attack Rate at Moderate Altitude. *Proceeds of the Sixth International Hypoxia Symposium*. Lake Louise, Canada.

Dean, A., Yip, R., & Hoffmann, R. (1990). High incidence of mild acute mountain sickness in conference attendees at 10 000 foot altitude. *Journal of Wilderness Medicine*, *1*, 86–92.

Farthing, M. J. (1991). Review article: prevention and treatment of travelers' diarrhoea. *Alimentary Pharmacology and Therapeutics*, *5*, 15–30.

Fatemian, M., Kim, D. Y., Poulin, M. J., & Robbins, P. A. (2001). Very mild exposure to hypoxia for 8h can induce ventilatory acclimatisation. *Pflugers Archiv*, *441*, 840–843.

Federal Aviation Administration. (2005). *Guide for Aviation Medical Examiners* version IV [Electronic version]. Washington, DC: Author.

Fowler, B., Prlic, H., & Brabant, M. (1994). Acute hypoxia fails to influence two aspects of short-term memory: implications for the source of cognitive deficits. *Aviation, Space & Environmental Medicine*, *65*, 641–645.

Geha, R. S. & Meltzer, E. O. (2001). Desloratadine: A new, nonsedating, oral antihistamine. *Journal of .Allergy and Clinical Immunology*, *107*, 751–762.

Gonzalez, M. A. & Estes, K. S. (1998). Pharmacokinetic overview of oral second-generation H1 antihistamines. *International Journal of Clinical Pharmacology and Therapeutics*, *36*, 292–300.

Harris, G., Cleland, J., Collie, A., Bennell, K., & McCrory, P. (2004). Computerised cognitive assessment is more sensitive than written tests at 5100 altitude above sea level. *High Altitude Medicine and Biology*, *5*, 471–509.

Herxheimer, A. & Petrie, K. J. (2005). *Melatonin for the prevention and treatment of jet lag*. The Cochrane Library [4]. Chichester, UK, John Wiley & Sons Ltd.

Herxheimer, A. & Waterhouse, J. (2003). The prevention and treatment of jet lag. *British Medical Journal*, *326*, 296–297.

Honigman, B., Theis, M., Koziol-McLain, J., Roach, R., Yip, R., Houston, C. et al. (1993). Acute Mountain Sickness in a general tourist population at moderate altitudes. *Annals of Internal Medicine, 118*, 587–592.

Hornbein, T. F. (1992). Long term effects of high altitude on brain function. *International Journal of Sports Medicine, 13 Supplement 1*, S43–S45.

Jason, G. W., Pajurkova, E. M., & Lee, R. G. (1989). High-altitude mountaineering and brain function: neuropsychological testing of members of a Mount Everest expedition. *Aviation, Space & Environmental Medicine, 60*, 170–173.

Joint Aviation Authorities Europe (2005). Joint Aviation Authorities Manual of Civil Aviation Medicine [Electronic version]. Hoofddorp, The Netherlands: Author.

Kamali, F. (1994). No influence of ciprofloxacin on ethanol disposition. A pharmacokinetic-pharmacodynamic interaction study. *European Journal of Clinical Pharmacology, 47*, 71–74.

Kay, G. G. & Harris, A. G. (1999). Loratadine: a non-sedating antihistamine. Review of its effects on cognition, psychomotor performance, mood and sedation. *Clinical & Experimental Allergy, 29 Supplement 3*, 147–150.

Kay, G. G. & Quig, M. E. (2001). Impact of sedating antihistamines on safety and productivity. *Allergy and Asthma Proceedings, 22*, 281–283.

Kennedy, R. S., Dunlap, W. P., Banderet, L. E., Smith, M. G., & Houston, C. S. (1989). Cognitive performance deficits in a simulated climb of Mount Everest: Operation Everest II. *Aviation, Space and Environmental Medicine, 60*, 99–104.

Krop, H. D., Block, A. J., & Cohen, E. (1973). Neuropsychologic effects of continuous oxygen therapy in chronic obstructive pulmonary disease. *Chest, 64*, 317–322.

Leifflen, D., Poquin, D., Savourey, G., Barraud, P. A., Raphel, C., & Bittel, J. (1997). Cognitive performance during short acclimation to severe hypoxia. *Aviation Space & Environmental Medicine, 68*, 993–997.

Lezak, M. (1995). *Neuropsychological Assessment.* (3rd ed.) Oxford: Oxford University Press.

Lieberman, P., Protopapas, A., & Kanki, B. G. (1995). Speech production and cognitive deficits on Mt. Everest. *Aviation Space & Environmental Medicine, 66*, 857–864.

Matarazzo, J. D. & Herman, D. O. (1984). Base rate data for the WAIS-R: test-retest stability and VIQ-PIQ differences. *Journal of Clinical Neuropsychology, 6*, 351–366.

Nelson, M. (1982). Psychological testing at high altitudes. *Aviation Space & Environmental Medicine, 53*, 122–126.

Nelson, T. O., Dunlosky, J., White, D. M., Steinberg, J., Townes, B. D., & Anderson, D. (1990). Cognition and metacognition at extreme altitudes on Mount Everest. *Journal of Experimental Psychology-General, 119*, 367–374.

Newsom-Davis, I. C., Lyall, R. A., Leigh, P. N., Moxham, J., & Goldstein, L. H. (2001). The effect of non-invasive positive pressure ventilation (NIPPV) on cognitive function in amyotrophic lateral sclerosis (ALS): a prospective study. *Journal of Neurology, Neurosurgery and Psychiatry, 71*, 482–487.

Ohrt, C., Richie, T. L., Widjaja, H., Shanks, G. D., Fitriadi, J., Fryauff, D. J. et al. (1997). Mefloquine compared with doxycycline for the prophylaxis of malaria in Indonesian soldiers. A randomized, double-blind, placebo-controlled trial. *Annals of Internal Medicine, 126,* 963–972.

Overbosch, D., Schilthuis, H., Bienzle, U., Behrens, R. H., Kain, K. C., Clarke, P. D. et al. (2001). Atovaquone-proguanil versus mefloquine for malaria prophylaxis in nonimmune travelers: results from a randomized, double-blind study. *Clinical Infectious Diseases, 33,* 1015–1021.

Petrie, K., Dawson, A. G., Thompson, L., & Brook, R. (1993). A double-blind trial of melatonin as a treatment for jet lag in international cabin crew. *Biological Psychiatry, 33,* 526–530.

Regard, M., Landis, T., Casey, J., Maggiorini, M., Bartsch, P., & Oelz, O. (1991). Cognitive changes at high altitude in healthy climbers and in climbers developing acute mountain sickness. *Aviation Space & Environmental Medicine, 62,* 291–295.

Regard, M., Oelz, O., Brugger, P., & Landis, T. (1989). Persistent cognitive impairment in climbers after repeated exposure to extreme altitude. *Neurology, 39,* 210–213.

Rendi-Wagner, P. & Kollaritsch, H. (2002). Drug prophylaxis for travelers' diarrhea. *Clinical Infectious Diseases, 34,* 628–633.

Roche Pharmaceuticals (2005). Lariam Data Sheet. Electronic Medicines Compendium [On-line]. Retrieved on November 4, 2005, from http://emc.medicines.org.uk/

Satish, U., Streufert, S., Dewan, M., & Voort, S. V. (2004). Improvements in simulated real-world relevant performance for patients with seasonal allergic rhinitis: impact of desloratadine. *Allergy, 59,* 415–420.

Schlaepfer, T. E., Bartsch, P., & Fisch, H. U. (1992). Paradoxical effects of mild hypoxia and moderate altitude on human visual perception. *Clinical Science (London), 83,* 633–636.

Schlagenhauf, P., Lobel, H., Steffen, R., Johnson, R., Popp, K., Tschopp, A. et al. (1997). Tolerance of mefloquine by SwissAir trainee pilots. *American Journal of Tropical Medicine and Hygiene, 56,* 235–240.

Schoene, R. B. (2001). Limits of human lung function at high altitude. *Journal of Experimental Biology, 204,* 3121–3127.

Shamiss, A., Atar, E., Zohar, L., & Cain, Y. (1996). Mefloquine versus doxycycline for malaria prophylaxis in intermittent exposure of Israeli Air Force aircrew in Rwanda. *Aviation Space & Environmental Medicine, 67,* 872–873.

Stuiver, P. C., Ligthelm, R. J., & Goud, T. J. (1989). Acute psychosis after mefloquine. *Lancet, 2,* 282.

The British Medical Association. (2004, May). *The impact of flying on passenger health.* London: Board of Science and Education.

Townes, B. D., Hornbein, T. F., Schoene, R. B., Sarquist, F., & Grant, I. (1984). Human cerebral function at extreme altitude. In J.West (ed.), *High altitude and man* (pp. 31–36). Bethesda: American Psychological Society.

Vuurman, E. F., Rikken, G. H., Muntjewerff, N. D., de Halleux, F., & Ramaekers, J. G. (2004). Effects of desloratadine, diphenhydramine, and placebo on driving performance and psychomotor performance measurements. *European Journal of Clinical Pharmacology, 60*, 307–313.

Wu, X., Li, X., Han, L., Wang, T., & Wei, Y. (1998). Effects of acute moderate hypoxia on human performance of arithmetic. *Space Medicine & Medical Engineering (Beijing), 11*, 391–395.

Chapter 4

Psychological Problems Among Passengers and On-Board Psychiatric Emergencies

Graham Lucas and Tony Goodwin

Introduction

Passenger numbers including long-haul flights are increasing. Pre-existing mental and or physical illness may be exacerbated impairing fitness to fly. Passengers who have specially modified facilities in their homes assume that airlines can accommodate accordingly, without discrimination (Goodwin, 2000).

Psychological Aspects of Air Travel

Cabins are pressurized to the equivalent of 6–8,000 feet (1,950–2,400m altitude). The lowered air pressure reduces the oxygen supply to tissues, especially cerebral and cardiac as well as causing internal gaseous expansion by some 30 percent. Apart from occasional transient ear or sinus pain, the otherwise healthy passenger is not affected. However, times of meals or of routine medication can be disrupted. Hypoglycaemia can be mistaken for aggression, intoxication, or even psychiatric illness.

The mental and physical components of health together with personality influence the behavior of passengers between themselves and towards the crew. The pre-selected crew, trained in human factors, relate to randomly reacting and interacting passengers. Interfacing with demanding, aggressive, or disturbed passengers can be stressful and could compromise safety, the primary role of cabin crew. Unfulfilled expectations cause anger, especially if the advertised level of services does not materialize. Delays can be extremely irritating. Safety depends on harmony between all those on board the aeroplane, an enclosed isolated community. Issues affecting flight and cabin crew members are similar to those described for passengers, but due to rostering can be exacerbated by the effects of multiple time zone changes, particularly east to west changes on successive trips often with minimum rest between. Ultra long haul operations currently up to 18 hours non stop with some carriers are soon to be increased to 20–21 hours. This will compound psychosocial and physical stressors such as dehydration and hypoxia.

Flight crew members, in particular, are often reluctant to discuss stress, anxiety or depression fearing stigmatization, and licensing issues. Cabin crewmembers are particularly vulnerable to acute stress reactions and even posttraumatic stress disorder (PTSD), following verbal or physical abuse (see the discussion of PTSD by Chung in this text). They may share passenger anxiety regarding turbulence and unusual noises. While serious in-flight psychiatric emergencies are relatively infrequent, varying levels of aviaphobia are common, requiring support, reassurance, explanations, and understanding (see a detailed account of the nature and treatment of fear of flying, by Iljon Foreman, Bor, and van Gerwen in this text).

Personality encompasses life-long traits dictating how an individual responds to a given situation, interacts and is then self perceived. A typical example is selfishness with hand luggage. Its weight, quantity, bulk, and how it is stowed can give offence. Ensuing resentment fueled by alcohol could ultimately erupt into air rage, easily mistaken for terrorism, and readily capitalized by the media.

Any personality trait which significantly impairs functional effectiveness can constitute a neurosis, the commonest being anxiety, depression, somatization, hypochondriacal, and obsessional states. All these can cause demanding behavior. Up to 90 percent of air passengers are said to have a degree of flying related anxiety or *fear of flying*. Also, there are the significantly rarer but more severe psychotic illnesses such as schizophrenia, manic depression, and the drug induced psychoses. Features of these may include impaired insight and judgment, delusions, hallucinations, impulsivity, and disruptive behavior.

Pre and in-flight stressors, particularly for the first time flier, include uncertainty, reason for the journey, delays on way to airport, at check-in, departure, and separation from luggage. Anticipatory anxiety exacerbated by media scares about deep vein thrombosis (DVT), terrorism, or recent accidents increases vulnerability. It may also reflect emotionally-tinged journeys, either religious (Hajj, Lourdes), for medical treatment, bereavement, or even deportation or repatriation under section (Gordon & Goodwin, 2004).

Physical illness is psychologically stressful, compounding inevitable uncertainty and loss of control, even more so in the elderly, who may already be prone to a degree of confusion. For them, in particular, supplemental oxygen is both psychologically and physiologically beneficial due to the generally adverse effects of hypoxia.

Passengers with cardio respiratory and cerebro vascular illness, diabetes, Irritable Bowel Syndrome (IBS), prostatism and menorrhagia due to limited toilet facilities, also face problems and musculo skeletal pain, arthritis and obesity impair mobility, increasing trapped feelings. Contact lenses can be uncomfortable due to dryness, relieved by artificial tears. Vertigo, tinnitus, impaired hearing, hyperacusis can all be aggravated, as can a variety of skin conditions. Hearing aids can magnify background noise alarmingly.

Safety demonstrations, compulsory seat belt restraint, restricted spaces, unexplained noises on take off or landing, air turbulence, unexplained 'fasten seat belt signs', long toilet queues, provocative or disruptive behavior, can both irritate and threaten, even more so since the hijacking of September 11, 2001. Articles entitled *One Million Good Reasons to Fly* and *Happy Landings for Flying Phobics* are few. Men are often less able than women to admit fear, reacting with aggression in an attempt to regain control. Flying entailing total loss of control even of the air breathed.

Passengers with Psychiatric Problems

Cabin crew training varies widely between airlines with reference to mental health. Understandably, there is often an emphasis on physical wellbeing and how to deal with physical health emergencies which are more common than those of psychiatric origin. However, health is defined as "mental, physical and social well-being" (World Health Organization. [WHO], 1980). Each of these components is relevant in caring for the airline passenger. Effective reassurance for aviaphobic or cardiac related anxiety depends on crew training allied to personality, communication skills, and sensitivity. Predictably, all of these vary, notwithstanding conscientiousness and devotion to duty. Moreover, only recently have skilled dedicated healthcare professionals come to appreciate the importance of whole person and environment health, so relevant in flying.

Handling of an in-flight incident, be it interpersonal, health or aeroplane related, entails a coordinated approach by flight and cabin crews. Inevitably this is hampered by locked flight deck policy.

Sleep

Psychiatric morbidity can predispose to, and be aggravated by sleep deprivation. Stress, anxiety and depression are common causes, and the nightmares of posttraumatic stress disorder can contribute to insomnia. Long-haul flying can profoundly disturb sleep with its time-zone changes, jet lag, and often unsocial hours of departure and arrival (see more on jet lag in Waterhouse et al., in this text).

Anxiety

Anxiety is characterized by an over-reaction to manageable situations, worrying about possibilities rather than probabilities, dwelling on *worst case* scenarios and having a need for control, surrendered in flight. It is aggravated by infants crying, seats being kicked, or suddenly tilting. Turbulence may be perceived as the pilot transiently losing control.

Paranoid people thrown into a potentially threatening milieu can become suspicious, authoritarian, and perceive unintended criticism. These, plus the obsessional and hypochondriacal travelers, demand extra attention from cabin crew, cooperation from other passengers, and can react adversely if thwarted.

Pathological anxiety may involve lifts or heights, as well as fear of flying. Suffice to say that prophylactic or emergency use of a short acting anxiolytic such as Diazepam 1–10mg is effective. However, it is only indicated when all behavioral methods have proved ineffective. An intramuscular injection of diazepam is justifiable to avert a dangerous crisis, but it can only be administered by a healthcare professional. Fear of flying responds to Cognitive Behavioral Therapy (CBT), but a prolonged course may be required, although quicker improvement may be achieved. Bor's synthesis of cognitive behavioral therapy with brief solution-focused therapy meets time constraints; however, selection of candidates for such a program is key (Bor & van Gerwen, 2003).

Panic Disorder

Defined as two or more panic attacks, panic disorder is a separate entity, differing from generalized anxiety. It is associated with an awareness of autonomic nervous system activity. Severity varies, sometimes constituting an emergency.

Panic attacks are sudden, unexpected episodes of heightened anxiety lasting for seconds or minutes, causing fear due to misinterpreting palpitations and chest tightness as a heart attack, or hyperventilation as suffocation. Previously experienced in-flight stresses such as vomiting, or observing another passenger vomit, may trigger panic. Those who have had one panic attack during a flight are sensitized in subsequent flights, embarrassment compounding the situation. Autonomic activity is associated with the release of adrenaline. Features may include dizziness, sweating, palpitations, hyperventilation, feelings of unreality, loss of control, trembling, shaking, headache, hyper arousal or *going mad*. Catastrophizing about such symptoms creates a vicious circle.

Depression

There may be obvious psychosocial triggers such as bereavement, breakdown of a relationship, or job loss, but often there is no clear cause. It can occur in up to 25 percent of those with physical illness, and can lower the pain threshold. It is described by Cantopher as a common illness affecting strong people (Cantopher, 2003), and it is common throughout life. Pre-flight assessment should include objective accounts from key relatives, friends, and general practitioner if necessary. Features may include spontaneous weeping, withdrawn behavior, low mood, irritability, guilt, suicidal thoughts, impaired energy/cognitive function/ sleep/appetite and libido, loss of self-confidence, indecision, isolation. Excessive use of alcohol, caffeine and nicotine are counter productive short-term coping strategies, with adverse longer term effects. In flight inability to smoke causes tension, usually relieved by Nicorette or other nicotine-replacement therapy. Anxiety and depression frequently coexist. Both are common and frequently undiagnosed, therefore untreated.

Post-natal depression (PND) follows approximately 10 percent of otherwise uneventful pregnancies and deliveries, puerperal *psychosis* only affecting 0.1-0.2

percent. PND may reflect poor social support, but can occur in favorable circumstances. Features include low mood and guilt provoked by comments such as "Aren't you lucky to have such a lovely baby?" Feelings of inadequacy, irritability, tiredness, loss of appetite, libido, and self-confidence are common. Occasionally, there may be transient thoughts of harming the infant. Suicidal thoughts are uncommon (Cox et al., 1987). It is widely recognized that air travel with an infant or child is potentially stressful, even in first class when feeling well. Women who wish to breastfeed need tactful help and comfortable facilities. When available, it may be possible to offer a more secluded seat.

Psychoses

Media coverage usually involves manic-depressive illness, schizophrenia or drug induced psychosis. The malaria prophylactic mefloquine (Lariam) occasionally causes psychosis. Most patients are effectively stabilized by anti-psychotic medication and community monitoring and are capable of flying. However, a report on medication regime should be carried. The treating psychiatrist confirms whether a traveling companion, or even a psychiatric nurse escort, is indicated.

Embarrassment to the patient, or apprehension in other passengers, is to be avoided. Before embarking, compliance with effective medication regime is essential. In flight it should be taken at *the home time* rather than at that of the long haul destination. If agitated during flight, the dosage should be increased, with ground-based advice being obtained if necessary. If pre-flight assessment identifies active or incipient psychosis, travel should be postponed to allow for treatment. If travel is essential, then a specialized repatriation carrier service is mandatory to ensure appropriate equipment and psychiatric nursing staff. The following medication may then be considered; lorazepam 1–4mg, or haloperidol (10–30mg) per 24 hours. If indicated procyclidine (5–15mg)/24 hours for extra-pyramidal signs. In an emergency, these drugs can be injected intramuscularly, but only by a healthcare professional.

The Elderly Passenger

It is now common for the elderly to fly recreationally, but sometimes also for distressing reasons. Impaired cognitive function is aggravated by hypoxia, restricted mobility or dehydration. Confusion is aggravated by anxiety and depression. Dementia occurs in some 10 percent of people over 65. A Mini-Mental State Examination score of less than 23 is said to signify dementia (Folstein et al., 1975). *Pseudo-dementia* is depression mimicking dementia, usually eminently treatable with anti-depressants. An escort is often necessary to negotiate the airport maze, facilitated by a wheelchair.

Air Rage

There is nothing specific about air rage, which is basically resentment and anger within a very confined space and in close proximity to other people with no opportunity for escape or even avoidance. This differentiates it from any other anger, be it *workplace, trolley, pool*, or even *road* rage, where alcohol abuse is less common, and escape is feasible. In road rage, witnesses and victims are protected to some extent by their own car and may be able to drive away, not feasible for the airline passenger (*Problems of Passenger Interference*, 1998). It can be triggered and compounded by physiological factors. Those feeling threatened can become aggressive in self-defense, and verbal abuse can easily escalate into physical violence.

Passenger perception of this phenomenon is influenced by high profile media cases. Busy crew may miss early signs of abnormal, irrational behavior prior to a crisis erupting. However, passenger behavior is the key to safety, and a sympathetic, firm, but non confrontational approach is essential.

Excessive alcohol is one of the commonest precipitating causes of disruptive behavior. Its management should be regarded as a practical rather than essentially a clinical problem.

Available data confirms that seriously disruptive behavior is not a widespread problem on board UK aircraft, whereas low level antisocial behavior is identified (Bor, 2003). This constitutes stress to passengers, cabin, and flight crew.

Air rage is one of the most dramatic emergencies which may seriously affect others in proximity and compromise the aeroplane's safety. A study of all in-flight calls for ground based physician consultations to MedAire. Matsumoto and Goebert (2001) found that of 1,375 consultations, 3.5 percent (48) were psychiatric, 90 percent of these presenting as acute anxiety and 69 percent requiring assessment on arrival; three cases necessitated diversion and landing. The value of rapid-onset anxiolytics for on-board medical kits was recognized. Only two psychiatric calls involved psychosis; none involved "air rage." However, in another study 15 percent of emergencies were psychiatric (Rayman, 1998). 75 percent of disruption has been found to originate in economy class, 17 percent in business and 8 percent in upper (Anglin, Neves, Giesbrecht, & Kobal-Mathews, 2003; United Kingdom Department of Transport, 2005).

A spectrum of disruptive behaviors is manifested by passengers in flight. To date, systematic research is limited. However, identifiable causes include alcohol consumption, mental instability, and environmental stress (Anglin et al., 2003).

Pop stars, sports, and media personalities while attracting publicity, often escape prosecution. The captain can authorize all reasonable measures to counteract such behavior. These include restraint and recruiting help from other passengers, except flight deck crew, to ultimately disembark the passenger, and delivering him or her into custody is necessary. Police can deal with smoking or drunkenness under the Public Order Act 1986.

Fine (2002) described contributory factors including alcohol 58.25 percent, no smoking/nicotine craving or withdrawal 19.9 percent, arguments with attitude 15.05 percent; "mentally disturbed" 7.2 percent; Drugs (licit/illicit) 2.4 percent.

Psychological and physiological risk factors include fear of flying, particularly anxiety, panic, claustrophobia, proximity to strangers, disputes, lack of exercise, restrictive clothing, as well as hypoxia, dehydration, and the alcohol altitude syndrome (for additional insights on the bearing of physical factors on psychological functioning see chapter by Richards, Cleland, and Zuckerman).

Other Factors Contributing to Air Rage

Concourse queues, flight delay, lack of information, carry on (hand) luggage, seating assignment disputes, unsatisfactory food/complaints (Source: US Government Agencies), quality of in-flight movie, poor ventilation, changing nature of passengers and "lager louts" have all been found to contribute to air rage.

Fine (2002) analyzed 168 cases of passenger rage obtained from the Federal Aviation Administration. The primary factors identified were: alcohol intoxication, nicotine withdrawal, disputes with flight attendants, and psychiatric morbidity.

Bor's review (2003) suggests the problem is becoming less common because of stricter enforcement of rules, custodial sentences and self restraint since 9/11. The risk of any such incident being misinterpreted as terrorism is recognized, as is the possibility of vigilante response.

Abuse of Alcohol or Other Drugs in Flight

Alcohol is our favorite mood-altering drug. The vast majority of passengers use alcohol appropriately. Moreover, for many of those with fear of flying, its anxiolytic effect can be beneficial. Anxious, phobic or unhappy fliers drink alcohol as a universal tranquilizer, however, its excessive use is hazardous. Some airlines actually encourage drinking by advertising "Best Bar in the Sky" whilst some charter airlines profit from its sales. Alcohol is often consumed between check-in and boarding, and excessively so if the flight is delayed. Holiday mood, free alcohol and peer pressure all increase consumption. There are various patterns of drinking: occasional, binge, heavy and continuous due to psychological and or physical dependence. Occasional or binge drinkers often pose more of a threat in-flight due to unpredictable behavior and are less likely to have considered modification of intake or to have sought advice or treatment. The long-standing guidelines for pilots requiring minimum abstinence of eight hours "from bottle to throttle" might be considered for passengers prior to check in plus random breathylizing. Comparable restrictions are now accepted at sporting events, involving fewer risks. Despite much health "education," the prevalence of excessive drinking continues to increase, vested interests being relevant in the flying industry worldwide; but it is postulated that "dry" airlines do have fewer disruptive behavior incidents.

However, separating alcohol from other forms of drug abuse is illogical, because all such substances affect the brain. In sufficient quantity, behavior, mood, and level of consciousness are altered. Cognitive function and sleep are adversely affected. Confusion and disinhibition predispose to impulsivity and even violence. Substances may interact with each other and with prescribed and over-the-counter medication, including hypnotics and analgesics. Alcohol increases the diuretic and dehydrating effects of caffeine. Cannabis and other centrally acting stimulant drugs such as cocaine and ecstasy can cause or exacerbate psychotic illness with rapid onset. The physical and emotional stress of flying can increase vulnerable personality traits and underlying psychiatric morbidity. These are compounded by substance abuse, and nicotine withdrawal. Smokers subjected to obligatory airline-imposed nicotine withdrawal can become irritable, and nicotine replacement with therapy should always be available and recommended. Alcohol is often used inappropriately to counteract nicotine withdrawal effects.

Incidents of disruptive behavior are of significance, although constituting a very small proportion of total passengers carried world wide. The USA reported 125 cases of disruptive behavior in 1994 and 300 cases in 1998. In 1998, 43.4 percent of the cases of disruptive passengers in the United Kingdom (UK) were related to alcohol misuse (UK Department of Transport, 2005).

Airline Pilots Association Conference 1997 (USA) reported that 25 percent of passenger disruptive behavior and misconduct was alcohol related, in UK 42 percent. Therefore it appears that some 25–45 percent of cases are alcohol related.

Alcohol is more likely to predispose to other triggers rather than being the specific cause of disruptive behavior. In-flight alcohol compounds the effects of that consumed before boarding, particularly when taken on an empty stomach.

Prevention

It is important to identify those whose mental and or physical health may be adversely affected by air travel, or whose ill health could adversely affect the flight. Therefore, ahead of departure, passengers are required to inform airlines of any significant illness. The airline's medical adviser can then request clarification or further information from the passenger's family doctor or specialist, often by completion of the internationally recognized "MEDIF" form. Predictably, apart from flying-related anxiety, psychiatric illness is often not declared. Therefore, pre-flight identification of personality characteristics and underlying mental as with physical illness is fundamental, but rarely possible. Ground staff has a vital initial screening role, significant behavioral aspects being communicated to flying crew and flight dispatcher. A decision is then made regarding whether there is a specific psychiatric contraindication to flying, and if not, whether psychotherapeutic or medicinal intervention is necessary prior to flight.

Unfortunately, this was not achieved with a BA Nairobi flight in 2000 when a schizophrenic passenger attacked the flight deck crew. Significantly, concern had been expressed during an earlier connecting flight and during check-in.

When mental health impairment is declared pre-flight, assessment of mental state should include behavior, orientation, mood, cognitive function, thinking, perception, judgment, and suicidal risk. Whatever the underlying diagnosis, if the prospective passenger is stabilized and compliant with recommended psychotropic medication, there may be no specific psychiatric contraindication to flying independently. However, written details should be carried of anti-psychotic and other medication regimes together with enough medication for the journey in the hand luggage, advice on any dosage modification of routine medication, and of the indications for "as required" medication and of a clinical contact. Ground-based medical advice can be accessed when indicated. Appropriate transfer of care on arrival should be organized well ahead of the flight. Last-minute arrangements are potentially unsatisfactory.

Planning is the key, requiring notification ahead of flight. The airline's Special Assistance Department should report to their medical adviser, who accesses details from the GP, who can give prophylactic advice on coping with physical/psychiatric morbidity. As with the GP, psychiatrists should liaise with airport medical services provided they have formal permission. An escort who may be a friend, nurse or doctor may be indicated to supervise and to ensure administration of medication as required.

1. Prophylactic beta blocker or diazepam 1–5mg can be effective when a flight has to be made urgently. However, the use and effects of such medication should be "rehearsed." Notifying vulnerability on booking and at check-in facilitates the early establishment of a "therapeutic alliance" with the cabin crew who can initiate essential informal assessment. This ensures soonest identification of significant change in behavior.
2. Considering the relevance of alcohol induced disruptive behavior, its pre and in flight monitoring has to be considered, although highly contentious. Nevertheless, those whose behavior is inappropriate at check in or boarding may constitute a safety risk and consideration given to exclusion from the flight may be indicated. It seems logical that passengers should be liable to breathylization as are the flight crew. If such sanctions were included, in routine boarding instructions, it would be reassuring for responsible passengers (Anglin et al., 2003). Air rage is preventable. Its main triggers are alcohol consumption, mental instability, and environmental stress. Therefore perhaps there is a case for reducing alcohol availability in airports and aeroplanes.
3. Individuals with a known history of uncontrollable anger and violence should be required to undergo anger management and social skills training to facilitate self assertion without aggression; again, very difficult to implement. However, police exclusion already in place of undesirable football followers is relevant in this context.

4. Travel fatigue has two components; stress from physical and psychological aspects of the flight itself and that from the need to reset the biological clock (Waterhouse, Reilly, & Edwards, 2004). Advice on sleep hygiene including stimulant, diuretic, and dehydrating effects of caffeine should be given. Rehydration, gentle in-flight exercise, avoidance of heavy meals, and alcohol are also important.
5. Clear communication and the captain's reassurance, and explanation of reasons for delay in take off, or any relevant in-flight changes including weather details and of the nature of unexpected noise.

Psychiatric/Medical Emergencies

Air rage is the most dramatic, but there are a number of other flight related situations affecting an individual and potentially all on board the aeroplane – an isolated community in the sky. These include an initial, or recurrence of a panic attack, particularly in the first time flier, alcohol intoxication, drug induced or other acute psychoses, and confusion in the elderly.

Disruptive behavior of whatever origin in flight is extremely hazardous, and its prompt and appropriate management is essential for safety and to prevent its ripple effect. It is dangerous to mistake an essentially physical condition for a psychiatric crisis or vice versa, therefore, cardio respiratory illness must always be excluded, as well as alcohol intoxication, drug induced or other acute psychoses, and confusion in the elderly. Therefore, the possibility of any form of organic or underlying physical condition must always be considered. Curtis described a man of 39, incoherent, agitated who broke into the flight deck with an axe threatening the pilot. There had been headaches for one month, but no previous psychiatric history or criminal violence. He was delirious due to encephalitis. There was no prosecution (Curtis, 2000, October 20).

Managing In-Flight Psychiatric Emergencies

Inevitably, vital communication is increasingly difficult in larger aircraft such as the A380, as well as isolation of the flight deck. When dealing with the mentally unwell, simple clear language, and calm reassuring tones relieve mild anxiety or even a panic attack. Although psychiatric emergencies are rare, stress and anxiety are common. Uncertainty generates anxiety, hence the importance of clarifying any aspect of in-flight changes.

When a passenger becomes disturbed, disruptive or violent, separation is helpful when feasible, and reassurance and behavioral strategies are utilized. In addition, an in-flight call to a ground based physician may be indicated. Rayman (1998) described 15 percent of emergencies as psychiatric.

If an acute crisis persists, despite utilizing the full range of behavioral strategies, intramuscular diazepam administered by a healthcare professional can be invaluable. It is held in the onboard emergency medical kit.

Acute Stress Reaction can occur in response to exceptional physical and mental stress without any previous mental health impairment. It usually subsides within hours or days (WHO, 1990). Features include fear, anxiety, helplessness, anger, guilt, sadness, distressing thoughts, dreams, muscle tension, aches, dry mouth, and fatigue.

Crew or passengers who are directly or indirectly subjected to in flight verbal or physical violence can develop an acute stress reaction, and even PTSD.

Availability of individual or group "de-briefing" is indicated. However, it should not be imposed routinely following a flying related incident. The decision to access such an intervention should be the choice for each passenger or crew as for some it can be harmful. Therefore initially "basic psychological first aid" is preferable including reassurance and emotional support to confirm an understanding of the victim's feelings, attitude, mood, and behavior, plus explanation of the likely cause. Likewise, advice to avoid alcohol or illicit drugs to cope with the anxiety.

Due to enforced proximity, reinforcement of such in-flight traumas can persist throughout the flight even if the acute crisis is contained and satisfactorily resolved.

Posttraumatic Stress Disorder

PTSD involves a normal reaction to terrifying events causing personal or vicarious near-death experiences. It is characterized by re-experiencing the event, avoiding similar situations, and hyper arousal. Due to disturbing high profile media reports, minor flight events may seem life-threatening to the susceptible, thereby precipitating acute posttraumatic stress. This usually resolves spontaneously, but when it persists, PTSD can develop.

Also PTSD may ensue as a delayed response to a brief or prolonged stressful situation of an exceptionally threatening or catastrophic nature, likely to cause pervasive distress in almost anyone (WHO, 1990). In-flight incidents, particularly since 9/11 fulfill these criteria, hence the importance of passenger awareness and the availability of de-briefing should they wish to access. However, as with acute stress reaction, its automatic imposition is contraindicated and could have an adverse effect.

Cabin crew is a uniquely at-risk group in combating terrorism. Demographic characteristics and standardized questionnaires including the Postraumatic Stress Disorder checklist, the Psychotherapy Outcome Assessment, and Monitoring System-Trauma Version were sent to 26,000 American Airline cabin crew. Of the 2050 respondents, 18.2 percent probably had PTSD. Those living alone were 1.48 times more likely to have PTSD than those living with someone else. Age and years of service did not predict the diagnosis. Substance abuse was not endorsed as a coping strategy. Ongoing terrorist threats indicate the importance of availability for assessment and management of stress-related symptoms (Lating, Sherman, Everly, Lowry, & Peragine, 2004).

Predisposing factors to PTSD include vulnerable personality traits, previous psychoneurosis/in-flight crises, RTA, or childhood abuse. The National Institute for Health and Clinical Excellence having reviewed evidence base recommends trauma-focused cognitive behavioral therapy (TFCBT) and eye movement desensitization and reprocessing (EMDR) as the two first line treatments for PTSD (National Institute for Health and Clinical Excellence, 2005).

Conclusion

Passengers manifest a wide range of responses to flying related stress depending on personality, mental and physical health, age, and the specific circumstances. Since 9/11, the possibility of hijacking, increased security, and media reporting predisposes to anxiety which may be exacerbated by the events in the UK of 07/07.

The threshold is lower for what constitutes an emergency, and the behavior of those on board is inevitably subject to greater scrutiny, however, tactful and reassuring.

Nevertheless, air travel is still safer than that by road, rail or sea, due to the professionalism of ground staff, flight deck, and cabin crew. Their expertise prevents or diffuses the vast majority of critical incidents. Identification and sensitive, but appropriately assertive intervention, are key in resolving such potentially hazardous situations. Unacceptable confrontation is to be avoided.

Education about air travel is important from childhood onwards (Nicholson & Cummin, 2002). This ensures the important understanding of the whole process from checking in to landing.

Matsumo's and Goebert's (see Lucas, 2002) figures may reflect the fact that world media report in-flight psychiatric emergencies as violence threatening to compromise flight safety, and necessitating unscheduled diversion to off load, is initially dealt with legally. Subsequent diagnosis of acute mental ill health may not be reflected statistically.

Maintaining harmony is facilitated by an awareness of whole person health, and by achieving a balance between service, caring, and safety. The latter being the priority for all on board.

Interfacing with demanding and potentially aggressive passengers constitutes considerable stress for cabin and flight crew. Fortunately, this is now recognized in Crew Resource Management Training.

Anxiety is the commonest form of psychiatric morbidity, and it is generated by uncertainty. Nowhere is this more relevant than in an aeroplane, a very small isolated community of strangers, where control has been surrendered, even of the air you breathe.

References

Anglin, L, Neves, P., Giesbrecht, N., & Kobal-Mathews, M. J. (2003). Alcohol-related air rage: From damage control to primary prevention. *Journal of Primary Prevention, 23*(3), 283–299.

Bor, R. (2003). Trends in disruptive passenger behavior on board UK registered aircraft 1999-2003. *Travel Medicine and Infectious Diseases, 1*, 153–157.

Bor, R. & van Gerwen, L. (eds.) (2003). *Psychological perspectives on the treatment of fear of flying.* Hampshire: Ashgate.

Cantopher, T. (2003). *Depressive illness: The curse of the strong.* London: Sheldon Press.

Cox, J. L., Holden, J. M., & Sagovsky, R. (1987). Detection of postnatal depression: Development of the 10-item Edinburgh postnatal depression scale. *British Journal of Psychiatry, 150*, 702–706.

Curtis, K. (2000, October 20). *Experts agree encephalitis prompted Alaska Airlines attack.* Associated Press release 17:14 PDT San Francisco. Retrieved on November 4, 2005 from http://www.sfgate.com/cgi-bin/article.cgi?file=/news/archive/2000/10/20/state1517EDT0155.DTL

Fine, E. W. (2002). Air rage: Implications for forensic psychiatry. *American Journal of Forensic Psychiatry, 23*, 1, 29–44.

Folstein, I., Folstein, S. E., & McHugh, P. R. (1975). Mini mental state: A practical method of grading the cognitive state of patients for the clinician. *Journal of Psychiatric Resources, 12*, 189–198.

Goodwin, T. (2000). In-flight medical emergencies: An overview. *British Medical Journal, 321*, 1338–1341.

Gordon, H., & Goodwin, T. (2004). Air travel by passengers with mental disorder. *Psychiatric Bulletin, 28*, 295–297.

Lating, J. M., Sherman, M. F., Everly Jr., G. S., Lowry, J. L., & Peragine, T. F. (2004). PTSD reactions and functioning of American airlines flight attendants in the wake of September 11. *Journal of Nervous Mental Diseases, 192* (6), 435–444.

Lucas, E. G. (2002). Mental health: Pre-flight and in-flight. In A. N. Matsumoto, K. & Goebert, D. (2001). In-flight psychiatric emergency. *Aviation, Space and Environmental Medicine, 72*, 919–923.

National Institute for Health and Clinical Excellence (2005). *Post-traumatic stress disorder: The management of PTSD in adults and children in primary and secondary care.* London: Department of Health.

Nicholson, A. N. & Cummin, A. R. C. (eds.) (2002). *Aviation medicine and the airline passenger.* London: Arnold Publishing.

Problems of passenger interference with flight crews and a review of H.R. 3064, the Carry-on Baggage Reduction Act of 1997: Hearing before the Subcommittee on Aviation of the Committee on Transportation and Infrastructure, House of Representatives, One 105th Cong., 2 (June 11, 1998) (testimony of Captain Stephen Luckey, Chairman, National Security Committee, Airline Pilots Association).

Rayman, R. (1998). In-flight medical kits. *Aviation Space and Environmental Medicine, 69*, 1007–1010.

United Kingdom Department of Transport (2005). Disruptive passengers on board UK aircraft: April 2004–March 2005. Retrieved on November 4, 2005 from http://www.dft.gov.uk.

Waterhouse, J., Reilly, T., & Edwards, B. (2004). The stress of travel. *Journal of Sports Sciences, 22*, 946–966.

World Health Organization (1980). *Health aspects of well-being in working places* [Euro Reports & Studies 31]. Report on a WHO working group, Prague, September 8–20, 1979.

World Health Organization (1990). *ICD-10 classification of mental and behavioral disorders*. Geneva: Churchill Livingstone.

Chapter 5

The Nature, Characteristics, Impact, and Personal Implications of Fear of Flying

Elaine Iljon Foreman, Robert Bor, and Lucas van Gerwen

Holidays out of this world will soon be available, going to one of the countless worlds out there. Space is no longer the Final Frontier, and only the domain of the star ship *Enterprise*. Science fiction has now evolved into reality. Holidays which boldly go where no man has gone before, are waiting to be reached for in the sky. On 4 October 2004, the Ansari X Prize of one million dollars was won by *Space Ship One*, the first privately funded space craft to reach Space (65 miles) twice within a fourteen-day period. Richard Branson stated that he expects 3,000 people could fly on *Space Ship One* within five years. Yet there are a large number of people for whom, even were finances not a consideration, would be unable to take advantage of these exciting new developments in aviation history – even a short domestic hop is too terrifying to contemplate.

Implications

The consequences of a fear of flying can be far reaching. It can limit the person's professional opportunities, affect leisure options, disrupt personal relationships as one partner may decide to take holidays without the other; and of course prevent people from participating in life's core rituals, such as attending a wedding or funeral. The problem can therefore have a profound impact on professional, social and family life, and can substantially affect marital or relationship satisfaction because a fear of flying hampers or restricts one partner's freedom of movement (Iljon Foreman, 2003; van Gerwen & Diekstra, 2000) and therefore destabilizes relationships (Bor, 2003). For those who manage to fly in spite of their fears, Greist and Greist (1981) report that 20 percent of fearful flyers who travel by air use alcohol or sedatives to cope with severe anxiety, and this finding is reported within the vast majority of the literature reviewed. Individuals affected by fear of flying can be divided into three sub-groups – those who avoid all flights, those who restrict flying to an absolute minimum, and experience considerable discomfort prior to and/or during each flight, and those who show continuous mild or moderate apprehension about flying, but do not avoid it, even though it remains an unpleasant experience (Ekeberg, Seeberg, & Ellertsen, 1990).

Substantial costs are incurred through the fear of flying. In 1982, average revenue loss for the airline industry through fear of flying was estimated at $1.6 billion (Roberts, 1989). Looking at it from the perspective of the sufferer, the cost to the individual for lost productivity and opportunities is perhaps incalculable (Wiederhold, Gevirtz, & Spira, 2001). These substantial costs may account, to some degree, for why, despite the high levels of reported fear cited in the studies of prevalence, apart from the automobile, air travel remains the most popular form of transport. The increasing popularity of the latter is illustrated in that overseas flights by British residents rose from 5.9 million in 1971, to 16.5 million in 1986 (Iljon Foreman & Iljon, 1994). By 2001, the UK Civil Aviation Authority reported that 180 million passengers used UK airports. These figures are reflected internationally, with a report from the European Airline Association stating that two billion people worldwide traveled by air in the year 2000. Looking at the recent growth in air travel, Henderson (2002) of the European Airline Association (EAA) states that in the last 40 years, the European air travel market has doubled in size five times – that is to say in 2000 it was 32 times bigger than in 1960. The last doubling cycle, to the year 2000, took 10 years, the previous one, 12 years. While there were three earlier periods of marked downturn, at the time of the Gulf Wars in 1999 and 1991, and in 1986 following the Chernobyl disaster and the bombing of Tripoli, in these cases, growth quickly resumed, and the lost market was regained. Any study of air travel cannot fail to refer to the shocking events of September 11, 2001. In the following 24 weeks, EAA carriers lost over a quarter of their North Atlantic traffic, and more than 10 percent of their short-haul, with overall volume down 15 percent compared to the previous year. By February 2002 North Atlantic air travel had marginally improved, but was still 10 percent down, while European routes gradually produced a small growth for the first time, and even Asia-Pacific routes rose from approximately a quarter down to near-previous year levels. It has been predicted that in the coming two decades, the number of passengers will double.

Prevalence

Looking at those who are reluctant to fly, estimates indicate that up to 40 percent of the population report a fear of flying (Bor & van Gerwen, 2003). Epidemiological studies have reported a wide range of point prevalence, from 10–25 percent of the population (Dean & Whitacker, 1982), 10–30 percent (van Gerwen & Diekstra, 2000), with the disorder being found by Frederickson, Annas, Fischer and Wik (1996) to be twice as common in women as in men. A recent study in Germany revealed that 15 percent of the population have a fear of flying, and that an additional 20 percent are apprehensive while flying (Muhlberger et al., 2001; Institute fur Demoskopie Allensbach, 1995). Another recent study in the Netherlands revealed that 16 percent of the adult population refuse to fly because of a fear of flying (van Gerwen, 2004). It has been reported by Capafons, Sosa, and Vina (1999) that 45–50 percent of the population suffer anything from a slight discomfort or apprehension to a very intense fear, and about 10 percent suffer from such a high degree of fear or anxiety that they totally avoid flying. Despite the belief that if someone is afraid of doing something, they will avoid it, Greco (1989) cited a study indicating that one in every four flyers shows a significant degree of fear or anxiety. This high prevalence is paralleled by the report of van Gerwen (2004) who suggests that 30 percent of the Dutch population are afraid of flying. In a recent survey of travelers in Scotland, McIntosh, Swanson, Power, Raeside, and Dempster (1998) found that 40 percent of their sample were worried by take off and landing. The high prevalence seems to be an international phenomenon although hardly any research from resource-poor countries has been published in the literature. People who can be affected by a fear of flying range from those who have never flown before, to frequent flyers, and also include both civilian and military aircrew (Carr, 1978; Dyregrov, Skogstad, Hellesoy, & Haugli, 1992; Tempereau, 1956).

Nature and Characteristics

In the early years, a reluctance to fly was regarded as a normal human attitude, and not as evidence of a mental disorder (Jones, 2000). On the contrary, those who wanted to fly were regarded as the ones whose sanity might be doubted. Fear of flying can be a symptom that may be a product of an acute or posttraumatic stress disorder, a generalized or phobic anxiety disorder or part of some other major or minor psychiatric condition. Fear of flying is classified in the Diagnostic and Statistical Manual of Mental Disorders (DSM-IV) (American Psychiatric Association, 1994) as a specific phobia that arises in a defined situation. It is characterized by a marked, persistent, excessive fear that is precipitated by the experience or immediate prospect of air travel. Exposure to this phobic stimulus almost invariably provokes an anxiety response – sometimes to the point of a panic attack – which the individual recognizes as unreasonable, and which produces significant interference or distress. It is in a different category to the other non-situational phobias, which are categorized as social or agoraphobias. In the previous edition of DSM, (DSM III-R) (American Psychiatric Association, 1987), fear of flying was classified as a simple phobia, more akin to spider phobia, or needle phobia. There is some concern among clinicians that the existing diagnostic and classification systems such as DSM IV and ICD-10 do not appropriately acknowledge the diverse nature, aetiology and types of fear of flying. It is a heterogeneous, not a unitary phenomena. Considerable thought therefore needs to be given to the classification within the upcoming DSM-V.

The nature of the avoidance behavior can be very subtle, and not easily distinguishable by others who do not share the fear. On board avoidance can be apparent in a special preference for a particular seat. Some people wish to be near the front, or near the window, while others try and sit as far as possible from the window. Certain people always try and book an aisle seat, or to be by the emergency exit ostensibly for additional legroom, but really to facilitate a quick escape. Some people with flight anxiety are afraid to walk about or even move because they fear they will unbalance the plane, and sometimes they literally close their eyes and block their ears to try and be as minimally aware as possible of the flight. Others are utterly alert and focus on every change in movement or noise, and experience matching high levels of muscle tension. Aside from the avoidance strategies, most people with a fear of flying will not eat on the flight, and tend to retreat into themselves. These behaviors also apply to crew who fear flying, when they are traveling as passengers.

The heterogeneity of this disorder is further illustrated by the finding that people suffering from a fear of flying quite frequently report the presence of other psychological difficulties. Wilhelm and Roth (1997) reported that 44 percent of their sample had met criteria for current panic disorder with agoraphobia, or had done so in the past. This group was more concerned with internal or social anxiety. Those who were phobic in their study did not have this worry, but both groups were equally concerned regarding the worry about external danger. Iljon Foreman and Borrill (1994) found that 60 percent of their sample reported "other fears," besides that of flying. These included a diverse range such as a fear of spiders, swimming, being outdoors, enclosed spaces, heights, falling, social embarrassment, crowds, collapsing, and a fear of "being under some one else's control." In a sample of fear of flying patients, 39 percent were reported to have an additional diagnosis of "personality pathology" (van Gerwen, Delorme, van Dyck, & Spinhoven, 2003). Furthermore, the association between fear of flying and other psychiatric disorders was found in 46 percent of travelers with a fear of flying who had other phobias; 33 percent present with agoraphobia, and 25 percent with claustrophobia (Dean & Whitaker, 1982).

It is generally agreed that fear of flying is not a unitary phenomena, but consists of various underlying fears. These may meet the criteria of identifiable psychological disorders, or may just be factors which apply only to the situation of air travel. Six separate fear categories have been identified, in the following order of significance: crashing 52 percent, heights 23 percent, confinement 18 percent, instability 11 percent, panicking 5 percent and lack of control 5 percent (Howard, Murphy, & Clarke, 1983). Dean and Whitaker (1982) reported that out of 562 fearful flyers they surveyed 29 percent had a fear of dying, 24 percent a fear of heights, and 7 percent feared bad weather. The heterogeneous nature of fear of flying has also been emphasized by Beck, Emery, and Greenberg (1985) who highlighted the following additional fears expressed by people who sought treatment for fear of flying: fear of suffocation due to deprivation of air, subjective tension and loss of control, crashing and death, loss of control in social situations, vomiting or fainting and the subsequent humiliation, agoraphobia, being trapped in an enclosed space, being separated from a caretaker and experiencing a serious disorder such as a heart attack. It has been suggested that people are less bothered by a fear of heights, the plane crashing, or even dying, than they are about experiencing negative feelings, perceived loss of control, and what others may think of them (Heller, 1993).

Specific phobias can be grouped into three categories – situational phobias, animal phobias, and mutilation phobias (Rosenhan & Seligman, 1989). The difficulty with classifying fear of flying as a specific phobia is that this does not clarify what it is about flying which frightens the person. It could as easily fit into the situational as the "mutilation" categories, depending on what it was that the person most feared.

As one person seeking help for their fear of flying succinctly put it: "Flying? That doesn't bother me at all! CRASHING? Now THAT bothers me!!"

Based on clinical assessment and standard questionnaires, analysis of a group of fearful fliers found that the fears could be divided into two separate sub-categories, with a few people experiencing fears from both categories (Iljon Foreman & Borrill, 1994). The first group reported fears that concerned a loss of internal control. These subsumed all those fears relating to social anxiety, panic disorder, claustrophobia, and agoraphobia. In this first group, the person fears some form of internal catastrophe, where in some way they will go out of control. They are unable to employ the strategy of escape, and therefore remain terrified of the frightening prospect. The second group report a fear of a loss of external control – something happening to the plane. The latter fear encapsulates heights, turbulence, bad weather, and all of the precursors to crashing. The third and smallest group have both fears – a loss of internal and of external control.

This finding is consistent with that of McNally and Louro (1992), who examined fear of flying, agoraphobia and simple phobia in 34 people. They concluded that the distinguishing features were that danger expectancies (loss of external control) motivate flight avoidance in simple phobia, whereas anxiety expectancies (loss of internal control) motivate flight avoidance in agoraphobia. It is therefore not so much panic per se that differentiates simple phobias from agoraphobia, but rather the fear of panic (Goldstein & Chambless, 1978). This view is supported by the findings of Wilhelm and Roth (1997) who found that those with panic disorder and agoraphobia were more concerned with internal, or social anxiety, while simple phobics did not have this worry, but all were equally concerned regarding external danger. Likewise, Howard, Murphy, and Clarke (1983) found that specific flying phobics fear crashing, while agoraphobics fear panic attacks.

As the next chapter on the treatment of fear of flying discusses (see chapter in this text by Iljon Foreman, Bor, & van Gerwen), the treatment of fear of flying can generalize to other phobias (van Gerwen & Diekstra, 2000; Iljon Foreman & Borrill, 1994). Many participants reported that they had successfully used therapeutic techniques learned in the treatment of their fear of flying to conquer other difficulties. Other studies have also reported this (Botella, Villa, & Banos, 1999; Denholtz, Hall, & Mann, 1978; Iljon Foreman, 2003). However, as is frequently the case with complex psychological problems, the opposite has also been reported. Scrignar, Swanson, and Bloom (1973) reported that if the patient had multiple anxieties, or a pre-existing psychiatric history, the poorer the prognosis. Likewise, van Gerwen et al. (2003) contend that in general, it has been presumed that treatment outcome is negatively influenced by the presence of personality pathology.

What is feared?

One of the most detailed studies of what patients most fear is that of van Gerwen et al. (1997) who carried out an analysis of data from 419 patients referred to their

Netherlands-based treatment agency because of fear of flying. They examined the nature of the fear and found that female patients stated "fear of being involved in an accident" as their primary fear, while males stated "not being in control." Using the conceptualization of Iljon Foreman and Borrill (1994), both genders can be seen to be reporting the same thing – a fear of an external loss of control. When considering the second reason given for the fears, women reported a "fear of confined spaces/claustrophobia" (lifts, underground travel, tunnels), while men cited a "fear of losing control over themselves" (crying, fainting, going mad, heart attack or heart palpitations). Once again, rather than this reflecting gender differences, both of these fears reflect the same category of a fear of a loss of internal control. Thus, both males and females report a loss of external control as their primary fear, and a loss of internal control as the second reason for their fear. The phobic fears that are most specifically associated with high levels of flight anxiety are claustrophobia, fear of water and fear of heights. This is to be expected as these are central to most people's experience of flying – one is in an enclosed space, we often fly over water and height or elevation is axiomatic to flight.

While there is consensus as to the heterogeneous nature of the fear of flying, there is no agreement as to the nature of the underlying and the associated fears (Moller, Nortje, & Helders, 1998). But what specifically is a fear of flying? The person's decision to fly, or not to fly seems to reflect a key factor in the way judgments are made. As emotional beings, we formulate responses on the basis of the perceived risks, and not the true risk of an activity. Most fearful fliers are aware that one has a greater chance of dying in a car crash on the way to the airport than during the flight itself. The following table compiled by Greco (1989) on the probability of a person coming to harm in different situations makes for fascinating reading.

As can be seen, one has a greater statistical chance of dying if one avoids flying and stays at home, than if one were to take the flight and be killed in a plane crash.

Table 5.1   Danger of Flying, Reported by Mode

Danger of flying in relation to other modes of transport or situations in the U.S.A.		
Mode of transport/ situation	Number of deaths per year in U.S.A.	Comparative safety of airline travel
Car	45,000	29 times safer
Walking/being a pedestrian	8,000	8 times safer
Staying at home	20,000 accidental deaths	18 times safer
Working on the job	11,000 accidental fatalities	10 times safer
Homicide by spouse or relative	7,000 homicides	6 times safer
Bus		4 times safer per mile
Train		4 times safer
Boating		8 times safer

Source: U.S.A. Department of transport Document. Greco (1989)

Given the odds of 1 in 14 million to win the UK National Lottery, it is sobering to realize that one is more likely to be dead by the end of the week, than to have won the lottery.

Causes

Despite a lack of consensus among researchers and clinicians, many have attempted to grapple with the underlying causes of the fear. Early writings on fear of flying from 1920–66 emphasize the internal, unconscious processes and mechanisms behind the fears. Morgenstern (1966) asserts that it is a reflection of the pervasive dualism and of man's feelings when neurotic illness causes the metamorphosis of an intense need to fly into an equally strong dread of flight. Freud's (1960) theory also suggests an underlying cause, whereby the aeroplane itself might represent the displacement of a strong figure of threat or desire in the person's internal world. Separation anxiety is given as an underlying cause (Shneck, 1989). From an early behavioral perspective, other authors have proposed that the fears can be seen as a conditioned response to an aversive experience (Watson & Rayner, 1920). Taking the cognitive behavioral perspective, a pattern of avoidance behavior is set up which reinforces the anxiety and prevents the possibility of testing, and invalidating, the feared predictions of future catastrophes (Greenberger & Padesky, 1995). Building on this view, Wilhelm and Roth (1997) propose that their results support a vulnerability-stress model, with flying phobia developing in people who were more susceptible to events that had little impact on non phobics. They suggest that flight phobia begins for many people with a rise in anxiety while flying, either triggered internally, or by a transitory overreaction to a minor external event. This results in direct conditioning of a phobic response to flight stimuli. They further propose that specific vulnerabilities of various kinds present at phobia onset may have promoted this process, and add that cognitive biases could have played an important role particularly in the initial progression and maintenance of the phobia. Williams (1982) proposed that fear of flying actually represents a difficulty in communication. The phobic person, he hypothesized, expresses a different message by refusing to fly. Examples include the child who does not want to return to boarding school, or the partner who resents being repeatedly uprooted to follow their spouse's promotion trail.

From another perspective, Bakal (1981) highlighted the nature of the language in use at some airports – last and final call for a particular flight, a flight terminating at its final destination, – the airport terminal and the departure lounge. He reported that it is interesting that correspondence with a number of the major airlines, in which the improper terminology was pointed out, resulted in appreciative letters from all except one. Despite this apparent appreciation, he reported that in over a year, no changes in terminology had been noted.

Air Crew Fear of Flying

It is not only passengers who can have a fear of flying, but also the air crew themselves. There is a need to differentiate between passengers and crew who have a fear of flying. Thus far in this chapter, the conceptualization and treatment methods for fear of flying has applied to passengers. There are clear medical guidelines as to how fearful crew members should be managed (see also the next chapter on treatment of fear of flying in this text). While it may seem counter-intuitive to consider the possibility of aviation personnel suffering from a fear of flying, the literature and the authors' experience reveals that this can indeed be the case.

As long ago as 1919, Anderson coined the term "aeroneurosis," to refer to the development of the reluctance to fly in World War I pilots. He noted that the disorder could be precipitated by the strain of learning to fly, the psychological trauma of a plane accident, (which was a frequent occurrence) or the physiological and psychological stress of combat flying. In the same year, Gotch explored the aetiology of aeroneurosis, suggesting that mental and physical exhaustion, disease, conduct disorders, traumatic physical experiences and malingering could lead to aeroneurosis. The disorder has been given a range of descriptive names, some less than complementary. They include: War Neuroses, Lack of Moral Fibre, Fear of Flying, Aviation Phobias, Flying Phobia, Aerophobia, and Aviaphobia.

It is important to discriminate between those studies which have examined fear of flying in military aircrew during wartime and peace, and civil aviation crew. Perhaps unsurprisingly, the incidence of fear of flying in Aitken, Daly, Lister, and O'Connor's (1971) study revealed higher rates of fear of flying in bomber crews, and the prevalence was closely related to higher casualty rates. They suggest that flying phobia in military air crew has been well documented; after all, air crew have chosen an occupation in which there is constant danger. It is not easy to define what brings about or triggers anxiety or a phobia in air crew who face real danger in their work. Anxiety can only be considered neurotic when previously it was absent and when it is not present in healthy colleagues. Conversely, avoidance need not be neurotic when it results from an intellectual decision after consideration of the risks.

Despite strenuous efforts to select only those who will succeed in training, O'Connor (1970) reported that 25 trained air crew were permanently taken off flying in the RAF every year as a result of psychiatric illness, and this despite a selection process which rejected 98 percent of initial applicants. An American study (Benson, 1985) reported that there is an overall 30 percent attrition rate from the US Naval flight training programme.

Even under non-combat conditions, fighter pilots have been reported to have an accident rate 8 to 20 times greater than pilots of transport planes (Bucove & Maioriello, 1970). In the two years before their study, in June 1965, no fighter pilot was judged to have a fear of flying at the base studied. The massive American involvement in Vietnam began in the summer of 1965, and all of the fighter pilots who had not yet flown in Vietnam could be expected to be assigned there after completion of duty at the base. All the fliers knew pilots who had been killed or captured in Vietnam.

The vociferous attacks against US military involvement in Vietnam in general, and the use of air power in particular, both on moral and strategic grounds, had been well publicized and discussed at the base. It is reported that a history of fear of heights was present in each of the fliers. However, a fear of heights was not deemed to be a useful predicting factor in fear of flying, as a random sample of seven first lieutenant and captain pilots revealed four had a history of fear of heights. Bucove and Marioriello (1970) conclude that the symptoms were not to be understood and treated as illness, but were better understood as a communication: "I don't want to fly jet fighters anymore," and that these symptoms had their own reward. It is interesting to compare this to the assertion by Marks, Yule, and De Silva (1995) that with reference to malingering and compensation neurosis, the literature suggests that compensation claims do not generally lead litigants to exaggerate symptoms, and symptoms do not suddenly remit once claims have been settled. This is supported by McCarthy and Craig (1995), who state that malingering or dissembling for secondary gain is seemingly rare.

Further confirmation of development of fear of flying in non-combat military personnel is provided by Fowlie (1999), who found that 40 percent of Royal Air Force officers who had survived ejection from an aircraft subsequently experienced prolonged emotional disturbances. 28 percent flew again, despite having significant fears, apprehensions, resentment, and anger. Appropriate counseling and psychological support can reduce prolonged emotional morbidity. Those flying ejection-seat aircraft are trained to cope with the obvious dangers of combat, but are also exposed to daily risks in peacetime while operating sophisticated high performance aircraft, frequently at low level. Fear of flying may not necessarily be linked to combat duties, as is demonstrated by Jones (1986). He asks the thought-provoking questions: "What motivates people to do something as dangerous as flying?," and "Why, after flying successfully for a number of years, does an accident-free aviator unexpectedly become unbearably anxious about flying?" He continues when one asks a flier "What do you think about the dangers of flying?" the most likely answers are "Well, you can get killed just crossing the street," or "You can get killed driving on the freeway." He ponders whether young pilots may be truly fearless because they do not understand the dangers of flight, or because they can consider them only as abstractions. He suggests that as these realities are brought home to them early in their flying career, through their own near misses, or through the deaths of their friends in aircraft accidents, the fears become part of their emotional lives and must be dealt with differently. Those pilots must move from the fearlessness of those who do not understand the truth of the matter to the courage of those who understand it well and who choose even so, to continue to fly. Jones (1986) proposes that most experienced pilots deal with their feelings about these real dangers by coping mechanisms which usually include a combination of denial, humour, suppression, intellectualization, and rationalization. He then continues to be provocative by proposing that "Perhaps one should not ask 'Why do some fliers become afraid to fly?' But rather 'Why are not ALL fliers afraid to fly!'" When pondering "Why do they do it" he concludes that one cannot underestimate the joy of flight. This joy does not apply to the passenger who is carried, rather to the flier who flies. Jones refers to the wish to achieve a sense of mastery of the sky above – to achieve freedom in space, and to grasp power over time itself.

It is not only military pilots who can suffer from a fear of flying. The authors' experience has shown that sufferers can include the range of aircrew – pilots working for an airline, general aviation pilots, and also cabin crew. Marks, Yule, and De Silva (1995) describe the responses and psychological sequelae for cabin crew following a disaster in which 47 people were killed, and 74 of the 79 survivors were seriously injured. The highest levels of distress were reported by the three most senior members of cabin crew with the most responsibility on board, who had also suffered the most severe physical injury. Aircraft crashes differ from natural disasters in that they are often caused by human error. Survivors may have to contend with feelings of anger and a sense of futility stemming from the fact that a potentially preventable disaster has occurred (Butcher & Dunn, 1989). The literature reflects this, indicating that there is an increased likelihood for the development of posttraumatic stress disorder following man-made versus natural disasters (Iljon Foreman, 2004). In the case of the airline crash, airline staff may have to cope with the additional stress of working in an environment in which blame resides. This suggestion is supported by Fowlie (1999) who found that there were a number of factors associated with an adverse emotional consequence for pilots who ejected from their aircraft.

In one study, it was found that following the accident, all crew members reported developing a fear of flying as well as of other modes of transport, particularly when being a passenger (Marks, Yule, & De Silva, 1995). Most reported being unsure of their future career prospects due to their physical injuries and fear of flying. All had accepted psychological help arranged by the airline after the crash, but generally perceived the company to be unsupportive and not to have understood the stress they were under. They felt this had exacerbated their difficulties. Once again, psychological interventions with fighter pilots have reflected this. Fowlie (1999) stressed the importance of fighter pilots getting the understanding and support from their close and senior colleagues as well as from clinicians. At follow up, Marks, Yule, and De Silva (1995) reported that five of the crew involved in the incident had left their original employer and had accepted jobs with different airlines. They suggest that crew form a special sub-population among the survivors of a crash, and that it is important that their responses be especially targeted and studied. Follow up at 18 months indicated that the crew still had all the major symptoms of posttraumatic stress disorder, and this points to the potential for the psychological effects of the experience to become chronic, despite the therapeutic input provided, in this and similar groups. The need for specialist psychological support for crews in such disasters as this at an early stage has been made (see Chapter 7 by Chung in this text) both in the short term as well as up to many months, even years, after the event. Crew involved in adverse incidents may express anger at the assumption that because they were members of staff, they should be able to cope with minimum distress. The presence of both practical and emotional support from employers and recognition of the traumatic effects of being involved in a life threatening event, therefore, seem important in facilitating recovery.

At an individual case study level, within clinical practice, crew members may present for therapy to overcome a fear of flying. In one case, a female flight attendant had recently had her first child. Shortly after returning to work, her aircraft was involved in what she understood to be a "close call." She began to ruminate on what could have happened, if the danger had not been averted. The more she thought about it, the more she found herself alarmed and distressed that her child could grow up not knowing his mother, should harm befall her. Her concentration became adversely affected, as did her ability to work, and by the time of referral, she had been medically grounded. Additionally, many presentations of cabin crew seeking help for their fear of flying have reported the fear developing soon after they had a child, but without any report of a traumatic air-related incident. Another crew member who sought psychological help for the development of a fear of flying reported the onset of the fear as following the death of a friend and colleague in an air disaster. The underlying fear was of the "it could have been me" variety, and ruminations started to preoccupy the person to the point where they could not perform their duties at work.

Emerging evidence from studies of crew members, who present with a fear of flying, points to the presence of co-morbid psychiatric problems. In one recent report of fear of flying among 1,101 Spanish air crew between 1985–2000, the researchers reported that two thirds of those who were treated for fear of flying were cabin attendants, while a third were pilots. In more than half of the total number of cases of fearful fliers, there was a pre-existing psychiatric disorder (e.g. depression, anxiety, and personality disorder). In this sense, fear of flying can be viewed as a symptom or outcome of another psychological problem (Medialdea & Tejada, 2005).

With regard to the treatment of aircrew, this too largely parallels the civilian findings to be reported later in this chapter. Actual exposure to flying is usually necessary for aircrew to recover from anxiety associated with flight (McCarthy & Craig, 1995), a view entirely consistent with the clinical experience of the authors of this chapter. McCarthy and Craig state that the treatment of the underlying psychological problem may be sufficient for the crew member to return to full flying duties.

A history of childhood phobias, presence of other adult phobias or family history of flying trauma suggests a greater likelihood of aircrew to develop flying phobia. Careful identification of vulnerable aircrew can help to prevent the onset of fear of flying if co-factors that increase susceptibility are taken into consideration at times of routine medical screening for licensing. Once vulnerable crew have been identified, supportive care could be offered to prevent the onset of symptoms, particularly if the crew members are to be posted abroad, or have worries concerning personal relationships which may be linked to the development of fear of flying. Any intervention should be aimed at reducing the general level of arousal by cognitive behavior therapy, pharmacotherapy, psychotherapy, or marital counseling. Information could be given about the arousing effect of ambiguous visual cues, and its influence on somatic symptoms, recommending control of breathing to reduce possible hyperventilation as a self-operated procedure. This could be related to improving the general level of tension, fatigue, and mood, and overall social adjustment. With further reference to preparatory training prior to difficulties arising, Butcher and Dunn (1989) recommend specific pre-accident training for cabin crew to enable them to perform and recover to a higher level in the event of a disaster. These views are also supported by those of Timm (1977), although sadly are seldom implemented. He provides a brief discussion concerning the potential usefulness of routinely incorporating relaxation exercises into flight training as a preventative approach to anxiety and stress related illness frequently found in aviators.

Conclusion

One is left to consider the case of the flier who talks with regret of symptoms – "I'd like to fly, but..." Jones (1986) suggests one may wonder whether the person is sick, or irrational, or malingering, or has he simply experienced an "acute onset of reality?"

With the increasing importance of commercial aviation, Strongin (1987) asserts that passengers' fear of being flown has been added to the list of disorders bearing the fear of flying label. He further states that the psychology of the passenger's fear may somewhat resemble that of the flier's, but that the implications and motivations of passengers differ from those of fliers and require a separate review. While there are certain differences in those suffering from a fear of flying, dependent on being civilian passengers, being military or civilian aircrew, or crew traveling as either crew or passengers, there are some common elements, events and personality factors present for those suffering from a fear of flying across all three groups, with certain elements that are unique to just one individual.

Though there is not yet a consensus on the cause or the nature of fear of flying, given the enormous negative impact this problem can have, it is heartening to know that a number of highly effective treatments have been developed. These are presented in the next chapter on the psychological treatment of fear of flying (in this text). So those currently suffering from "high anxiety" may yet be freed from their terror in the skies.

References

Aitken, R.C.B., Daly, R.J., Lister, J.A., & O'Connor, P.J. (1971). Treatment of flying phobia in aircrew. *American Journal of Psychotherapy*, 25, 4, 530–542.

American Psychiatric Association (1987). *Diagnostic and statistical manual of mental disorders* (3rd edn., revised) Washington, DC: Author.

American Psychiatric Association (1994). *Diagnostic and statistical manual of mental disorders* (4th edn.) Washington, DC: Author.

Anderson, H.G. (1919). *The medical and surgical aspects of aviation*. London: Oxford University Press.

Bakal, P.A. (1981). Hypnotherapy for flight phobia. *The American Journal of Clinical Hypnosis*, 23, 4, 248–251.

Beck, A.T., Emery, G., & Greenberg, R. (1985). *Anxiety and its disorders: A cognitive perspective*. New York: Basic Books.

Benson, J.W. (1985). Psychological differences noted in aircrew members. *Aviation, Space and Environmental Medicine*, 56, 3, 238–241.

Bor, R. (ed.) (2003). *Passenger behaviour*. Ashgate: Aldershot.

Bor, R. & van Gerwen, L.J. (eds.) (2003). *Psychological perspectives on fear of flying*. Ashgate: Aldershot.

Botella, C., Villa, H., and Banos, R. (1999). The treatment of claustrophobia with virtual reality: Changes in other behaviors not specifically treated. *CyberPsychology and Behavior*, 2, 135–141.

Bucove, A.D. and Maioriello, R.P. (1970). Symptoms without illness: Fear of flying among fighter pilots. *Psychiatric Quarterly*, 44, 1,125–142.

Butcher, J.N. & Dunn, L.A. (1989). Human responses and treatment needs in airline disasters. In R. Gist & B. Lubin (eds.) *Psychosocial aspects of disaster* (pp. 86–119). New York: Wiley.

Capafons, J. I., Sosa, C.D., and Vina, C.M. (1999). A reattributional training program as a therapeutic strategy for fear of flying. *Journal of Behavior Therapy and Experimental Psychiatry*, 30, 259–272.

Carr, J.E. (1978). Behaviour therapy and the treatment of flight phobia. *Aviation, Space and Environmental Medicine*, 49, 115–118.

Dean, R.D. & Whitacker, K.M. (1982). Fear of flying: Impact on the U.S. travel industry. *Journal of Travel Research*, 21, 7–17.

Denholtz, M.S., Hall, L.A., & Mann, E. (1978). Automated treatment for flight phobia: A three and a half year follow up. *American Journal of Psychiatry*, 135, 1340–1343.

Dyregrov, A., Skogstad, A., Hellesoy, O.H., & Haugli, L. (1992). Fear of flying in civil aviation personnel. *Aviation, Space and Environmental Medicine*, 63, 9, 831–838.

Ekeberg, O., Seeberg, I., & Ellertsen, B.B. (1990). A cognitive behavioural treatment program for flight phobia with 6 months and 2 years follow up. *Norsk Psykiatrisk Tidsskrist*, 44, 365–374.

Fowlie, D.G. (1999). Emotional consequences of ejection, rescue and rehabilitation in royal air force aircrew. *Human Performance in Extreme Environments*, 4, 1, 119–122.

Frederickson, M. Annas, P., Fischer, H., & Wik, G. (1996). Gender and age differences in the prevalence of specific fears and phobias. *Behavior Research and Therapy*, 34, 1, 33–39.

Freud, S. (1960). *The standard edition of the complete psychological works of Sigmund Freud*. The Hogarth Press: London.

Goldstein, A.J. & Chambless, D.L. (1978). A reanalysis of agoraphobia. *Behavior Therapy*, 9, 47–59.

Gotch, O.H (1919). The aeroneurosis of war pilots. In: *The medical and surgical aspects of aviation* (pp. 109–149). London: Oxford University Press.

Greco, T.S. (1989). A cognitive-behavioural approach to fear of flying: A practitioner's guide. *Phobia Practice and Research Journal*, 2, 1, 3–15.

Greenberger, D. & Padesky, C. (1995). *Mind over mood*. Guildford Press: New York.

Greist, J.H. & Greist, G.L. (1981). *Fearless flying: A passenger's guide to modern air travel*. Chicago: Nelson Hall.

Heller, R.F. (1993). Overcoming the fear of flying. In W. Dryden & L.K. Hill (eds.), *Innovations in rational emotive therapy* (pp. 238–252). Newbury Park: Sage Publications.

Henderson, D. (2002). Uncertainty and the outlook for world air travel – Europe and the North Atlantic. Paper given at the FAA Commercial Aviation Forecast Conference, Washington.

Howard, W.A., Murphy, S.M., & Clarke, J.C. (1983). The nature and treatment of fear of flying: A controlled investigation. *Behaviour Therapy*, 14, 557–567.

Iljon Foreman, E. & Borrill, J. (1994). Long-term follow-up of cognitive behavioural treatment for three cases of fear of flying. *Journal of Travel Medicine*, 1, 1, 30–34.

Iljon Foreman, E. & Iljon, Z. (1994). Highwaymen to hijackers: A survey of travel fears. *Travel Medicine International*, 12, 4,145–152.

Iljon Foreman, E. (2003). Just plane scared? An overview of fear of flying. In R. Bor (ed.) *Passenger behaviour* (pp. 45–59). Aldershot: Ashgate.

Iljon Foreman, E. (2003). Putting fear to flight: Cases in psychological treatment. In R. Bor & L. van Gerwen (eds.) *Psychological perspectives on the fear of flying* (pp. 229–244). Aldershot: Ashgate.

Iljon Foreman, E. (2004). The airline's response to disaster: Psychology and post-traumatic stress. MSc. Aviation Psychology Module, City University, London/Dubai.

Jones, D.R. (1986). Flying and danger, joy and fear. *Aviation, Space and Environmental Medicine*, 57, 2, 131–136.

Jones, D.R. (2000). Fear of flying: No longer a symptom without a disease. *Aviation Space and Environmental Medicine*, 71, 4, 438–440.

Marks, M., Yule, W., & De Silva, P. (1995). Post traumatic stress disorder in airplane cabin crew attendants. *Aviation, Space and Environmental Medicine*, 66, 3,128–132.

McCarthy, G.W. & Craig, K.D. (1995). Flying therapy for flying phobia. *Aviation, Space and Environmental Medicine*, 66, 12, 1179–1184.

McIntosh, I.B., Swanson, V., Power, K., Raeside, F., & Dempster, C. (1988). Anxiety and health problems related to air travel. *Journal of Travel Medicine*, 5, 198–204.

McNally, R.J. & Louro, C.E. (1992). Fear of flying in agoraphobia and simple phobia: Distinguishing features. *Journal of Anxiety Disorders*, 6, 4, 319–324.

Medialea, J. & Tejada, F. (2005). Phobic fear of flying in aircrews: Epidemiological aspects and comorbidity. *Aviation, Space and Environmental Medicine*, 76, 566–568.

Moller, A.T., Nortje, C., & Helders, S.B. (1998). Irrational cognitions and the fear of flying. *Journal of Rational-Emotive and Cognitive Behavior Therapy*, 16, 2, 135–148.

Morgenstern, A.L. (1966). Fear of flying and the counterphobic personality. *Clinical Aviation and Aerospace Medicine*, 37, 404–447.

Muhlberger, A., Hermann, M.J., Wiedemann, G., Ellgring, H., & Pauli, P. (2001). Repeated exposure of flight phobics to flights in virtual reality. *Behavior Research and Therapy*, 39, 1033–1050.

Roberts, R.J. (1989). Passenger fear of flying: Behavioral treatment with extensive in vivo exposure and group support. *Aviation, Space and Environmental Medicine*, 60, 342–348.

Rosenhan, D.L. & Seligman, M.E. (1989). *Abnormal psychology*. New York: Norton.

Scrignar, C.B., Swanson, W.C., & Bloom, W.A. (1973). Use of systematic desensitisation in the treatment of airplane phobic patients. *Behavior Research and Therapy*, 11, 129–131.

Shneck, J.M. (1989). Separation anxiety and fear or avoidance of flying. *Journal of Clinical Psychiatry*, 50, 12, 474.

Strongin, T.S. (1987). A historical review of the fear of flying among aircrewmen. *Aviation, Space and Environmental Medicine*, 58, 3, 263–267.

Tempereau, C.E. (1956). Fear of flying in Korea. *American Journal of Psychiatry*, 113, 218–223.

Timm, S.A. (1977). Systematic desensitisation of a phobia aviation. *Aviation, Space and Environmental Medicine*, 48, 370–372.

van Gerwen, L.J. (2004). Fear of flying: Assessment and treatment issues. Dissertation Leiden University. Asten: Skyline Asten b.v., The Netherlands.

van Gerwen, L.J., Delorme, C., Van Dyck, R., & Spinhoven, P. (2003). Personality pathology and cognitive behavioral treatment of fear of flying. *Journal of Behavior Therapy and Experimental Psychiatry*, 14, 171–189.

van Gerwen, L.J. & Diekstra, R.F.W. (2000). Fear of flying treatment programs for passengers: An international review. *Aviation, Space and Environmental Medicine*, 71, 4, 430–437.

van Gerwen, L.J., Spinhoven, P., Diekstra, R.F.W., & Van Dyck, R. (1997). People who seek help for fear of flying: typology of flying phobics. *Behavior Therapy*, 28, 237–251.

van Gerwen, L.J., Spinhoven, P., Diekstra, R.F.W., & Van Dyck, R. (2002). Multi-component standardised treatment programmes for fear of flying: Description and effectiveness. *Cognitive and Behavioral Practice*, 9, 2, 138–149.

Watson, J.B. & Rayner, R. (1920). Conditioned emotional reactions. *Journal of Experimental Psychology*, 3, 1–14.

Wiederhold, B.K., Gevirtz, R.N., & Spira, J.L. (2001). Virtual reality exposure therapy vs imagery desensitisation therapy in the treatment of flying phobia. In G. Riva & C. Galimberti (eds.) *Towards cyberpsychology* (pp. 253–272). IOS.

Wilhelm F. H. & Roth, W.T. (1997). Acute and delayed effects of alprazolam on flight phobics during exposure. *Behavior Research and Therapy*, 35, 9, 831–841.

Williams, M.H. (1982). Fear of flight: Behavior therapy versus a systems approach. *The Journal of Psychology*, 111, 193–203.

Chapter 6

Flight or Fright? Psychological Approaches to the Treatment of Fear of Flying

Elaine Iljon Foreman, Robert Bor, and Lucas van Gerwen

> Flying was my biggest fear, but there were lots of others to keep it company; my encounters with tube trains, chocolate, coffee, lifts, strong odours, nutmeg, cheese, crowded spaces, cinemas...made me feel as though I was facing death or insanity. Now that I am conquering these phobias, the seemingly uncontrollable fear is dissipating and in its place is an overwhelming euphoria and renewed confidence at being able to regain my life, to embrace it instead of holding it at arms length. I used to worry about doing almost everything, now all I worry about is doing as many things as I possibly can.
>
> Emma 2005

Unaware of the above, some health professionals may dismiss the significance of a patient presenting with a fear of flying and question the appropriateness of using limited resources to treat the problem. A fear of flying affects between 10 and 40 percent of air travelers (see chapter on nature of fear of flying in this text) and, contrary to popular wisdom, wreaks havoc in people's personal and professional lives. It may also signal the presence of related psychological problems (e.g. depression, phobias, anxieties, relationship, and developmental problems) that may require further assessment and treatment. The outrageous terrorist attacks on September 11, 2001 may have contributed to an increase in the number of people suffering from this fear, and could well, counter-intuitively, affect the number seeking help to overcome the problem – many who were considering undertaking therapy could now assume that it is out of the question.

Up until 2003, anyone with an interest in studying fear of flying and its psychological treatment would have to search through published material from disparate sources, such as psychology, aviation, medical, and specialist mental health journals and books. The psychological treatment of fear of flying is not the preserve of any one discipline. The publication of the first comprehensive textbook *Psychological Perspectives on the Treatment of Fear of Flying* by Robert Bor and Lucas van Gerwen (2003) sought to remedy this. The current chapter provides a contemporary discussion and review of approaches to the psychological treatment of fear of flying.

Air travel is increasingly more commonplace and nearly 2 billion passengers annually travel on commercial airlines worldwide. In the early days of commercial flight, air travel was considered an adventure, risky and at times even dangerous. It was accessible mostly to the rich and famous, and while quicker than most other forms of transportation, by today's standards, it was a lengthy ordeal. The last quarter of the twentieth century was characterized by mass jet-powered air travel where almost anywhere on earth could be reached in less than 24 hours. Modern air travel is safe, efficient and accessible, though arguably less romantic and adventurous than in preceding years. Particularly in the early years, air travel presented physical and psychological challenges to the passenger, and the first *air stewardesses* recruited by the fledgling airlines all had a background in nursing. However, this requirement has changed since the altitudes at which most large commercial jets fly render most passengers less susceptible to motion sickness and pressurized cabins remove much of the discomfort to the inner ears traditionally associated with the early days of air travel (Bor, 2003). While air travel has certainly become far safer over the years, fear associated with flying remains among the most common phobias and anxiety states for which patients seek psychological intervention. For some, the fear is so overwhelming that the problem remains untreated as they avoid seeking professional help and even avoid flying altogether (Iljon Foreman & Borrill, 1993). This is unfortunate because the success rates for treating fear of flying and most other phobias are encouragingly high, as will be demonstrated in this chapter.

Fear of flying is classified in psychiatric terms in the *Diagnostic and Statistical Manual of Mental Disorders* (*DSM-IV*) of the American Psychiatric Association (1994) as a specific phobia, characterized by a marked, persistent, excessive fear that is precipitated by the experience or immediate prospect of air travel. An anxiety response (e.g. avoidance, increased heart rate, adrenalin release or even panic attacks) is invariably provoked by exposure to this phobic stimulus. Interestingly, the individual generally recognizes that their response is disproportionate, though this understanding may only serve to intensify the distress and add a measure of shame about having the problem. Many sufferers report symptoms of anxiety and distress long before they set out for the airport. Indeed, this sort of anticipatory anxiety might be triggered at the time of planning the trip, booking the ticket, packing, or en route to the airport. For some travelers this is their first experience of a phobia, while for as many as 46 percent they will have previously experienced phobias that were triggered by other, often related, situations. The most common of these are agoraphobia and claustrophobia.

It is generally agreed among psychologists that a fear of flying is not a single, unitary problem, but comprises several underlying fears including a fear of crashing, heights, confinement, instability, and lack of control (Iljon Foreman, 2003; van Gerwen, Spinhoven, Diekstra, & van Dyck, 1997). Some people can develop a secondary fear associated with flying. They fear having a phobia and have undue concern that this might signal the onset of a more serious and pervasive psychological problem with even more worrying consequences. For all these reasons, a full and detailed assessment of the patient must first be made before treatment commences.

Failure to carry out a general mental health assessment, even if briefly, could result in other problems being overlooked, with obvious consequences for the efficacy and duration of the psychological treatment (Bor, 2004).

The essentials of the initial assessment include finding out:

1. When the problem first started, and what might have triggered this. Careful attention should be paid to life stage transitions (e.g. shortly after the birth of a child, or following a bereavement or other separation), and also to distressing experiences associated with previous air travel (e.g. excessive, enduring turbulence, bumpy landings or perceived emergencies).
2. How the individual has coped with or reacted to air travel in the past. A fear of flying might have been preceded by many years of stress-free air travel. This foundation may enable a better understanding of the nature of the person's fears, from which to build their coping skills and confidence.
3. How the problem affects the individual's life and routines, as well as their relationships and career.
4. Whether co-factors relating to their psychological state (e.g. depression, anxiety, etc.) may have triggered, exacerbated or caused the fear of flying. It may then be necessary to address these additionally.
5. Whether anyone else in the family has a fear of flying or suffers from other phobias or fears. A positive family history of similar problems is common. It is also helpful to discover whether, and if so how, that individual learned to cope with or even overcome the problem.
6. What steps the individual has already taken to cope with or overcome the problem. Some solutions (e.g. self-medication, reliance on alcohol or avoidance) serve to maintain or even exacerbate the problem and may even give rise to new problems.
7. The extent of the person's motivation to overcome the problem. While seeking treatment may signal some motivation, it is necessary to determine whether this is sufficient. Motivation is directly related to efficacy and long-term outcome of treatment.
8. What might happen to the person in different areas of their life, or in relation to their self-concept, if psychological treatment proved unsuccessful? It is always necessary to consider the implications or possible consequences of failure, as this is a real possibility for a small proportion of those who undergo treatment.
9. Results of self-report questionnaires for the assessment of fear of flying can be useful in some cases. Several inventories have been developed: the Flight Anxiety Situations questionnaire (FAS), which assesses flying-related anxiety experienced in different situations, and the Flight Anxiety Modality questionnaire (FAM), which measures symptom modalities in which anxiety in flight situations is expressed. These questionnaires are already widely used and the psychometric properties of the inventories have proved to be excellent (van Gerwen, Spinhoven, van Dyck, & Diekstra, 1999). Besides being

available in English and Dutch versions, they have also been translated into French, German, Hebrew, Italian, Korean, Portuguese, Spanish, and Swedish. The questionnaires can be useful as pre- and post-treatment and follow-up measurements. Using the same questionnaires makes it easier to compare studies from different countries (van Gerwen, 2004).

Considering the preferred psychological treatment option, the approach or method used links to the therapist's conceptualization of the fear of flying. The psychodynamic perspective has its origins in Freudian psychoanalysis and posits that this fear has its origins in unconscious processes, most likely linked to problems with early childhood development and primitive needs and drives. A fear of flying within this theoretical model is viewed as a symptom of *deeper* problems, such as a fear of reduced control, hostility towards a parent figure or even a symptom of a fear of attachment. The approach requires the patient to gain insight into these processes. Treatment often takes many years. The aim is to bring the unconscious into consciousness, and then process this in a new and more productive manner. Psychodynamic therapy and psychoanalytic approaches to treatment are demanding intellectually, emotionally, and financially. The fact that more than the *symptom* of the fear of flying is the object of treatment makes it difficult to evaluate outcomes. Interestingly, the treatment of choice for a fear of flying before the mid-1960s was psychodynamic psychotherapy, although the success rate of treatment was seldom more than 18 percent (Carr, 1978).

The development of behavioral therapies in the 1960s and 1970s radically changed approaches to the treatment of a whole range of psychological problems, phobias and fears included. Refinements of behavioral therapy continued apace and cognitive behavioral therapy (CBT) is now the dominant model for the psychological conceptualization of fear of flying and its treatment. There have been additional developments in therapy, including virtual reality exposure therapy, but all share a core underpinning with CBT. More significantly, the success rate for the treatment of fear of flying with this approach is now between 70–96 percent (Bor, Parker, & Papadopoulos, 2000; van Gerwen et al., 2002).

The essential ingredients of the CBT approach are as follows:

1. The person is helped to identify the signs and symptoms of anxiety and panic, and to cope with them as a first step to ultimately overcoming them.
2. Cognitive restructuring is taught to help utilize logical thinking, and is combined with some information describing the basics of aerodynamics, principles of flight and safety issues in aviation.
3. Exposure to the feared situations is encouraged, ideally in actual flight, but sometimes under simulated conditions, such as a cabin-flight-simulator, or employing computerized technology.

The common goal of all therapies for fear of flying is to intervene in the internal representations of reality that prove to be non-functional with respect to the required adaptation to the environment (Vincelli, 1999). CBT proposes that it is not events per se, but rather a person's interpretation of them that is responsible for the production of feelings such as anxiety and problems such as phobias. Where individuals suffer from anxiety, it is hypothesized that the interpretations relate to an exaggerated perception of danger. It is not only the external events that can be seen as a source of danger, but also internal events such as the physiological symptoms of anxiety themselves. There is also a reciprocal relationship between the external event and the perception of danger, such that once individuals have labeled a situation as dangerous, they tend to selectively scan and interpret situations in ways that augment their sense of being in danger. Specific techniques to modify cognitions and thus affect the interaction between thoughts, feelings, and behaviors form the predominant core of the cognitive behavioral approach.

The range of different psychotherapeutic approaches can make even the experienced clinician wonder whether all are equally effective. The range described in the literature includes varied CBT programs, psychoanalytic therapy, systemic therapy, hypnosis, virtual reality, re-attributional training, systematic desensitization, stress inoculation training, coping self-talk, cognitive preparation, flooding, implosion, *in vivo* exposure and relaxation training which have all been described in the literature (Beckham et al., 1990; Capafons, Sosa, & Vina, 1999; Denholtz & Mann, 1975; Haug et al., 1987; Roberts, 1989; Rothbaum, Hodges, & Kooper, 1997). The dilemma of whether to combine different treatment approaches or identify which one works best for whom will be discussed in more detail later in this chapter.

CBT as an umbrella terminology can be confusing, covering a bewildering, often extremely heterogeneous, range of treatment techniques. As an example, both progressive muscle relaxation and cognitive restructuring can be called CBT, yet the former focusses almost entirely on physical symptoms, and the latter on cognitions. The term "evidence-based psychological treatment" is more appropriate for describing the large number of treatments that have been shown to be effective for treating anxiety-related problems (Antony, 2002). In a comparison between treatments, Howard et al. (1983) examined systematic desensitization, flooding, implosive therapy, and relaxation training, and reported that all were equally effective compared to no treatment control. Notwithstanding the development of virtual reality methods of treatment, *in vivo* exposure is still considered the *gold standard* of effective treatment of phobias (Muhlberger et al., 2001).

One of the comparatively recent developments in the field of CBT has been the use of computer-assisted treatment interventions. Two forms have predominated – that of Computer Assisted Exposure (CAE) and Virtual Reality (VR), otherwise know as Virtual Reality Exposure Therapy (VRET). The former presents a series of visual images on a computer screen, which can be accompanied by an audio soundtrack. This approach has largely been developed by Bornas, Tortella-Feliu, Llabres, and Fullana (2001). They highlight the difficulty and expense of *in vivo* flight exposure, which has daunted many researchers and therapists (Rothbaum et al., 1996). There are a number of advantages to computer-assisted therapy programs compared with VR – the technology is more widely available, and less expensive than VR, and can nevertheless be very realistic. The programs can be sub-divided into their various components relatively easily to meet the idiosyncratic needs of each patient. Patients with their own computers can also practice at home.

In a comparison of Computer Assisted Exposure (CAE) with a more traditional fear of flying multi-component program, the results suggested that CAE is indeed a better clinical choice than traditional information and relaxation components (Bornas et al., 2001). Treatment was also shorter and more cost effective. The authors reported that the average treatment duration was 5, 50-minute sessions of CAE for patients to feel ready to take a flight. The therapist went on the plane with the patients for the test flight, though the reasons for this are unclear. It is suggested that perhaps CAE suits some sub-types of flight phobia but not others and that there might be differences in responses between those with panic disorder, agoraphobia, and claustrophobia. It is also added that in further research, consideration should be given to the limitations regarding comparability of volunteers recruited through newspaper adverts to a clinical population.

A detailed review of another computer-assisted technique is provided by Krijn et al. (2004). They refer to Virtual Reality Exposure Therapy (VRET) rather than the previously more frequently utilized term *Virtual Reality*. VRET integrates real-time computer graphics, body tracking devices, visual displays and other sensory input devices to immerse patients in a computer-generated virtual environment, which, it is suggested, may be a possible alternative to standard *in vivo* exposure. VRET has been found to be effective for participants with fear of heights and of flying, whereas for other phobias, research to date is not conclusive. VRET also still needs to be assessed as a stand-alone treatment, and not a component of a wider treatment package. A recent review confirms the greater effectiveness of VRET over imaginal exposure (Emmelkamp, 2003). If a virtual environment could elicit fear and activate the anxiety-provoking structure, it could function as an alternative mode to induce exposure. As in the case of exposure *in vivo*, or in imagination, the information that disconfirms anxiety beliefs could be experienced, and habituation may occur. In the treatment of fear of flying, the advantages of VRET over standard exposure therapy are said by VRET proponents to be considerable. VRET is cost effective given that components of the flight can be repeated endlessly in the therapist's office, and different weather conditions can be simulated in seconds.

Potential drawbacks of VRET are discussed by Muhlberger et al. (2001). They found that only 53 percent of the flight phobics reported intense fear during the VR exposure while a proportion reported some fear. However, the strongest criticism made to date is that there is a fundamental difference between any Virtual Reality exposure and the real world – there is no way the simulated aeroplanes in the former could fall out of the sky, while as for the latter, the opposite can occur, although thankfully very rarely.

An important issue is whether the effects of VRET generalize to the world outside the laboratory. Only a few researchers have included behavioral avoidance tests in their studies, which was usually done only once in one specific situation. Although follow-up results are promising, these are based on self-report rather than on formal behavioral tests. It is only when creating virtual worlds that are copied from real world situations that an actual comparison between the effectiveness of both exposure programs can be made. A last drawback is the substantial number of drop-outs among participants in some VRET studies. While VRET may be safer than exposure *in vivo*, because the whole situation can be controlled easily, the counter-argument is that the whole point of *in-vivo* is its very uncontrollability.

In VRET, unexpected factors that can confound the treatment are few and far between. A clinical vignette serves to illustrate the perhaps counter-intuitive effect of an uncontrolled in vivo event. In one group treatment session, the first author was on a scheduled flight, with four patients as well as other regular passengers. Several of the small group had only boarded the flight with considerable reluctance. As the plane was taxiing, a loud bang was heard. The pilot slowed to a stop and investigated what had happened. A burst tire was discovered. Passengers were bussed to the terminal, and given the option of boarding a different flight, or waiting for their plane to be repaired. The psychological therapy treatment group unanimously chose the latter course of action, and 90 minutes later re-boarded their plane – with considerably less reluctance and anxiety than at the first attempt.

It has been suggested by Borkovec and Sides (1979) that vividness of imagery and not relaxation per se may be a critical element of laboratory exposure therapy. This also resonates with the findings of Bornas et al. (2001) who suggested that the information/relaxation component did not add anything to therapeutic efficacy. Likewise it has been shown that a particular group of fearful flyers gained little from information alone, whether it be information on aviation or on anxiety (Verschragen, 2000). *In vivo* exposure seems to offer significant benefits to the treatment process as real life cannot be switched off when it starts to feel intolerable. Through VR, one can re-expose the patient to the particular part of the experience that is most troublesome for them – e.g. take offs – but this, as mentioned, is an inadequate form of intervention as it is impossible for a VR session to really *crash*. Even though one can include vestibular *clues* such as motion and vibration in the course of VR, there will always be limits on actual fear as experienced by the patient. It is conceded that real life has been, and still is, the gold standard for treatment efficacy.

A study by Maltby, Kirsch, Mayers, and Allen (2002) compared VRET to other types of therapy and showed that the results of VRET for fear of flying are promising. However, VRET is not yet routinely used in actual clinical practice, though some clinical practices may include this component in their standardized fear of flying treatment program in the years ahead (van Gerwen, Diekstra, Arondeus, & Wolfger, 2004). Comparing the results of a Cognitive Behavioral Group Treatment (CBGT) (van Gerwen, Spinhoven, & van Dyck, (pending)) with the VRET study of Maltby et al. (2002), the latter demonstrated clinically significant change on three scales, with an average of 46 percent of VRET study participants meeting criteria for clinically significant change, while 76.7 percent of CBGT group met criteria for clinically significant change.

A word of caution must be introduced at this point regarding the employment of new therapeutic tools. It is important to note that while technological aspects of virtual reality and simulation treatment allow portrayal of images in a manner that is provocative, interesting, and engaging, it is primarily the skills of the therapist that allow achievement of the therapeutic results.

Further treatment innovations are Newman's (1999) palmtop computers and Hoffman's (1999) use of psycho-physiological data. They may improve a patient's ability to cope with extreme anxiety, and the prompting capability may improve adherence to treatment. Hoffman argues that accurate psycho-physiological feedback can increase the efficiency of treatment, which might additionally reduce the length and cost of treatment.

Given that, to date, a single effective treatment component responsible for improvement in all cases has yet to be established, there is no consensus within the published literature as to whether to employ a "multifaceted package of interventions" (Sidley, 1990). A trial by Wilhelm and Roth (1997) raises concerns about the efficacy of the use of combinations in the treatment of fear of flying. They found that using a combination of medication and cognitive behavioral interventions produced a poorer result than the cognitive behavioral treatment alone. At the first International Fear of Flying Conference held in New York in 1996, a range of treatment programs was described, from one to twelve sessions. While most included relaxation training and breathing techniques, not all did so, and the review by Antony (2002) questions the necessity and indeed clinical efficacy of these techniques, as well as other aspects of the multi-modal treatment package. Fueling the debate over multi-modal or individually tailored treatments, van Gerwen et al. (2002), describe a study in which patients were non-randomly assigned to either a one or two day treatment program. Given that flying phobics are a heterogeneous group, they propose that a multimodal treatment program seems appropriate to help patients who have different mechanisms and backgrounds that underlie their fear of flying. However, an alternative viewpoint can be considered, which examines the need to tailor the particular treatment intervention specifically to each individual patient, and not to utilize any unnecessary aspects of the *multi-modal* package.

Considering established treatment programs for fear of flying, rather than individual research studies which are of a more academic nature, Jones (2000) highlights the concerns raised by van Gerwen and Diekstra (2000) that there could be a "one size fits all" approach. In addition, concerns are expressed that the treatment of fear of flying may be undertaken by anyone, whether or not trained as a therapist, whether or not licensed to treat patients. In the last 15–20 years, a substantial number of airlines (at least 36 at last count) have participated in or initiated courses on prevention or reduction of flight anxiety.

The most parsimonious treatment study published to date appears to be that of Ost, Brandberg, and Alm (1997). A single three-hour session of massed treatment, including a return domestic flight was compared to five sessions of exposure and cognitive restructuring for 28 randomly assigned patients. The former group were more successful immediately post treatment. At one year follow up, there was a reduction in the number who took the behavioral test with immediate post treatment results of 93 percent of the one session group and 79 percent of the five-session group falling to 64 percent of both groups. The patients studied fulfilled the *DSM-IV* criteria for specific phobia, but were excluded if they had other psychiatric problems requiring immediate treatment. It is thus unclear whether people with panic disorder, claustrophobia and social phobia were included in the study, and therefore one cannot tell if this treatment is of the "one size fits all" variety.

Given the differences between courses outlined by van Gerwen and Diekstra (2000) above, and the research study format used by Ost et al. (1997), which is not offered as an ongoing clinical intervention, as are the courses of Iljon Foreman and van Gerwen, the structure of the Freedom to Fly course (Iljon Foreman, 2003) is worth noting. It consists of three stages: a telephone assessment, one session in the Consulting Room, and then a week later, a return short-haul flight on a scheduled flight.

The telephone assessment enables assessment of suitability – there may be some people for whom it seems that the treatment is unlikely to be of benefit. It is clearly best that they do not join the course in the first place – for their sake, for that of the therapist, and for the others in the group. Once accepted onto the course, people are seen either individually, or as a maximum of four per group. The first session involves taking a detailed history. An explanation for the development of the fear for each individual is explored, and the nature of anxiety explained. The maintaining factors are considered, and clients are encouraged to test out their fears regarding the consequences of their anxiety. The second session involves meeting at the airport, and taking a scheduled return short-haul flight. Technical information provided is minimal. The rationale is that one does not necessarily need to understand how something works in order to be comfortable in using it. Many people who have a fear of flying will nevertheless happily use other forms of travel that have equal disaster potential, which they have neither any idea of how it works, nor any control over it, and yet not feel at risk (Wilhelm & Roth, 1997).

The results of Iljon Foreman and Borrill's (1994) study indicate that employing the conceptualization of the two types of fears – a "loss of internal control" (subsuming panic disorder, social anxiety, claustrophobia, and agoraphobia) and "loss of external control" (subsuming fears of heights, turbulence, and all the elements which ultimately can be reduced to a fear of crashing) can enable a treatment program to be successfully employed in which one size does indeed fit all.

No clinician or researcher would ever be likely to claim a 100 percent success rate, however. The next challenge is to examine in more detail both people for whom treatment of whichever form has been successful, and also those who could be considered as *failures*. The latter terminology may be seen as somewhat harsh, and alternative terminology that published articles have employed are "negative outcomes, therapeutic drop outs, change resistant, treatment resistant, lack of success," and finally people who indicate a "lack of treatment related progress." Published research has often been criticized for a reluctance to examine the group showing "lack of treatment related progress." This must be viewed in the context of 8 percent of British clinical psychologists being responsible for 50 percent of the published work. Understandably, perhaps, the choice is to focus on the factors implicated in *success*, especially as negative findings are substantially less likely to be accepted for publication. The modal number of journal articles and conference papers published by British clinical psychologists is zero, and it is therefore important to bear in mind that published articles are likely to be atypical of normal clinical practice. The Ost et al. (1997) study is an example of this. Discussion with the clinician concerned indicated that such an intensive therapeutic treatment intervention would be unlikely to be offered on a regular ongoing basis of a standardized course, as the delivery of something so intense would probably lead to therapist "burn out" in the longer term. The challenge therefore is one of trying to apply research findings in a meaningful way to normal clinical practice. Concentrating on both the negative as well as the positive research findings, it should be possible to identify the critical factors that make for therapeutic efficacy.

What factors are responsible for successful treatment outcome? It has been argued that relying on acceptance-based strategies may be a more effective way of dealing with psychological distress, rather than control-orientated strategies (Antony, 2002).

He also suggests that perhaps too many techniques are taught to patients. He asserts that overloading patients with treatment techniques is a feature of many CBT treatments. Patients run the risk of learning a little bit about a lot of interventions, without mastering the most important or relevant techniques for that individual. Rather than using a large number of interventions for all patients with a given diagnosis, the challenge in the next few years will be to identify which patients are likely to benefit from which interventions.

A shortcoming of the empirically supported treatment movement has been the tendency to focus on diagnostic entities rather than on symptoms or core dimensions. For example, treatments have been developed for particular conditions such as panic disorder, social anxiety disorder, and generalized anxiety disorder, rather than the key features which comprise these disorders. A disadvantage of developing a single treatment for a given *DSM-IV* disorder is the need to include strategies that target all possible components of the disorder, even if they are not relevant for a given patient.

An alternative approach would be to identify the core dimensions that are relevant to a particular patient and to choose treatments that target those dimensions. In anxiety disorders, a number of dimensions exist that cut across disorders, including the presence of fear, anticipatory anxiety, worry, situational avoidance, avoidance of thoughts and feelings, interoceptive anxiety (i.e. anxiety sensitivity), compulsive rituals and overprotective behaviors. These symptoms are moderated by such factors as skills deficits, family issues, life stress and medical complications. To improve treatment outcome for a given individual, it is important to be able to measure the most salient symptoms and to select the most appropriate intervention for those symptoms, regardless of the diagnosis. For example, exposure to arousal sensations is likely to be helpful to any individual who fears these symptoms, regardless of whether the fear occurs in the context of panic disorder, social anxiety disorder, or a specific phobia of enclosed places, of which the aeroplane is a prime example – not only does one feel trapped – one really is.

One very important consideration for the treatment of fear of flying is therefore to consider optimal group size, so that all concerns are appropriately targeted, but participants are not subjected to information and techniques that are of no relevance to them, may be confusing, and perhaps even counter-therapeutic for their difficulties. Courses described have ranged from groups of two to 200 people. Clearly there are clinical implications for this. At the second World Congress on Fear of Flying in Vienna, it was not possible for the participants to reach agreement on the maximum group size, or the optimum number of sessions, and therefore a range of options continues to be available at the present time (van Gerwen, Diekstra, Arondeus, & Wolfger, 2004).

Conclusion

Those involves in aviation have witnessed enormous growth and massive developments in little more than 100 years since the Wright Brothers made their first flight. It is not surprising that human evolution has not kept pace with these developments, producing in some a fear of flying, since, to the species, it can appear safer being self propelled in the terrestrial domain. An increasing number of fear of flying treatment programs are being developed although almost all of these are confined to Western and developed countries. The insights and approaches that have been described in the literature are derived from clinical and academic research. The aim of psychological treatment is to understand, and to enable people to overcome their fear of flying, which is a highly debilitating and distressing disorder. An additional benefit of such treatment is the generalizing of improvement to other fears and difficulties. Findings to date indicate that there is a wealth of therapeutic options both currently on offer, and increasingly being refined. For those who currently are suffering from a fear of flying there is the ever increasing possibility that with access to one of the many courses or therapists available, they will no longer be grounded by fear – and after that – well, the sky is no longer the limit.

References

American Psychiatric Association (1994). *Diagnostic and statistical manual of mental disorders* (4th edn.). Washington, DC: Author.

Antony, M.M. (2002). Enhancing current treatments for anxiety disorders. *Clinical Psychology, Science and Practice*, 9, 1, 91–94.

Beckham, J.C., Vrana, S.R., May, J.G., Gustafson, D.J., & Smith, G.R. (1990). Emotional processing and fear measurement synchrony as indicators of treatment outcome in fear of flying. *Journal of Behavior Therapy and Experimental Psychiatry*, 21, 3, 153–162.

Bor, R. (ed.) (2003). *Passenger behavior*. Alershot: Ashgate.

Bor, R. (2004). *Anxiety at 35,000 feet: An introduction to clinical aviation psychology*. London: Karnac.

Bor, R., Parker, J., & Padadopoulos, L. (2000). Psychological treatment of a fear of flying: a review. *Journal of the British Travel Health Association, 1*, 21–26.

Bor, R., & Van Gerwen, L. (eds.) (2003). *Psychological perspectives on the treatment of fear of flying*. Aldershot: Ashgate.

Borkovec, T.D. & Sides, J.K. (1979). The contribution of relaxation and expectancy to fear reduction via graded, imaginal exposure to feared stimuli. *Behavior Research and Therapy, 17*, 529–540.

Bornas, X., Tortella-Feliu, M., Llabres, J., & Fullana, M.A. (2001). Computer-assisted exposure treatment for flight phobia: A controlled study. *Psychotherapy Research, 11*, 3, 259–273.

Capafons, J. I., Sosa, C.D., & Vina, C.M. (1999). A re-attributional training program as a therapeutic strategy for fear of flying. *Journal of Behavior Therapy and Experimental Psychiatry, 30*, 259–272.

Carr, J.E. (1978). Behavior therapy and the treatment of flight phobia. *Aviation, Space and Environmental Medicine, 49*, 115–118.

Denholtz, M.S. & Mann, E.T. (1975). An automated audiovisual treatment of phobias administered by non-professionals. *Journal of Behavior Therapy and Experimental Psychiatry, 4*, 111–115.

Emmelkamp, P.M. (2003). Behavior therapy with adults. In M. Lambert (ed.) *Handbook of psychotherapy and behavior change* (5th ed., pp. 393–446). New York: Wiley.

Haug, T., Brenne, L., Johnson, B.H., Bentzen, D., Gotestam, K., & Hugdahl, K. (1987). A three system analysis of fear of flying: a comparison of a consonant vs. a non-consonant treatment method. *Behavior Research and Therapy, 25*, 187–194.

Hoffman, S.G. (1999). The value of psychophysiological data for the cognitive behavioral treatment of panic disorder. *Cognitive and Behavioral Practice, 6*, 244–248.

Howard, W.A., Murphy, S.M., & Clarke, J.C. (1983). The nature and treatment of fear of flying: A controlled investigation. *Behavior Therapy, 14*, 557–567.

Iljon Foreman, E. (2003). Just plane scared? An overview of fear of flying. In R. Bor (ed.) *Passenger behavior* (pp. 45–59). Aldershot: Ashgate.

Iljon Foreman, E. (2003). Putting fear to flight: Cases in psychological treatment. In R. Bor & L. Van Gerwen (eds.) *Psychological perspectives on the fear of flying* (pp. 229–244). Aldershot: Ashgate.

Iljon Foreman, E. & Borrill, J. (1993). Plane scared – brief cognitive therapy for fear of flying. *Scottish Medicine, 13*, 4, 6–8.

Iljon Foreman, E. & Borrill, J. (1994). Long-term follow-up of cognitive behavioral treatment for three cases of fear of flying. *Journal of Travel Medicine, 1*, 1, 30–34.

Iljon Foreman, E. & Iljon, Z. (1994). Highwaymen to hijackers: A survey of travel fears. *Travel Medicine International, 12*, 4,145–152.

Jones, D.R. (2000). Fear of flying: No longer a symptom without a disease. *Aviation Space and Environmental Medicine*, *7*, 1, 438–440.

Krijn, M., Emmelkamp, P.M.G., Olafsson, R.P., & Biemond, R. (2004). Virtual reality exposure therapy of anxiety disorders: a review. *Clinical Psychology Review*, *24*, 259–281.

Maltby, N., Kirsch, I., Mayers, M., & Allen, G. (2002). Virtual reality exposure therapy for the treatment of fear of flying: A controlled investigation. *Journal of Consulting and Clinical Psychology*, *70*, 1112–1118.

Muhlberger, A., Hermann, M.J., Wiedemann, G., Ellgring, H., & Pauli, P. (2001). Repeated exposure of flight phobics to flights in virtual reality. *Behavior Research and Therapy*, *39*, 1033–1050.

Newman, M.G. (1999). The clinical use of palmtop computers in the treatment of generalised anxiety disorders. *Cognitive and Behavioral Practice*, *6*, 222–234.

Ost L.G., Brandberg, M., & Alm, T. (1997). One versus five sessions of exposure in the treatment of flying phobia. *Behavior Research and Therapy*, *35*, 1, 987–996.

Roberts, R.J. (1989). Passenger fear of flying: behavioral treatment with extensive in vivo exposure and group support. *Aviation, Space and Environmental Medicine*, *60*, 342–348.

Rothbaum, B.O., Hodges, L., Watson, B.A., Kessler, G.D., & Opdyke, D. (1996). Virtual reality exposure therapy in the treatment of fear of flying: a case report. *Behavior Research and Therapy*, *34*, 477–481.

Rothbaum, B.O., Hodges, L., & Kooper, R. (1997). Virtual reality exposure therapy. *Journal of Psychotherapy Practice and Research*, *6*, 3, 219–226.

Sidley, G.L. (1990). Brief clinical reports: A multi-component intervention with a lady displaying an intense fear of flying - a case study. *Behavioral Psychotherapy*, *18*, 307–310.

van Gerwen, L.J. (2004.) Fear of flying: Assessment and treatment issues. Dissertation Leiden University. Asten: Skyline Asten b.v., the Netherlands.

van Gerwen, L.J., Spinhoven, P., Diekstra, R.F.W., & van Dyck, R. (1997). People who seek help for fear of flying: Typology of flying phobics. *Behavior Therapy*, *28*, 237–251.

van Gerwen, L. J., Spinhoven, P., van Dyck, R., and Diekstra, R. F. W. (1999). Construction and psychometric characteristics of two self-report questionnaires for the assessment of fear of flying. *Psychological Assessment*, *11*, 2, 146–158.

van Gerwen, L.J. & Diekstra, R.F.W. (2000). Fear of flying treatment programs for passengers: An international review. *Aviation, Space and Environmental Medicine*, *71*, 4, 430–437.

van Gerwen, L.J., Spinhoven, P., Diekstra, R.F.W., & van Dyck, R. (2002). Multi-component standardised treatment programmes for fear of flying: Description and effectiveness. *Cognitive and Behavioral Practice*, *9*, 2, 138–149.

van Gerwen, L.J., Spinhoven, Ph., & van Dyck, R. (2005). Behavioral group treatment for fear of flying: A randomized controlled trial. Manuscript submitted for publication.

van Gerwen, L.J., Diekstra, R.F.W., Arondeus, J.M., & Wolfger, R. (2004). Fear of flying treatment programs for passengers: An international update. *Travel Medicine and Infectious Disease*, *2*, 27–35.

Vincelli, F. (1999). From imagination to virtual reality: The future of clinical psychology. *Cyber Psychology and Behavior*, *2*, 3, 241-248.

Verschragen, M.J. (2000). A study on the effect of information provided to people with fear of flying. Unpublished master thesis, Leiden University.

Wilhelm F. H. & Roth, W.T. (1997). Acute and delayed effects of alprazolam on flight phobics during exposure. *Behavior Research and Therapy*, *35*, 9, 831–841.

Chapter 7

Posttraumatic Stress Reactions Following Aircraft Disasters

Man Cheung Chung

Introduction

Experiencing a disaster means the emergence of at least two basic features, namely, the disaster and the victims involved, either of which can be categorized in different ways. An aircraft disaster, our main concern in this chapter, can be categorized as a type of technological disaster in that it results from failure of manmade products (Weisaeth, 1994). The victims involved can be categorized into six types: 1. primary victims (people who have had maximal exposure to the disaster); 2. secondary victims (the grieving relatives and friends of the primary victims); 3. tertiary victims (the rescue and recovery personnel who have been involved in a disaster rescue operation); 4. quaternary victims (the altruistic community which has a need to give aid, the convergers of various kinds in the community and organizations in the community which might have contributed to the disaster by any act of omission, commission or shared responsibility); 5. quinternary victims (people who have developed pathological behavior as a result of the disaster); 6. sesternary[1] victims (people who, but for chance, would have been primary victims, or people who have led others to take a course of action which results in their becoming primary victims) (Taylor & Frazer, 1981).[2] Much of what we know about the psychological distress that victims of aircraft disasters experience can be conceptualized within the above victim typology.

---

1 The numbering sequence of victims has been established in earlier literature (see Taylor & Frazer, 1981).

2 For some researchers, what's so important about Taylor and Frazer's category is that it has made us realize the range of people who can be affected by a disaster, many of whom are seldom mentioned, if at all, in disaster research. Of course, there are other ways of categorizing types of victim. For example, Figley and Kleber (1995) have simplified the definition by simply looking at primary victims and secondary victims. While they define primary victims in more or less the same way as Taylor and Frazer, their definition of secondary victims merges Taylor and Frazer's definitions of secondary and tertiary victims.

Following exposure to an aircraft disaster, victims may manifest posttraumatic stress reactions. The main purpose of this chapter is to review existing studies that explore such reactions. The review will be organized according to the foregoing victim typology. The chapter will end with some descriptions on the literature focussing on crisis interventions and psychological debriefing. At the outset, though, a few words need to be said about two things: one is why I use the term posttraumatic stress "reactions" in this chapter and the other is what the official definition of posttraumatic stress disorder is. One reason for using the term posttraumatic stress reactions in this chapter is to capture the different research aims of existing studies on aircraft disasters. As one will see in what follows, some studies explored posttraumatic stress reactions with a view to a diagnosis of posttraumatic stress disorder following aircraft disasters, in addition to making the association between posttraumatic stress symptoms and health problems. Some studies, however, focussed on posttraumatic stress reactions in terms of the association between posttraumatic stress symptoms and health problems but not the diagnosis. Another reason for using the term posttraumatic stress reactions is to capture the idea that when one examines posttraumatic stress disorder or symptoms, one is often examining a disorder or a group of symptoms that exist in conjunction with other aspects of psychiatric co-morbidity and other disorders (Solomon et al., 1991). The findings of large scale population surveys have indeed confirmed that in comparison with non-victims, victims of disasters are more likely to develop psychiatric co-morbidity such as anxiety, depression, obsessive-compulsive disorder, dysthymic disorder, schizophrenia, panic disorder, social phobia, substance abuse, and personality disorder (Davidson et al., 1991; Helzer et al., 1987; Keane & Wolfe, 1990; Norris et al., 2002).

In terms of the official definition of posttraumatic stress disorder, to put this simply, according to the fourth edition of the *Diagnostic and Statistical Manual of Mental Disorders* (*DSM-IV*, 1994), posttraumatic stress disorder is composed of three classic symptoms, namely, re-experiencing, avoidance, and hyperarousal. Re-experiencing causes victims to have recurrent and intrusive distressing recollections (e.g. images, thoughts or perceptions) of the disaster. They could also have recurrent distressing dreams of it, act or feel as if the disaster were recurring, and experience intense emotional upset (e.g. feeling scared, angry, sad, guilty) or physiological reactivity (e.g. sweating, heart beating fast) when reminded of the disaster. Victims could also experience avoidance symptoms in that they find themselves trying to avoid thoughts, feelings or conversations associated with the disaster, and activities, places or people which may arouse recollections of it. They could also become incapable of recalling an important aspect of the disaster, lose interest or participate much less often in important activities, feel distant or cut off from others and experience a restricted range of affect (e.g. unable to have loving feelings) and feel as if their future hopes and plans will not be fulfilled. They could also experience hyperarousal symptoms in that they have difficulty falling or staying asleep and difficulty concentrating on, for example, conversations with others or retaining information that they have read. They could feel irritable or have fits of anger, experience hypervigilance (i.e. overly alert) and exaggerated startle responses (i.e. jumpy or easily startled). These symptoms could trigger a great deal of distress or impairment in social, occupational or other areas of functioning (e.g. sex life).

When considering a diagnosis of posttraumatic stress disorder, the research suggests that one should base his or her decision on the duration of the symptoms (re-experiencing, avoidance, and hyperarousal). For example, as part of the diagnosis of posttraumatic stress disorder, the symptoms need to persist for more than one month. If the duration of these symptoms is less than three months, the individual would be considered to have acute posttraumatic stress disorder. However, if the duration is more than three months, the individual would be considered to have chronic posttraumatic stress disorder. If, however, the duration of these symptoms does not exceed one month, researchers are inclined to agree that the victim is suffering from acute stress disorder. Immediately after the disaster, victims experience dissociative, intrusive, avoidance, anxiety, and hyperarousal symptoms. These symptoms should last for at least two days and must affect victims' social or occupational functioning and make it difficult for them to engage in some necessary tasks (DSM-IV, 1994). Acute stress disorder is thought to be a clinical entity that is separate from posttraumatic stress disorder.

Primary Victims: People on Board

It is often difficult to conduct studies of survivors of aircraft accidents, since in many cases the primary victims (i.e. the passengers and air crew) perish in the event. Despite this fact, researchers have found sufficient numbers of individuals who have not only survived, but are willing to share their thoughts and feelings about these traumatic events. Studying them has given us a glimpse of the immediate reactions that these primary victims displayed during the disaster.

For example, in 1985, a Boeing 737 was waiting to take off on the runway at Manchester International Airport, United Kingdom. Suddenly, fire broke out in the aircraft and killed 54 people, while more than 80 people survived (Lee, 1989). It has been reported that as the fire gradually spread inside the aircraft, some people were confused, extremely anxious or screamed hysterically. Others remained reasonably calm; some were stunned, dazed, bewildered, became submissive and dependant, or felt a need to follow instructions or orders, even though these orders might not have been appropriate (e.g. they were told to stay in their seats with their seat belts fastened). Upon their rescue, they were severely shocked and sedated; they showed no emotions but acted passively. Meanwhile, they felt a compulsion to talk about what had happened. One would speculate that talking about what happened was a way of coping. By reliving the traumatic experience, one became subsequently desensitized to it. Some of these victims, on the other hand, coped by blaming others and finding a scapegoat. However, they subsequently developed paranoid attitudes about, for example, those who asked them to stay in their seats. Some experienced survivor guilt in that they had survived while others had died (Rawlins, 1985).

Two studies with small sample sizes have recorded some anecdotal evidence showing the range of posttraumatic stress reactions that the primary victims of those on board experienced. One focusses on a disaster in 1977 in which a KLM 747 and a Pan Am 747 collided, costing the lives of 580 people. Fortunately, 64 people survived, 8 of whom were recruited for the study. The other focusses on an aircraft that crashed and subsequently exploded on the ground immediately after take off in 1997, resulting in 23 dead and 29 seriously injured. Ten survivors were assessed. The combined studies reported the posttraumatic stress symptoms of uncontrollable emotions (particularly anxiety and rage), sleep disturbance and nightmares or daydreams related to the disaster, avoidance behavior, numbness and derealization, depersonalization with auditory hallucinations, and major depression. In addition to posttraumatic stress disorder, some victims were diagnosed as suffering from acute stress disorder and suffered from peritraumatic dissociative symptoms. There was evidence to suggest the long-lasting effect of the trauma in that one year later, three victims from the collision study suffered from a major physical illness associated with the trauma, and psychic numbing, disbelief, phobia, mood swings, depression, and psychosomatic problems (Birmes et al., 1999; Perlberg, 1979).

After some aircraft accidents, in which all of the persons on board have survived, there are few if any serious injuries to primary victims. However, despite the apparent reduction of bodily trauma, personal experiences of terror remain, especially after surviving a crash landing. In 1984, the East Tennessee State University basketball team and supportive personnel experienced a crash-landing. They found that twelve days after the accident, most of the victims (93 percent of 30 males studied) experienced hyperalertness (82 percent), decreased concentration or memory (79 percent), intrusive thoughts (71 percent), irritability (71 percent), sleep disturbance (68 percent), a high degree of physical pain (65 percent), feeling detached or estranged (54 percent) and dreams about the accident (50 percent). One year later, these posttraumatic stress symptoms were still prevalent among three to five victims, despite the fact that over time, the overall impact of the crash-landing upon the victims as a group declined, along with their levels of anxiety and depression, and their psychiatric and posttraumatic stress symptoms (Sloan, 1988).

Research has also focussed on the traumatic experience of ejection following an aircraft accident. Again, although those on board survived the accident, they nevertheless experienced life-threatening terror. Based on the experience of military aircrew who survived an accident by ejection, 40 percent of the Royal Air Force ejectees developed fits of anger, apprehension, anxiety, transient confusional states, paranoid disturbance, nightmares, flashbacks of the accident, and fear of being entangled in a crashed aircraft. They also felt exhausted due to their intense preoccupation with the accident, and their desire to return to flying had been diminished. Most of these ejectees did not think that their training had equipped them to cope with the experience of ejection (Fowlie & Aveline, 1985; Aveline & Fowlie, 1987). However, helicopter air crew who somehow managed to escape a helicopter crash, thought that their training had helped them to remain calm, relaxed, and confident and be in control during the accident (Hytten, 1989). Interestingly, there was evidence showing the emergence of traumatic growth or positive change in that several of the foregoing victims, after the accident, re-prioritized what was important in their lives. They came to the conclusion that spending time with family was more important than serving the air force (Aveline & Fowlie, 1987; Fowlie & Aveline, 1985) (see Linley & Joseph, 2004 for a recent review on traumatic growth or positive change).

Primary Victims: People in Community

Primary victims of aircraft disasters are not restricted to only those on board. Indeed, recent studies have focussed on people in the community as primary victims of aircraft disasters. For example, studies have been carried out to look at the posttraumatic stress reactions of the people in the community who were exposed to two aircraft disasters. One involved an Air Force jet fighter that crashed into a Ramada Inn in Indianapolis costing the lives of ten people in 1987. Within 4 to 6 weeks, 22 percent of the hotel employees were found to suffer from posttraumatic stress disorder as a result of the crash. Fifty-four percent fulfilled the criteria of a psychiatric disorder. The other disaster involved a Boeing cargo plane that crashed into two high-rise blocks of flats in Bijlmermeer, Amsterdam, in 1992, costing the lives of 43 people. Six months after the disaster, 26 percent of the community residents from one study suffered from posttraumatic stress disorder associated with material damage and loss of loved ones or friends; 44 percent had partial posttraumatic stress disorder (i.e. people only partially met the full set of diagnostic criteria for posttraumatic stress disorder). The forgoing studies showed a combination of symptoms among these community victims including extreme emotional upset (52 percent), depression (40 percent), hyperalertness (40 percent), intrusive thoughts (39 percent), generalized anxiety disorder (20 percent) and alcohol abuse (10 percent) (Carlier & Gersons, 1995, 1997; Smith et al., 1990).

To shed further light on the experiences of these primary victims, one study described the traumatic phases that the victims of the Bijlmermeer crash had gone through. Within the first 24 to 26 hours, community residents experienced an immediate shock phase. During this phase, they felt numb about the disaster and found themselves wanting to talk about it. A few days later, they experienced the cry out phase in which their feelings of shock and numbness began to be replaced by strong emotions and a growing desire to describe their traumatic experiences. During this phase, they also felt helpless and became angry and aggressive towards the authorities that were thought to be responsible for the disaster. At the same time, they experienced positive feelings resulting from people's closeness or bonding together. This phase lasted for about two weeks. Then, these victims began to experience acute posttraumatic stress symptoms alongside depression, anxiety, and agitation. They then reached the final phase, namely the intermediate posttraumatic period that signalled the beginning of disillusionment. During this phase, the positive feelings of closeness disappeared. People continued their lives with posttraumatic stress symptoms and had to face up to their losses (Gersons & Carlier, 1993).

The Lockerbie, Scotland disaster that occurred in 1988, is another example of primary victims of aircraft disasters who were not on board at the time of the event. In this disaster, Pan Am Flight 103 exploded in mid-air. Consequently, wreckage, burning aviation fuel, and mutilated human bodies fell onto the town of Lockerbie, causing death to eleven community residents. One study showed that 72 percent of a cohort of 66 residents had or had had posttraumatic stress disorder; 15 percent suffered from panic disorder; 12 percent anxiety disorder and 29 percent major depression (Brooks & McKinlay, 1992). The long-term traumatic effects upon these primary victims were also estimated. Thirty-six months after the disaster, the percentage of community residents suffering from posttraumatic stress disorder fell to 48 percent at follow-up. By way of contrast, the symptoms of those with severe or moderate posttraumatic stress disorder during the initial assessment were shown to persist through follow-up in these same individuals. On the other hand, depression increased from 28 percent at the initial assessment to 36 percent at the follow-up assessment (Scott, Brooks, & McKinlay, 1995).

To compare the posttraumatic stress reactions of elderly and younger victims in Lockerbie, one study showed reactions to be quite similar. They showed similar levels of intrusive thoughts and avoidance behavior. The posttraumatic stress reactions of the elderly were associated with personal loss, friends' injury, and witnessing human remains. However, this was not the case for the younger victims. The elderly victims suffered from higher rates of major depression than the younger victims. Loss of material possessions, personal loss or witnessing of human remains was not associated with depression among the elderly victims. However, material loss was associated with depression for the younger victims. The elderly victims experienced social dysfunction more than the younger victims (Livingston et al., 1992).

Focussing on the long-term traumatic effects upon the elderly victims, 36 months after the disaster, one study showed that the elderly cohort showed no significant changes in terms of proportion of psychiatric cases. However, the impact of the disaster seemed to have reduced to a significant degree in terms of the overall traumatic impact of the disaster, their somatic problems, anxiety and social dysfunction. The prevalence of posttraumatic stress disorder among the elderly also fell from 74 percent to 16 percent. While posttraumatic stress was associated with the witnessing of gruesome sights, loss of a partner or destruction of personal property during initial assessment, these associations were not established during follow-up. However, severe depression did not appear to decline 36 months after the disaster (Livingston, Livingston, & Fell, 1994).

Secondary Victims: The Grieving Relatives of the Primary Victims

Systematic studies looking at posttraumatic stress reactions among secondary victims (i.e. grieving families) of aircraft disasters are few and far between. Following the Boston air disaster in which 89 lives were lost, the bereaved relatives experienced traumatic reactions such as anger and bitterness towards the airline. These reactions were associated with finding someone to blame for the disaster and the lengthy legal battle for compensation in which these relatives had to engage. The relatives also experienced a powerful sense of grief and anxiety. These reactions were associated with their apprehension about viewing and identifying the bodies of their loved ones (Krell, 1974).

One study focussed on describing the formal professional support for bereaved families who had lost loved ones in the Delta Airlines jet disaster at Dallas-Fort Worth International Airport in 1985. One hundred and thirty-seven passengers lost their lives in that disaster. When the bereaved families arrived, they were provided with accommodation in a secluded hotel where they waited for victims' bodies to be retrieved and identified. The idea was that these relatives would stay in a protected, nurturing, and supportive environment where they would receive help from psychiatrists, nurses, hotel staff, airline representatives, and clergy. The families were then given opportunities whereby they could regress safely and allow themselves to satisfy their basic yearning for an idealized caregiver who would fulfil all their needs. Providing such an environment was thought to be important and useful as an intervention strategy for future disasters, especially those with a high number of casualties (Black, 1987).

Tertiary Victims: The Disaster Personnel

It is not difficult to imagine the degree of trauma to which disaster personnel are exposed. By engaging in body handling and recovery, for example, disaster personnel are at risk of developing posttraumatic stress disorder, especially during the six-month period following the disaster (e.g. Hershisher & Quarantelli, 1976; Jones, 1985; Miles, Demi, & Mostyn-Aker, 1984; Paton, 1989, 1990; Raphael, 1984). For example, following their work in recovering, identifying and handling bodies in the Mount Erebus aircraft disaster in 1979 (a DC10 aircraft crashed into the ice-covered slopes of Mount Erebus, killing the 257 on board), disaster personnel (53 percent) were found to experience moderately severe intrusive thoughts, nightmares, sleep difficulties, withdrawal, tension, depression, somatic problems, obsessive-compulsive symptoms, interpersonal sensitivity, and anxiety. One year later, there was evidence of improvement in that 80 percent of the personnel felt that they had successfully overcome their problems, while a small group (8 percent) felt the need to talk about their work with the disaster. However, 15 percent of the personnel still experienced flashbacks of the disaster occasionally; 5 percent were still angry about the aircrash and 4 percent experienced marital difficulties that were thought to be associated with the disaster (Taylor & Frazer, 1982).

Similarly, following their work in recovering, identifying and handling bodies during the Gander aircraft disaster (in 1985, a chartered airliner, carrying 8 aircrew and 248 US Army soldiers, took off from Gander, Newfoundland, but crashed into the forest at the end of the runway, killing all on board), the disaster personnel reported the extreme distress that they felt in viewing, smelling, and touching dead bodies. In particular, children's bodies or seemingly undamaged bodies or bodies which showed no obvious cause of death were just as distressing as burned and mutilated bodies. After all, the damaged bodies could be treated as non-human but undamaged ones could not. These disaster personnel reported a high degree of surprise and shock, despite their training and preparation for the disaster (Ursano and McCarroll, 1990).

While there are disaster personnel who are at the forefront of the recovery and identification of human bodies, there are disaster personnel who are assisting the families of the deceased. However, the degree of their posttraumatic stress cannot be understated. For example, it was reported that six months after the Gander disaster, family assistance officers experienced headaches (40 percent), nervousness or tenseness (33 percent), sleep disturbance (31 percent), general aches and pains (29 percent), common colds (25 percent), depressed mood (21 percent), and tiredness/lacking energy (20 percent). In general, these symptoms doubled in size one year later (Bartone et al., 1989).

McCarroll, Ursano, Wright, and Fullerton (1993) investigated the posttraumatic stress reactions of disaster personnel before, during and after working with dead bodies in different disasters, including the Gander aircraft crash in 1985 and the United Airlines flight 232 crash in 1989. Before exposure, the anticipated reaction to the dead, and the lack of information on the nature of the work caused the personnel a great deal of stress. They felt nervous because they were not sure what condition the bodies would be in or how difficult it would be to identify them. Some personnel believed that they should be told the worst before handling bodies, in order to minimize the surprises.

During the exposure, disaster personnel had to rely on different methods of helping themselves cope with the sensory stimuli of smelling and seeing the burned and mutilated bodies. These included, for example, masking the smell by burning coffee, smoking cigars, and using fragrances (e.g. peppermint oil and orange oil inside surgical masks). They also had to take frequent breaks or reduce visual contact with the bodies when they were handling them. They also tried to use humour in order to distance themselves emotionally from the bodies. They tried not to identify any similarities between themselves and the bodies. However, for many of these personnel, after having worked with the bodies for some time, they developed a tendency to think in terms of identification or similarity (Rosen, 1977). That is, they began to identify with the bodies and recognize similarities between the dead bodies and themselves (Ursano et al., 1988). They might feel that a particular body resembled themselves, a family member or a friend, and that it could have been them or one of their family members involved in the disaster. Some personnel would even feel that they knew the families of the dead. Someone involved with the Gander disaster reported feeling anxious when he realized he was lying in bed in the same position as one of the bodies he had worked on (Ursano & Fullerton, 1990).

After the exposure, instead of receiving professional counseling or psychiatric help, many disaster personnel felt that it was important to receive some form of post-disaster debriefing and for each to ask questions about the disaster. It was also important for them to feel appreciated for their work. At this stage, many of them experienced fatigue and irritability and felt the need to have time off work. Social support from spouses was important at this stage and alcohol use was common among these personnel (McCarroll et al., 1993).

Quaternary Victims: The Organization in the Community and the General Community Members

Large-scale aircraft disasters can have devastating effects on a community (see Ursano, McCaughey, & Fullerton, 1994; Wright & Bartone, 1994) and can create a group of quaternary victims who belong to an organization in the community to which the deceased belonged. For example, in 1989, seven aircrew and one passenger lost their lives when a C-141B Starlifter cargo jet from a US Air Force Base crashed as it was preparing to land during a storm. One study focussed on the posttraumatic reactions of 71 members who belong to the Air Force community. It showed that 67 percent of these members experienced intrusive thoughts about the crash and avoidant behavior. The latter were associated with high levels of depression and closer community ties. That is, the closer they felt the community ties to be, the greater intrusive thoughts, avoidance and depression they experienced (Fullerton et al., 1999).

Similarly, in the Gander disaster mentioned earlier, the quaternary victims include, for example, the commanders who mourned the loss of their troops and the soldiers who returned safely on other flights (Wright et al., 1990). Focussing on the soldiers, there are four phases through which it is thought soldiers progress. Phase 1 (1–6 weeks after the disaster) is called *numb dedication*. During this phase, soldiers experienced feelings of disbelief and denial, and interruption of sleep due to disturbing dreams. They also experienced posttraumatic stress symptoms of avoidance, hyperalertness, startle responses and survivor guilt. The phenomenon of guilt was particularly severe and persistent for the soldiers who had exchanged seats with those who were killed in the disaster. However, they detached themselves emotionally from the disaster and continued to behave like soldiers and concentrate on their mission.

Phase 2 (week 6–10) is called *anger betrayal*. During this phase, many of the soldiers were angry at the airline for safety violations and at the army who apparently did not thoroughly check the safety of the aircraft. Accompanying their anger was betrayal. They felt betrayed by the army because they thought the army would always ensure their safety and welfare. They then questioned the commitment of the army. While some solders managed to put aside their anger and invest their energy into caring for their dead colleagues' bereaved families, many soldiers used tranquilizers and alcohol to cope with their distress.

Phase 3 (week 10–20) is called *stoic resolve* in which many soldiers experienced a sense of relief when the last body had been recovered and identified. They subsequently felt that they could now get on with their lives by putting the disaster behind them and not holding on to their sadness and anger. During this phase, the disaster becomes a taboo. At this stage soldiers could still experience sleep disturbances and alcoholic problems.

Phase 4 (week 20–30) is called the *integration*, in which soldiers had come to terms with the loss and integrated it into their life experience. Consequently, some viewed their experience of the disaster as part of personal growth and learning. They now felt a renewed sense of hope, solidarity, and brotherhood within the army. They could now work with the relief soldiers and view them as members who can contribute to the army rather than as replacements of the dead. Nevertheless, some still experienced physical and emotional exhaustion (Bartone & Wright, 1990).

An additional subset of victims who could potentially belong to the group of quaternary victims is the airline employees, belonging as they do to an organization in the community to which the deceased belonged. These employees are involved in daily aircraft operation and could, following the disaster, experience a great deal of post-accident anxiety. These employees include pilots, designers of the machine, technicians who service and check the machine, tower controllers, flight instructors and the like. They may feel a sense of responsibility (Popplow, 1984). In general, following an air disaster, there are four common cognitive processes through which employees tend to progress. Going through these stages is meant to help victims recover a sense of control over the disasters.

Firstly, there tends to be a great deal of fear and worry experienced by these employees. They are fearful and worried about the safety of their own family members or friends who could have been on board the lost aircraft. Consequently, they are driven to find information concerning the health or safety of their family and friends. This fear, worry, and preoccupation with the safety of their family and friends reminds them of their own mortality. Secondly, these employees would engage in a problem-solving mode of thinking, trying to figure out what had happened. Often, their discussions turn into a debate about the probable causes, such as mechanical failure, pilot error and criminal act. Rumors or evidence that suggested that mechanical failure was the cause, would increase already high levels of anxiety about flying on this type of aircraft. The anxiety was only mildly relieved if the evidence led to a grounding of the type of aircraft until a fleet-wide inspection could be made. Despite this relaxation of anxiety, employees often share feelings of survivor guilt, because they survived the disastrous effects of mechanical failure while others did not. Rumors or evidence that suggested that pilot error was the cause, usually relieved anxiety regarding flight in the same type of aircraft, since it was not the aircraft, but the pilot at fault. Rumors or evidence that suggested that a criminal act was the cause increased within the employees a sense of frustration and anger. They would find themselves preoccupied with the notion of innocence and the issue of security.

The third cognitive process for these employees may involve the conscious experience and repeated fantasizing of the events that led to the destruction of the aircraft. These fantasies may be composed of the combination of known facts about the crash (e.g. the type of aircraft, weather conditions, the communication between the pilot and the control tower, etc.) and known or rumored personality traits of the employees on board the lost aircraft. If new information emerges about the crash from, for example, the recovered "black box" flight recorder, such new information would be incorporated into their fantasies. The fourth and final cognitive process is concerned with the decision of the employees to continue flying, to maintain the aircraft, or to provide services that keep others flying. The crash often makes people realize once again that life is important, a thing to be cherished, but also a thing to be carefully guarded in its fragile state. The decision to continue flying causes a dilemma for the employees. They are discouraged by family and friends to stop flying, but they also feel the need to renew their commitment to the flying industry. They alone can decide what is important for them and whether to continue in their jobs (Butcher & Hatcher, 1988).

The 1986 Challenger space shuttle explosion (the Challenger exploded in mid air shortly after take off) generated another group of quaternary victims. These were the general community members who came together and witnessed the explosion, in person or on television. Several studies focussed on the posttraumatic stress reactions of children living in different local communities. They found that following the explosion, children were experiencing feelings of shock or disbelief, sadness, anxiety and empathy toward the families of the deceased (Levitt & Leventhal, 1986). Male students were seemingly affected more than female students and appeared to recall the impact of the event more than female students (Gould & Gould, 1991). A sense of impersonal regret and attenuated affect was more characteristic of boys than girls (Wright et al., 1989). One year following the disaster, some children still felt concerned about the explosion and would often recall the impact of the disaster (Gould & Gould, 1991; Zeece, 1986).

Sesternary Victims: Community Residents Who, But For Chance, Would Have Been Primary Victims

We have recently examined the posttraumatic stress reactions of the sesternary victims of an aircraft disaster in the United Kingdom. One morning in 1994, a Boeing 737-2D6C was returning to Coventry, United Kingdom, having unloaded a cargo of live animals in Amsterdam. It lost control, struck a pylon and began to descend on a community called Willenhall. Fortunately, it did not crash into the community but clipped the roofs of two houses and crashed into a wooded area close to the edge of the community. Five people on board died, and hundreds of residents narrowly escaped death (HMSO, 1996). These community residents were sesternary victims who, but by chance, would have been primary victims. The posttraumatic stress reactions of these victims are substantial and cannot be undermined, despite the fact that no damage was caused to their property, nor injury sustained.

In our study, we examined posttraumatic stress reactions among 82 sesternary victims 6 months after the disaster. We found that 73 percent of them experienced severe intrusive thoughts about the disaster and avoidance behavior. They also experienced a great deal of general health problems. For example, anxiety was the most common problem, followed by somatic problems and social dysfunction. 57 percent of the victims were thought to be possible psychiatric cases. The more they experienced intrusive and avoidance symptoms following the aircraft disaster, the more they experienced somatic problems, anxiety, social dysfunction, and depression. We also found that there was a high degree of death anxiety among these victims. The more they experienced death anxiety, the more they experienced the impact of the disaster and health problems.

As part of the research program, we investigated the kinds of coping strategies that these victims used in order to cope with the posttraumatic stress resulting from the aircraft disaster. The results showed that distancing (e.g. they tried to forget the whole disaster) was used the most, followed by self-controlling (e.g. they tried to keep their feelings to themselves) and escape-avoidance (e.g. they avoided being with other people who also experienced the disaster). We also found that the victims who experienced avoidance behavior tended to be those who used the escape-avoidance coping strategy. Those persons who experienced intrusive thoughts about the disaster tended to be the same persons who used distancing and escape-avoidance coping strategies. Those who used distancing and escape-avoidance coping strategies tended to be those with health problems. Finally, we also examined the role that personality factors could play in the posttraumatic stress reactions of these victims. We found that neurotic and introverted victims tended to report intrusive thoughts about the disaster or display avoidance behavior (Chung, Easthope, Eaton, & McHugh, 1999; Chung, Easthope, Chung, & Clark-Carter, 1999; Chung, Chung, & Easthope, 2000; Chung et al., 2001).

Crisis Interventions and Psychological Debriefing for Victims of Aircraft Disasters

After having described posttraumatic stress reactions following aircraft disasters, attention has been drawn to the kinds of crisis interventions that can be implemented for different types of victims of aircraft disasters. Suggestions have been made that psychologists and other mental health professionals should work together to implement the crisis interventions and indeed the general disaster planning and response in airports (Anderson, 1988). There are different components to these interventions including mental health education and crisis hot lines (Williams, Solomon, & Bartone, 1988). Another important component is the establishment of psychiatric or mental health consultation teams (Butcher, 1980). For example, following the United Airlines flight 232 that crashed at Sioux City, Iowa in 1989, a consultation team was set up to: 1. provide consultation to disaster personnel on the mental health of the community; 2. provide direct psychiatric services or interventions to, for example, the high-risk groups; 3. train mental health personnel as consultants following disasters and; 4. develop and implement a research plan that would address both the immediate and long-term health consequences of the rescue work (McCarroll et al., 1992). There are important lessons that one needs to keep in mind as one sets up these teams. One is that the roles of the leadership and of the team members need to be clarified. It is also important to enable the team to mourn the people lost and to establish a clear plan of mental health services for both short-term and long-term management (Xenaki et al., 1991).

As part of the crisis interventions, suggestions have been made that debriefing should also be implemented for all victims, employees, and disaster personnel who are at risk of developing problems as a result of providing emotional assistance to the bereaved, and being involved in the rescue and recovery operation of the disaster (Butcher & Hatcher, 1988; Williams, Solomon, & Bartone, 1988). One form of debriefing is psychological debriefing which is a preventive strategy and aims to reduce the traumatic effect of long-term suffering among victims of disasters. We know from research that posttraumatic stress reactions can be long lasting. One study has estimated that the lifetime prevalence of posttraumatic stress disorder is 7.8 percent (Kessler et al., 1995). Co-morbidity items such as anxiety and depression would not likely disappear within a short period of time either. The idea is that psychological debriefing should help to reduce the traumatic effect. This, it is hoped, would reduce the lifetime prevalence of posttraumatic stress disorder.

Psychological debriefing comes in different forms, one of which is called Critical Incident Stress Debriefing (CISD) (Mitchell, 1983).[3] This strategy was first developed as a way of helping emergency services personnel. Now, it is being used for victims of other kinds. Based on one study looking at the medical workers involved in an aircraft crash in a 1989 air show in Germany, victims who came forward for debriefing tended to be those with high exposure to the disaster and female victims (Fullerton et al., 2000). According to the original protocol, CISD was thought to be best implemented between 24 and 72 hours after the disaster. However, it is now commonly thought that the time between the event and the debriefing needs to be somewhat flexible, but in all cases a trained mental health professional accompanied with support workers should carry out CISD. CISD can be a very emotionally demanding procedure lasting as long as two–three hours.

Very briefly, there are seven phases involved in CISD. The first phase is called the *introductory* phase in which the rules and rationale for doing psychological

---

3   There are other forms of psychological debriefing approaches that are somewhat different from CISD (e.g. see Raphael, 1986; Dyregrov, 1989 and the Multiple Stressor Debriefing Model (MSDM) (Armstrong, O'Callahan, & Marmar, 1991). Nevertheless, there are similarities between these approaches and CISD.

debriefing are disclosed to the victims. Victims are encouraged to talk and to express feelings about their experience of the disaster. The information disclosed during the session would be completely confidential. Phase 2 is called the *fact* phase in which the victims are asked to describe what has happened to them during the disaster. In describing the facts, victims can clarify for themselves the role that they played in the disaster, and clarify the way in which the disaster happened and any misunderstanding about what actually happened. Phase 3 is called the *thought* phase, where victims are encouraged to talk about the first or most significant thought during the disaster. Phase 4 is called the *reaction* phase during which victims are encouraged to talk about the component of the disaster which were the most distressing for them and which they found extremely difficult to deal with. The fifth phase is referred to as the *symptom* phase, in which victims are encouraged to talk about the kinds of symptoms, both psychological and physical, that they have experienced during or since the disasters. Phase 6 is the *teaching* phase, in which the facilitators of CISD make it clear to the victims that their psychological reactions are normal reactions towards an abnormal situation. The facilitators would also teach the victims some techniques that may be useful in coping with the traumatic stress. In turn, victims would hope to recover quickly or at least not to deteriorate further. The last phase is referred to as the *re-entry* phase, in which the facilitators would answer any questions which, in the victims' view, remain unanswered. The facilitators will also provide a summary of the debriefing session (Mitchell, 1983).

Although psychological debriefing is a widely used preventive strategy for disaster victims including aircraft disaster victims (in fact, one study has outlined the guidelines for conducting preventive mental health interventions using CISD for the aviation communities which have experienced fatal aircraft disasters (Cigrang, Pace, & Yasuhara, 1995)), what remains controversial is whether or not this preventive strategy is ultimately an effective one in reducing traumatic stress. Some researchers have found that CISD is effective in reducing the risk of posttraumatic stress related symptoms and psychological distress among disaster personnel such as emergency services personnel (Jacobs, Horne-Moyer, & Jones, 2004), rescue workers, police and disaster victim identification officers (Chemtob et al., 1997; Nurmi, 1999). Emergency services personnel and others such as welfare and hospital personnel found it valuable and helpful to be able to talk to someone about their experiences of the disaster in which they were engaged. They also thought that debriefing had increased their self-understanding (Robinson & Mitchell, 1993). Fire fighters have also reported crisis debriefing to be beneficial in reducing their level of stress (Regehr & Hill, 2000). Psychological debriefing was thought to be particularly effective in the short-term in reducing posttraumatic stress symptoms, in particular, intrusive thoughts (Humphries & Carr, 2001). Using a meta-analysis on adequately controlled, peer-reviewed journal articles and clinical proceedings, a conclusion was drawn that group psychological debriefings are effective in alleviating the effects of psychological distress and posttraumatic stress among emergency care providers (Everly, Boyle, & Lating, 1999).

However, some researchers believe that the effectiveness of CISD is in fact highly questionable, particularly among primary victims (e.g. victims of accidents or burns, interviewed while still in hospital) (Jacobs, Horne-Moyer, & Jones, 2004). According to one study on fire fighters, although those who received psychological debriefing shortly after their exposure to fire were, on the whole, less likely to develop acute posttraumatic stress symptoms, others developed delayed-onset posttraumatic stress and were more likely to have received debriefing than not (MacFarlane, 1988). Even if CISD is effective, it is no more than a brief psychoeducational intervention and does not represent anything special (Humphries & Carr, 2001). To echo this, one study showed that there was no difference in terms of posttraumatic stress symptom reduction between the fire fighters who were debriefed and those who had talked to their colleagues informally about their experience, despite the fact that those who had received debriefing seemed to have increased confidence (Hyteen & Hasle, 1989).

Casting further doubt onto the effectiveness of psychological debriefing, one study showed that on the basis of the analysis of fifteen empirical studies, psychological debriefing lacked empirical support for its effectiveness, in particular, psychological debriefing does not prevent posttraumatic stress disorder (Lewis, 2003). On the basis of randomized controlled studies (RCS), no evidence was found to suggest that psychological debriefing is an effective intervention in terms of preventing posttraumatic stress and psychological morbidity. Another study showed that debriefing made no difference in the subsequent psychiatric problems among army troops at nine-month follow up (Deahl et al., 1994). One other study showed no evidence that the debriefed helpers of disasters recovered quicker than those who were not. On the contrary, those who were not debriefed experienced a reduction in general health problems faster than the debriefed helpers (Kenardy et al., 1996).

Some researchers go so far as to argue that psychological debriefings are in fact harmful, and may even have an adverse effect on the victims (Kenardy & Carr, 2000). For example, one study examined the posttraumatic stress reactions of road traffic accident victims and found that the debriefed and non-debriefed groups did not show a significant reduction in symptoms. However, the debriefed groups were worse than the non-debriefed groups on some of the distress symptoms (Hobbs et al., 1996). Similarly, among burn trauma victims, at 13 month follow-up, victims who were debriefed experienced higher anxiety and depression symptoms than those who were not debriefed. The prevalence of posttraumatic stress disorder was also higher among those who were debriefed (Bisson et al., 1997). Attending crisis debriefing was also associated with the increase of intrusion among fire fighters, despite the fact that no significant association was found between attending crisis debriefing and depression (Regehr & Hill, 2000). To echo this finding, one study showed that disaster personnel who had received debriefing experienced significantly higher levels of intrusion and avoidance than those who had not (Griffiths & Watts, 1992). One other study also showed that eight months following the aforementioned Bijlmermeer aircraft disaster, the debriefed police officers involved in the rescue operation and the non-debriefed police officers did not differ significantly in posttraumatic symptoms. However, 18 months following the disaster, the debriefed police officers experienced significantly more disaster-related hyperarousal symptoms than the non-debriefed police officers (Carlier, Lamberts, van Uchelen, & Gersons, 1998).

In short, while psychological debriefing has been widely used by professionals as part of the crisis intervention for victims of disaster, including those involved in aircraft disasters, the scientific community remains sceptical on whether or not debriefing is an effective preventive strategy. More testing on this strategy must be done. There is very much a dearth of systematic studies that examine the acute and long-term effectiveness of debriefing (e.g. Avery et al., 1999; Gist, Lubin, & Redburn, 1999; Matthews, 1998; Raphael, Meldrum, & McFarlane, 1995; also see Raphael & Wilson, 2000).

Conclusion

This chapter reviewed some of the existing research on posttraumatic stress reactions resulting from aircraft disasters. Victims of disasters, particularly air disasters have been, by convention, separated into six categories, based on the psychological and psychiatric needs of each group as these needs become evident during post-event psychological debriefing or during prolonged psychotherapy. Although there might be numerous psychological issues present before and after an air disaster, I have focussed my attention on one of the more prevalent psychological outcomes of a disastrous event, that being posttraumatic stress disorder. I was principally concerned with the differentiated symptoms, based on proximity to the event.

The DSM-IV definition of posttraumatic stress disorder lacks relevance and specificity until it is used to diagnose victims of traumatic events. From primary to sesternary victims, the manifestations of psychological trauma differ and must be dealt with in a systematic and responsible fashion. I have drawn the reader's attention to a number of systematic means of after-event care, ending with a brief critique of those methods. Although we have anecdotal reports and formal studies of successes in psychological debriefings, it is still too early to categorically say that these debriefings actually help the victims. In fact, a growing number of studies suggest that psychological debriefings are ineffective. It is still too soon for a definitive response to this approach. Perhaps in future studies, as suggested by Ursano and others, we will pay more attention to the victim's predisposition to the effects of traumatic events as a beginning for diagnosis and treatment.

References

Anderson, T. (1988). An airport director's perspective on disaster planning and mental health needs. *American Psychologist, 43*, 9, 721–723.

American Psychiatric Association (1994). *Diagnostic and statistical manual of mental disorders* (4th edition). Washington, DC: American Psychiatric Association.

Armstrong, K., O'Callahan, W., & Marmar, C.R. (1991). Debriefing red cross disaster personnel: The multiple stressor debriefing model. *Journal of Traumatic Stress, 4*, 581–593.

Aveline, M.O., & Fowlie, D.G. (1987) Surviving ejection from military aircraft: Psychological reactions, modifying factors and intervention. *Stress Medicine, 3*, 15–20.

Avery, A., King, S., Bretherton, R., & Orner, R. (1999). Deconstructing psychological debriefing and the emergence of calls for evidence-based practice. *Traumatic Stress Points, 13*, 2, 6–8.

Bartone, P.T., & Wright, K.M. (1990). Grief and group recovery following a military air disaster. *Journal of Traumatic Stress, 3*, 4, 523–539.

Bartone, P.T., Ursano, R.J., Wright, K.M., & Ingraham, L.H. (1989). The impact of a military air disaster on the health of assistance workers. *The Journal of Nervous and Mental Disease, 177*, 6, 317–328.

Birmes, P., Arrieu, A., Payen, A., Warner, B.A., & Schmitt, L. (1999). Traumatic stress and depression in a group of plane crash survivors. *The Journal of Nervous and Mental Disease, 187*, 12, 754–755.

Bisson, J.L., Jenkins, P.L., Alexander, J., & Bannister, C. (1997). Randomised controlled trial of psychological debriefing for victims of acute burn trauma. *British Journal of Psychiatry, 171*, 78–81.

Black, J.W. (1987). The libidinal cocoon: A nurturing retreat for the families of plane crash victims. *Hospital and Community Psychiatry, 38*, 12, 1322–1326.

Brooks, N., & McKinlay, W. (1992). Mental health consequences of the Lockerbie disaster. *Journal of Traumatic Stress, 5*, 4, 527–543.

Butcher, J.N., & Hatcher, C. (1988). The neglected entity in air disaster planning. *American Psychologist, 43*, 9, 724–729.

Butcher, J.N. (1980). The role of crisis intervention in an airport disaster plan. *Aviation, Space and Environmental Medicine, 51*, 11, 1260–1262.

Carlier, I.V.E., & Gersons, B.P.R. (1995). Partial posttraumatic stress disorder (PTSD): The issue of psychological scars and the occurrence of PTSD symptoms. *The Journal of Nervous and Mental Disease, 183*, 2, 107–109.

Carlier, I.V.E., & Gersons, B.P.R. (1997). Stress reactions in disaster victims following the Bijlmermeer plane crash. *Journal of Traumatic Stress, 10*, 2, 329–335.

Carlier, I.V.E., Lamberts, R.D., van Uchelen, A.J., & Gersons, B.P.R. (1998). Disaster-related post-traumatic stress in police officers: A field study of the impact of debriefing. *Stress Medicine, 14*, 143–148.

Chemtob, C.M., Thomas, S., Law, W., & Cremniter, D. (1997). Postdisaster psychosocial intervention: A field study of the impact of debriefing on psychological distress. *American Journal of Psychiatry, 154*, 3, 415–417.

Chung, M.C., Chung, C., & Easthope, Y. (2000). Traumatic stress and death anxiety among community residents exposed to an aircraft crash. *Death Studies, 24*, 8, 689–704.

Chung, M.C., Easthope, Y., Eaton, B., & McHugh, C. (1999). Describing traumatic responses and distress of community residents directly and indirectly exposed to an aircraft crash. *Psychiatry: Interpersonal and Biological Processes, 62*, 125–137.

Chung, M.C., Easthope, Y., Chung, C., & Clark-Carter, D. (1999). The relationship between trauma and personality in victims of the Boeing 737-2D6C crash in Coventry. *Journal of Clinical Psychology, 55*, 5, 617–629.

Chung, M.C., Easthope, Y., Chung, C., & Clark-Carter, D. (2001). Traumatic stress and coping strategies of sesternary victims following an aircraft disaster in Coventry. *Stress and Health, 17*, 67–75.

Cigrang, J.A., Pace, J.V., & Yasuhara, T.T. (1995). Critical incident stress intervention following fatal aircraft mishaps. *Aviation, Space and Environmental Medicine, 66*, 9, 880–882.

Davidson, J.R.T., Hughes, D., Blazer, D.G., & George, L.K. (1991). Posttraumatic stress disorder in the community: an epidemiological study. *Psychological Medicine, 21*, 713–721.

Deahl, M.P., Gillham, A.B., Thomas, J., Searle, M.M., & Srinivasan, M. (1994). Psychological sequelae following the Gulf War: Factors associated with subsequent morbidity and the effectiveness of psychological debriefing. *British Journal of Psychiatry, 165*, 60–65.

Dyregrov, A. (1989). Caring for helpers in disaster situations: Psychological debriefing. *Disaster Management, 2*, 25–30.

Everly, G.S., Boyle, S.H., & Lating, J.M. (1999). The effectiveness of psychological debriefing with vicarious trauma: A meta-analysis. *Stress Medicine, 15*, 4, 229–233.

Figley, C.R., & Kleber, R.J. (1995). Beyond the "victim": Secondary traumatic stress. In R.J.Kleber, C.R.Figley, and B.P.R.Gersons (ed.) *Beyond trauma*. New York: Plenum Press.

Fowlie, D.G., & Aveline, M.O. (1985). The emotional consequences of ejection, rescue and rehabilitation in Royal Air Force Aircrew. *British Journal of Psychiatry, 146*, 609–613.

Fullerton, C.S., Ursano, R.J., Kao, T.C., & Bharitya, V.R. (1999). Disaster-related bereavement: Acute symptoms and subsequent depression. *Aviation, Space and Environmental Medicine, 70*, 9, 902–909.

Fullerton, C.S., Ursano, R.J., Vance, K., & Wang, L. (2000). Debriefing following trauma. *Psychiatric Quarterly, 71*, 3, 259–276.

Gersons, B.P.R., & Carlier, I.V.E. (1993) Plane crash crisis intervention: A preliminary report from the Bijlmermeer, Amsterdam. *Crisis, 14*, 3, 109–116.

Gist, R., Lubin, B., & Redburn, B.G. (1999). Psychosocial, ecological, and community perspectives on disaster response. In R.Gist and B.Lubin (ed.), *Response to disaster psychosocial, community and ecological perspectives on disaster response* (pp. 1–20). Ann Arbor, MI: Braun-Brumfield.

Gould, B.B., & Gould, J.B. (1991). Young people's perception of the space shuttle disaster: case study. *Adolescence, 26*, 102, 295–303.

Griffiths, J., & Watts, R. (1992). *The Kempsey and Grafton bus crashes: The aftermath*. East Lismore, Australia: Instructional Design Solutions.

Helzer, J.E., Robins, L.N., & McEvoy, L. (1987). Posttraumatic stress disorder in the general population. *The New England Journal of Medicine, 317*, 1630–1634.

Hershisher, M.R., & Quarantelli, E.L. (1976). The handling of the dead in a disaster. *Omega, 7*, 3, 195–208.

HMSO (1996) *Report on the Accident to Boeing 737-2D6C, 7T-VEE at Willenhall, Coventry, Warwickshire on 21 December 1994*. London: HMSO.

Hobbs, M., Mayou, R., Harrison, B., & Worlock, P. (1996). A randomised controlled trial of psychological debriefing for victims of road traffic accidents. *British Medical Journal, 313*, 1438–1439.

Humphries, C.L., & Carr, A. (2001). The short term effectiveness of critical incident stress debriefing. *Irish Journal of Psychology, 22*, 3–4, 188–197.

Hytten, K. (1989). Helicopter crash in water: Effects of simulator escape training. *Acta Psychiatrica Scandinavica, Supplement 355, 80*, 73–78.

Hytten, K., & Hasle, A. (1989). Firefighters: A study of stress and coping. *Acta Psychiatrica Scandinavica, Supplement 355, 80*, 50–55.

Jacobs, J., Horne-Moyer, H., & Jones, R. (2004). The effectiveness of critical incident stress debriefing with primary and secondary trauma. *International Journal of Emergency Mental Health, 6*, 1, 5–14.

Jones, D.J. (1985). Secondary disaster victims: The emotional effects of recovery and identifying human remains. *American Journal of Psychiatry, 142*, 3, 303–307.

Keane, T.M., & Wolfe, J. (1990). Comorbidity in posttraumatic stress disorder: an analysis of community and clinical studies. *Journal of Applied Social Psychology, 20*, 1776–1788.

Kenardy, J.A., & Carr, V.J. (2000). Debriefing post disaster: Follow-up after a major earthquake. In B.Raphael and J.P.Wilson (eds.), *Psychological debriefing: Theory, practice and evidence* (pp.174–181). Cambridge: Cambridge University Press.

Kenardy, J.A., Webster, R.A., Lewin, T.J., Carr, V.J., Hazell, P.L., & Carter, G.L. (1996). Stress debriefing and patterns of recovery following a natural disaster. *Journal of Traumatic Stress, 9*, 1, 37–49.

Kessler, R.C., Sonnega, A., Bromet, E., Hughes, M., & Nelson, C.B. (1995). Posttraumatic stress disorder in the national comorbidity survey. *Archives of General Psychiatry, 52*, 1048–1060.

Krell, G.I. (1974). Support services aid relatives of victims. *Hospitals JAHA, 48*, 56–59.

Lee, I. (1989). When an aeroplane catches fire: The Manchester International Airport disaster, 1985. In M.Walsh (ed.), *Disasters: Current planning and recent experience*. London: Edward Arnold.

Levitt, L., & Leventhal, G. (1986). Technological catastrophe: reactions to the space shuttle disaster. *Perceptual and Motor Skills, 63*, 670.

Lewis, S.J. (2003). Do one-shot preventive interventions for PTSD work? A systematic research synthesis of psychological debriefings. *Aggression and Violent Behavior, 8*, 3, 329–343.

Linley, P.A., & Joseph, S. (2004). Positive change following trauma and adversity: A review. *Journal of Traumatic Stress, 17,* 1, 11–21.

Livingston, H.M., Livingston, M.G., & Fell, S. (1994). The Lockerbie disaster: A 3-year follow-up of elderly victims. *International Journal of Geriatric Psychiatry, 9,* 989–994.

Livingston, H.M., Livingston, M.G., Brooks, N., & McKinlay, W.W. (1992). Elderly survivors of the Lockerbie air disaster. *International Journal of Geriatric Psychiatry, 7,* 725–729.

Matthews, L.R. (1998). Effect of staff debriefing on posttraumatic stress symptoms after assaults by community housing residents. *Psychiatric Services, 49,* 207–212.

McCarroll, J.E., Ursano, R.J., Fullerton, C.S., & Wright, K.M. (1992). Community, consultation following a major air disaster. *Journal of Community Psychology, 20,* 271–275.

McCarroll, J.E., Ursano, R.J., Wright, K.M., & Fullerton, C.S. (1993). Handling bodies after violent death: Strategies for coping. *American Journal of Orthopsychiatry, 63,* 2, 209–214.

McFarlane, A. (1988) The longitudinal course of most traumatic morbidity. The range of outcomes and their predictors. *Journal of Nervous and Mental Disease, 176,* 30–39.

Miles, M.S., Demi, A.S., & Mostyn-Aker, P. (1984). Rescue workers' reactions following the Hyatt Hotel disaster. *Death Education, 8,* 315–331.

Mitchell, J. (1983). When disaster strikes. The critical incident stress debriefing process. *Journal of the Emergency Medical Services, 8,* 36–39.

Norris, F.H., Friedman, M.J., & Watson, P.J. (2002). 60,000 disaster victims speak: part I. An empirical review of the empirical literature, 1981-2001. *Psychiatry, 65,* 3, 207–239.

Nurmi, L.A. (1999). The sinking of the Estonia: The effects of critical incident stress debriefing (CISD) on rescuers. *International Journal of Emergency Mental Health, 1,* 1, 23–31.

Paton, D. (1989) Disaster and helpers: psychological dynamics and implications for counselling. *Counselling Psychology Quarterly, 2,* 3, 303–321.

Paton, D. (1990). Assessing the impact of disasters on helpers. *Counselling Psychology Quarterly, 3,* 2, 149–152.

Perlberg, M. (1979). Trauma at Tenerife: The psychic aftershocks of a jet disaster. *Human Behavior, 8,* 49–50.

Popplow, J.R. (1984). After the fire-ball. *Aviation, Space and Environmental Medicine, 55,* 4, 337–338.

Raphael, B. (1984). *The anatomy of bereavement.* London: Hutchinson.

Raphael, B. (1986). *When disaster strikes.* London: Hutchinson.

Raphael, B., & Wilson, J.P. (2000). *Psychological debriefing: Theory, practice and evidence.* Cambridge University Press.

Raphael, B., Meldrum, L., & McFarlane, A.C. (1995). Does debriefing after psychological trauma work? *British Medical Journal, 310,* 1479–1480.

Rawlins, T. (1985). Survivors. *Nursing Times, November, 27*, 12–14.
Regehr, C., & Hill, J. (2000). Evaluating the efficacy of crisis debriefing groups. *Social Work with Groups, 23*, 3, 69–79.
Robinson, R.C., & Mitchell, J.T. (1993). Evaluating of psychological debriefings. *Journal of Traumatic Stress, 6*, 3, 367–382.
Rosen, V. (1977). Disorders of communication in psychoanalysis. In V.H.Rosen (ed.), *Style, character and language* (pp.127–150). New York: Jason Aaronson Inc.
Scott, R.B., Brooks, N., & McKinlay, W. (1995). Post-traumatic morbidity in a civilian community of litigants: A follow-up at 3 years. *Journal of Traumatic Stress, 8*, 3, 403–417.
Sloan, P. (1988). Post-traumatic stress in survivors of an airplane crash-landing: A clinical and exploratory research intervention. *Journal of Traumatic Stress, 1*, 2, 211–229.
Smith, E.M., North, C.S., McCool, R.E., & Shea, J.M. (1990). Acute postdisaster psychiatric disorders: Identification of persons at risk. *American Journal of Psychiatry, 147*, 2, 202–206.
Solomon, Z., Mikulincer, M., & Arad, R. (1991). Monitoring and blunting: implications for combat-related posttraumatic disorder. *Journal of Traumatic Stress, 4*, 209–221.
Taylor, A.J.W., & Frazer, A.G. (1981). *Psychological sequelae of operation overdue following the DC10 aircrash in Antarctica*. Victoria University of Wellington Publications in Psychology, No.27.
Taylor, A.J.W., & Frazer, A.G. (1982) The stress of post-disaster body handling and victim identification work. *Journal of Human Stress, 8*, 4, 4–12.
Ursano, R.J., McCaughey, B.G., & Fullerton, C.S. (1994) *Individual and community responses to trauma and disaster*. Cambridge: Cambridge University Press.
Ursano, R.J., & Fullerton, C.S. (1990). Cognitive and behavioral responses to trauma. *Journal of Applied Social Psychology, 20*, 1766–1775.
Ursano, R.J., & McCarroll, J.E. (1990). The nature of a traumatic stressor: Handling dead bodies, *Journal of Nervous and Mental Disease, 178*, 6, 396–398.
Ursano, R.J., Wright, K., Ingraham, L., & Bartone, P. (1988). *Psychiatric responses to dead bodies*. Proceedings of the American Psychiatric Association Annual Meeting (pp.132). Washington: American Psychiatric Press.
Weisaeth, L. (1994) Psychological and psychiatric aspects of technological disasters. In R.J. Ursano., B.G. McCaughey., & C.S. Fullerton (ed.) *Individual and community responses to trauma and disaster*. Cambridge: Cambridge University Press.
Williams, C.L., Solomon, S.D., & Bartone, P. (1988). Primary prevention in aircraft disasters. *American Psychologist, 43*, 9, 730–739.
Wright, K.M., & Bartone, P.T. (1994). Community responses to disaster: the Gander plane crash. In R.J. Ursano, B.G. McCaughey, & C.S. Fullerton (ed.) *Individual and community responses to trauma and disaster*. Cambridge: Cambridge University Press.

Wright, K.M., Ursano, R.J., Bartone, P.T., & Ingraham, L.H. (1990). The shared experience of catastrophe: An expanded classification of the disaster community. *American Journal of Orthopsychiatry*, *60*, 1, 35–42.

Wright, J.C., Kunkel, D., Pinon, M.F., & Huston, A.C. (1989). How children reacted to televised coverage of the space shuttle disaster. *Journal of Communication*, *39*, 2, 27–45.

Xenakis, S.N., Maury, J.L., Marcum, J.M., & Duffy, J.C. (1991). Consultation in the aftermath of an air tragedy. *Military Medicine*, *156*, 23–26.

Zeece, P.D. (1986) Young children's understanding of the shuttle disaster. *Journal of Psychology*, *124*, 5, 591–593.

PART 2
Psychological Processes Among Passengers and Crew

Chapter 8

Psychiatric Disorders and Syndromes Among Pilots

Jennifer S. Morse and Robert Bor

Introduction

The impact of psychiatric disorders on pilot performance must not be underestimated. The early diagnosis and treatment of psychiatric disorders among pilots is critical to maintaining aviation safety, and to reducing the effect of these disorders on the pilot's health and quality of life. At the outset, we acknowledge the difference between emotional upset that affects flight safety and emotional upset that has no effect on flight safety. In this chapter we focus more intently on individual emotional upset that influences flight safety, subtly or more expressively. Pilots with an identified psychiatric disorder must be grounded until fully recovered, and in some cases, a period of observation will be required prior to any consideration of a request to return to flying duties.

Psychiatric disorders are not uncommon among the general population but are still frequently under recognized. Due to extensive health history and medical screening prior to certification, pilots are generally considered to be at less risk for psychiatric disorders and psychologically better equipped to cope with stress. But pilots, despite their stereotypical image as "the right stuff," are biologically human and therefore prone to all the ailments that lay dormant within one's brain chemistry. It is paramount that medical and mental health personnel caring for pilots recognize that pilots, however resilient, are in no way immune to psychiatric disorders.

In this chapter the more common psychiatric disorders and syndromes will be reviewed, as well as treatment options and aeromedical disposition. This chapter will present the FAA requirements, which are similar though not identical to most other aviation authorities' aeromedical psychiatry requirements. Readers should check with their local aviation authorities for the unique and specific requirements for their country or region. For those readers who are involved in military aviation, please check to see if rules established by national authorities have been modified to support military aeromedical operational requirements.

The FAA Guide for Aviation Medical Examiners includes specific guidance regarding psychiatric diagnoses and aerospace medical disposition. It is not within the scope of this chapter to include all of that information, but readers can access the

Guide for Aviation Medical Examiners via the FAA website (www.faa.gov) for the latest updates.

Addressing Diagnoses of Relational Problems in Pilots

When relationship problems and pilot behavior appear to be linked, everyone should take notice, and in many instances interventions are necessary to ensure public safety. Studies by Jones, Katchen, Patterson, and Rea (1997) pointed to the need for such interventions when previously well-functioning and non-anxious pilots suddenly become overly concerned about flying duties. They insisted that family issues could very well be the underlying cause for this sudden behavior change and that these family issues must be addressed before allowing the pilot to resume.

If the pilot's emotional upset and subsequent poor performance is indeed linked to relationship problems, a clinician might very well understand his or her responsibility to contain the effect of that upset. However, refusing to allow a pilot to fly cannot be based merely on the presences of relational problems. Many pilots operate aircraft safely and effectively while dealing with relationship problems with spouses or life partners. Which then leads us to the question, "Why are these relationship problems suddenly affecting this pilot's professional behavior?"

As is widely understood by clinicians, the *Diagnostic and Statistical Manual of Mental Disorders*, fourth edition [DSM-IV-TR] (American Psychiatric Association, 2000) provides the mental health community with an orderly process for diagnosis and treatment of mental disorders. The *DSM-IV-TR* is also useful when a diagnosis is made difficult by other presenting issues. When a pilot fails to perform well and there is evidence that this failure was in part caused by relational problems with a spouse, the clinician can use V Codes and the multiaxial system to continue his or her assessment.

Under the circumstances of performance failures due, in part, to relational problems, a pilot might be grounded in the acute phase, depending on the severity of the symptoms. Once the condition is stable and resolved, a pilot can be returned to flying if no psychotropic medication were prescribed and used and if there was no associated disturbance of thought or history of recurrent episodes. If psychotropic medications were used for less than six months, the medications must have been discontinued for at least three months before the pilot can return to flying. Psychotropic medications are generally not required for these diagnoses, and the need for such medication should raise questions as to whether the diagnosis of V Code is correct.

It is the nature of the airline industry and military aviation that individuals spend time away from *home and hearth*, often for days, weeks, and months. Careers in commercial aviation and military aviation are inherently stressful for pilots. Although the experienced pilot has developed specific adaptive mechanisms to allow him or her to deal with the stresses inherent in flying, these same coping mechanisms may lead to difficulties in other settings. Several papers (Fine & Hartman, 1968; Reinhardt, 1970; Ursano, 1980) have addressed the tendency of pilots to avoid and deny their internal emotional life. Inner feelings tend to be perceived in much the same manner as external events would be perceived, which in affect diminishes introspection and increases the chances that a pilot will take steps to assuage internal conflict by adjusting his or her environment. Pilots often seek to *solve the problem* as an active coping mechanism, but when confronted with perceived failure to reach a career goal, such as airline industry layoffs, a death of a family member or close friend, or a spouse or partner who is leaving them, pilots may have few mechanisms available to deal with these issues. Retrospective studies have linked stress factors such as pilot career strain, financial setbacks, and interpersonal problems to aircraft mishaps (Alkov, Borowsky, & Gaynor, 1985; Little, Gaffney, Rosen, & Bender, 1990).

It has long been recognized that stable spouse and family relationships can act as a buffer against stress in the workplace; conversely discord in close relationships may intensify stress, leading to impaired work performance. Regrettably, where pilots are concerned, there are too few studies on the effect of the spouse or significant other on pilot behavior and more needs to be done in this area (Cooper and Sloan, 1985a, b). In the meantime, it is quite evident that spousal relationships help mitigate the effects of work stress and appear to reduce the likelihood of unsafe acts in flight. It follows that this issue needs to be promoted more in pilot training and should be studied more by those interested in how pilots daily cope with stress and personal relationships (Karlins, Koss, & McCully, 1989).

In their research on coping strategies of commercial airline pilots, Cooper and Sloan (1985a, b) discovered that overall mental ill-health has a very close association with lack of autonomy at work, fatigue, the inability to relax, and lack of sufficient social support (Sloan and Cooper, 1986). Raschmann, Patterson, and Schofield (1990) studied the psychosocial lives of pilots and noted that pilots who suffer from marital distress may be less able to concentrate effectively on their piloting duties and responsibilities. Based on these studies, it appears quite evident that the pilot who lacks sufficient emotional support from a significant life partner or spouse cannot cope with life events as well as pilots who are emotionally supported by positive and meaningful relationships among friends, partners, and spouses. Karlins et al. (1989) have helped us to appreciate that in an *airline marriage* the spouse can function as a very helpful social support system, thus aiding the pilot in dealing effectively with psychosocial stressors. Despite these studies that suggest that social and emotional support groups positively influence pilot behavior, we are still some distance away from fully appreciating the effect of life disruptions on pilots who frequently leave home to perform their flying duties.

Adjustment Disorders

Adjustment Disorders are generally considered one of the most common mental disorders diagnosed in pilots. Adjustment disorder (with its various subtypes) is defined as a maladaptive reaction(s) to an identifiable psychosocial stressor(s) that occurs within three months after onset of that stressor. The disorder is considered a transient minor disorder in the broader scheme of the various disorders and their severity. The definition of *maladaptive reaction* is potentially broad and systemically relative depending on the racial, ethnic, and cultural identifications of the patient and psychiatrist. Psychiatric symptoms result from the disruption of the individual's typical coping mechanisms, and frequently include anxiety, fatigue, depressed mood, inattention, indecisiveness, irritability, and decreased concentration. Relationship, occupational, career, and financial stressors are the most common predisposing stressors.

A common dilemma for medical personnel is the differentiation between the following: 1. a *typical* reaction to a stressor without evidence of functional impairment; 2. an adjustment disorder; and 3. a mood or anxiety disorder. Each of these categories has different implications for the pilot's ability to maintain medical certification.

Consider the following vignette: a 45-year-old pilot presents with a three-week history of insomnia requesting a temporary hypnotic. When questioned further, he becomes defensive, saying "Who wouldn't lose some sleep with the way things are going at work?" in reference to his employer's discussion of possible bankruptcy and pension cuts. In the face of the pilot's minimization and rationalization of his symptoms, a busy primary care physician might simply write a prescription for a sleep agent without further investigation.

The fact that the reluctant pilot has chosen to focus solely on the somatic symptom of sleep disturbance is generally merely the tip of the iceberg. As with any evaluation for mental health issues, the examiner must always remember to address the possibility of substance abuse. Increased use of alcohol or other substances in the face of life stressors is too common to be ignored. In this case, further questioning revealed that the pilot endorsed feeling *keyed up*, irritable with a short fuse, with frequent rumination about his finances even in the cockpit. The pilot denied panic attacks. The examiner determined that although the pilot's symptoms were not of the duration and severity of an anxiety disorder, the pilot's reaction to the stressor had resulted in impairment of the pilot's level of functioning.

Pilots with the diagnosis of an Adjustment Disorder should be grounded. Adjustment Disorders typically respond well to mental health counseling. If the symptoms have resolved without the use of medication and are not recurrent, a medical certificate can be issued. If psychotropic medications are required and used for less than six months, and the pilot has been symptom free and off medications for at least three months, a medical certificate can be issued. Clinical records and update must be submitted to the FAA. It is preferable that counseling or psychotherapy be provided by mental health personnel with knowledge and recognition of the unique demands of a career in the aviation environment.

Psychotic Disorders

Psychotic disorders include schizophrenia, schizophreniform disorder, schizoaffective disorder, delusional disorders, brief psychotic disorder with and without marked stressors, psychotic disorder not otherwise specified, substance-induced psychotic disorder, and psychotic disorder due to a general medical condition. Bipolar disorder and major depressive disorder with psychotic features may also present with psychotic symptoms. In addition, some personality disorders such as borderline or schizotypal may have intermittent psychotic symptoms which would be disqualifying. Psychosis presents as a gross impairment of an individual's ability to perceive reality even in the face of evidence to the contrary. Psychotic symptoms include hallucinations, delusions, and disorganized speech and behavior, to include grossly bizarre behavior.

It is vitally important for medical personnel to understand the importance of performing a complete diagnostic medical assessment in pilots presenting with psychotic symptoms. Common pharmacologic agents such as steroids, anticholinergic agents, cimetidine, indomethacin, antabuse, and INH have been associated with onset of psychosis. Medical conditions may present with psychotic symptoms, such as hypothyroidism (Heinrich & Graham, 2003), HIV, encephalitis, SLE, Addison's disease, or a brain neoplasm.

Psychotic disorders are considered permanently disqualifying by the FAA. Medical examiners should closely review any and all medical records in these cases. In cases where the psychotic symptoms have resolved, the etiology is clearly judged to have been a psychoactive agent (not illicit drugs) or a resolved general medical condition considered not likely to recur, and the pilot wishes to return to flying, aerospace medical examiners should call the FAA Regional Flight Surgeon (RFS) or the Aerospace Medical Certification Division (AMCD) to discuss these specific cases. These complex cases require prolonged observation of the pilot for one or more years as well as close review of family history before a decision would be made by the FAA regarding a return to flying. United States military pilots must request a waiver based on the individual military service aeromedical psychiatry guidance regarding substance-induced psychotic disorders, psychotic disorders due to a general medical condition, and brief psychotic disorders with marked stressors.

Mood Disorders

The Mood Disorders include such diagnoses as Major Depressive Disorder (MDD), Dysthymic Disorder, Depressive Disorder NOS, and Bipolar Disorder.

Major Depressive Disorder MDD is widespread. In the United States, 16.2 percent of the population has had MDD in their lifetime (Kessler et al., 2003). As of 2000, MDD ranked as the fourth leading cause of disability for women and seventh for men, and it is projected to be second only to cardiovascular diseases by the year 2020 (Ustun, Avuso-Mateos, Chatterji, Mathers, & Murray, 2004). Although it is not possible to determine the incidence of MDD in pilots because not all cases are reported by pilots, it is certainly a disorder that is not uncommon among pilots, particularly with the increasingly stressful occupational factors of flying.

Depressed pilots may experience both somatic and psychological symptoms, but pilots in general are more aware of their physical symptoms. Somatic symptoms may include: sleep disturbance, fatigue, loss of usual motivation and drive, decreased concentration, distraction, and change in appetite and possible weight change. Psychological symptoms typically include: depressed mood most of the time for at least two weeks, loss of interest in usual activities, psychomotor agitation or retardation, feelings of worthlessness or guilt, irritability, loss of joy or anhedonia, and suicidal ideation. It should be clear to the reader why a pilot would be disqualified from flying based on the above symptoms.

Unfortunately in some circumstances, flight surgeons, medical personnel, and mental health practitioners may collude with pilots to minimize the above symptoms as just stress and maintain that the pilot is safe to fly despite his depressive symptoms. This not only creates significant risk to aviation safety, but also may result in an increase in the duration and severity of a pilot's depression due to lack of appropriate treatment. In other situations, such as in a case of acute bereavement (without suicidal ideation or psychotic symptoms), the pilot may be incorrectly diagnosed with MDD and prescribed antidepressants, thus resulting in an unnecessarily prolonged grounding period. Clearly, the appropriate diagnosis should be made based on an objective assessment of the symptoms. Aviation medical examiners and medical personnel should always complete a medical assessment, as certain medical conditions may present with depressive symptoms, most commonly thyroid disorders such as hypothyroidism, and neurological disorders such as head trauma, multiple sclerosis, and dementia in the early stages. Some medications can also produce depressive symptoms such as steroids, indomethacin, and cardiac agents.

FAA standards state that depression is disqualifying as well as every medication that is used for the condition. In pilots without recurrent episodes or associated suicide attempts, generally the pilot may return to flying after individual case review by the FAA after the pilot is free of symptoms, used psychotropic medication for less than six months, and has not been taking psychotropic medications for at least three months. Treatment generally consists of psychotherapy and/or antidepressant medications.

Bipolar Disorder The diagnosis of Bipolar Disorder (manic-depressive disorder in laymen's terms) is made based on history of an episode of mania. Individuals diagnosed with Bipolar Disorder generally have a history of recurrent episodes of depression and mania. During an episode of mania, the individual is generally euphoric and grandiose, or may be extremely irritable for a period of three to seven days. Some of the symptoms include decreased need for sleep, hyperproductivity, pressured speech, risk-taking behaviors such as substance use, reckless driving, and hypersexuality. Manic episodes may include psychotic symptoms as well. Individuals with this disorder generally require long-term treatment with mood stabilizer medications and are at significant risk for abrupt onset of symptoms and for suicide despite ongoing treatment.

FAA standards state that a diagnosis of Bipolar Disorder is disqualifying. However, FAA guidelines state that some applicants diagnosed with Bipolar Disorder may be considered for an Authorization after individual case review when the symptoms do not constitute a threat to safe aviation operations.

Dysthymic Disorder and Depressive Disorder NOS These diagnoses are considered milder forms of depression as compared to MDD, but pilots should be grounded while symptomatic, and all pertinent clinical records submitted. By FAA standards, medical certificates can be issued only after the condition is resolved and stable, without associated disturbance of thought or recurrent episodes. If psychotropic medications were used for less than six months and have been discontinued for at least three months, medical certificates may be issued.

Anxiety Disorders

Anxiety disorders are the most prevalent psychiatric disorders in the United States and the world, and these disorders, although not as visible as a psychotic or mood disorder, can be quite disabling in the aerospace environment. This category includes Panic Disorder, Agoraphobia, Posttraumatic Stress Disorder (PTSD) and acute stress disorder, obsessive-compulsive disorder, generalized anxiety disorder, social and specific phobias, anxiety due to a general medical condition, and substance-induced anxiety disorder, and anxiety disorder not otherwise specified. All of these disorders have apprehension and impaired function in common (Levy, 2000). The presentation of anxiety disorders involves a complex interplay of genetic, biological, and stress factors. Anxiety disorders are stress dependent, and recent findings also indicate that stress produced by an event or by a persistent and chronic disorder is capable of causing secondary biological changes in specific brain structures (Axelson et al., 1993; Bremner, 2003). Anxiety can be associated with a fear of flying (see Chapters 5 and 6 on fear of flying by Iljon Foreman, Bor, and van Gerwen, in this text). The following is a brief overview of the more significant anxiety disorders.

Panic Disorder Panic disorder is diagnosed when an individual suffers at least two unexpected panic attacks, followed by at least one month of concern over having another attack. A panic attack is defined as the abrupt onset of an episode of intense fear or discomfort, which peaks in approximately 10 minutes, and includes at least four of the following symptoms: shortness of breath, dizziness, chest pain or tightness, tingling in the hands and around the mouth, nausea, sweating, a sense of doom, a fear of losing control, or a fear of death. Individuals with this disorder may also be prone to situationally predisposed attacks. The frequency and severity of the attacks varies from person to person, and a pilot might suffer repeated attacks for weeks, or (s)he might have short bursts of very severe attacks which could certainly impair their ability to function in the aviation environment. Certain medical disorders can mimic Panic Disorder, to include arrhythmias, supraventicular tachycardia, thyroid abnormalities, hypoglycemia, and seizure disorders.

Agoraphobia Agoraphobia often, but not always, coincides with Panic Disorder. Agoraphobia is characterized by a fear of having a panic attack in a place from which escape is difficult (such as a plane). Pilots with agoraphobia may develop a fixed route, or territory, from which they cannot deviate, and in some cases, traveling outside of what they consider their safety zones becomes impossible due to severe anxiety.

Generalized Anxiety Disorder The essential characteristic of Generalized Anxiety Disorder is excessive worry about everyday things present more days than not for at least six months, interfering with the individual's ability to function and concentrate. This constant worry affects daily functioning and can cause physical symptoms, such as muscle tension, sweating, nausea, gastrointestinal discomfort and/or diarrhea, fatigue, and jumpiness.

Obsessive-Compulsive Disorder Obsessive-Compulsive Disorder is characterized by uncontrollable obsessions and compulsions which the individual usually recognizes as being excessive or unreasonable. Obsessions are recurring thoughts or impulses that are intrusive or inappropriate and cause anxiety. Some common obsessions are: fear of contamination; extreme need for orderliness; aggressive thoughts or impulses; and persistent doubts about whether, for example, doors are locked or appliances turned off. Compulsions are repetitive behaviors or rituals performed to neutralize the anxiety caused by obsessive thoughts, but relief is only temporary. Common compulsions include: cleaning; checking; counting; and repeating. OCD would clearly interfere with a pilot's ability to concentrate.

Posttraumatic Stress Disorder (PTSD) PTSD may occur in pilots who have been exposed to a traumatic event involving either witnessing the death or severe injury to others, or the perceived threat of death or serious injury to themselves or others, and who responded to that exposure with fear, helplessness, or horror. Pilots are at risk for PTSD particularly following an aircraft mishap or near mishap, participation in a mishap investigation which may include body recovery or exposure to pictures of dead passengers or co-workers, or the death of a fellow pilot or aircrew (see also Chapter 7 by Chung in this text). Heightening anxiety in the wake of an airline disaster, terrorist attacks or terrorist threats increases anxiety for all airline passengers, and has a ripple effect on the crew. Psychological problems have been found in many survivors of air accidents (Lundin, 1995), and PTSD and depression were the most common psychological outcomes in the survivors of the Lockerbie disaster (Brooks & McKinlay, 1992). In a study of 175 Royal Air Force officers, for example, who survived ejection from their aircraft, 40 percent experienced prolonged psychological disturbance (Fowlie & Aveline, 1985). One study that directly addressed psychological reactions among flight crew to an accident suggested that commercial pilots suffer more stress than was previously thought in the aftermath of a major incident (Johnston & Kelly, 1988). However, a study that assessed the stress reactions of a submarine crew seven months after they had been forced to abandon their vessel in high seas after flooding and fire found that acute exposures of highly trained professionals to potentially fatal events may not result in high levels of posttraumatic symptoms (Berg, Grieger, & Spira, 2005).

It is important to realize, however, that not everyone exposed to a traumatic event will develop PTSD. Within the first month after such an event, individuals may experience some of the same symptoms of PTSD, but in the early phase this is considered a *normal* reaction to an *abnormal* situation or event. However, if the individual experienced dissociative symptoms at the time of the event, and then develops symptoms of PTSD within the first month, the diagnosis of Acute Stress Disorder should be considered. PTSD is discussed in more detail in Chapter 7 by Chung in this text.

For PTSD to be diagnosed, symptoms must be present for more than one month and be accompanied by a decline in the ability to socialize, work, or participate in other areas of daily functioning. To be diagnosed with PTSD, a pilot would need to experience symptoms from each of the four main criteria: 1. re-experiencing the event, which can take the form of intrusive thoughts and recollections, or recurrent dreams; 2. avoidance behavior in which the sufferer avoids activities, situations, people, and/or conversations which he/she associates with the trauma; 3. a general numbness and loss of interest in surroundings; this can also present as detachment; and; 4. hypersensitivity, including: inability to sleep, anxious feelings, overactive startle response, hypervigilance, irritability and outbursts of anger. The diagnosis of an anxiety disorder is disqualifying. Many anxiety disorders respond well to psychotherapy, although some individuals may require treatment with SSRIs, SNRIs, or anxiolytics such as benzodiazepines. In less severe anxiety disorders, the pilot may be able to return to flying after the symptoms have resolved, or if psychotropic medications were used for less than six months and the pilot has been symptom free and off medications for at least three months. However, FAA decisions could vary depending on the type of anxiety disorder, its severity and prognosis (for example, a pilot who experienced panic attacks in the aircraft). Readers are encouraged to review such cases with their local Regional Flight Surgeons or aviation authorities.

Substance-Related Disorders: Alcohol and Drugs

Substance dependence is generally defined by the FAA as a condition in which a person is dependent on a substance (other than tobacco or caffeine beverages) as evidenced by: a. increased tolerance; b. manifestation of withdrawal symptoms; c. impaired control of use; or d. continued use despite damage to physical health or impairment of social, personal, or occupational functioning. 'Substance' includes alcohol, other sedatives and hypnotic; anxiolytics; opioids; central nervous system stimulants such as cocaine, amphetamines and similarly acting sympathomimetics;

hallucinogens; phencyclidine (PCP) or other similarly acting arylcyclohexylamines; cannabis; inhalants; and other psychoactive drugs or chemicals.

Substance abuse is generally defined by the FAA as: a. the use of a substance in the last two years in which the use was physically hazardous (e.g. DUI or DWI) if there has been at any other time in the pilot's history an instance of the use of a substance also in a situation in which the use was physically hazardous; b. the pilot has received a verified positive drug test result under an anti-drug program of the Department of Transportation or one of its administrations; or c. the Federal Air Surgeon finds that an applicant's misuse of a substance makes him or her unable to safely perform the duties or exercise the privileges of the airman certificate applied for or held.

Substance dependence and substance abuse are disqualifying and require an FAA decision before a medical certificate can be issued. Pilots with a diagnosis of substance dependence must submit clinical evidence, satisfactory to the Federal Air Surgeon, of recovery, including sustained total abstinence from the substance(s) for not less than the preceding two years, to be considered for a medical certificate.

Foremost among psychiatric disorders that present among pilots are alcohol abuse and dependence. Alcohol primarily affects the central nervous system and leads to impairment of reaction time, reasoning, balance and coordination, speech, and the long-term effects of dementia, impaired memory, rigidity of thought, and problems in relationships, both socially and occupationally. Seizures can occur with alcohol dependence. Absenteeism may be the first obvious signal to colleagues at work and to management that there may be a serious problem. Those who suffer from alcohol dependence are more likely to be in accidents generally, and drunk-driving accidents in particular.

Alcohol and drug misuse may be a co-factor in other psychiatric disorders, including adjustment disorders, anxiety disorders such as panic disorder and posttraumatic stress disorder, and mood disorders such as depression. Onset of alcohol or drug misuse may begin with the break-up of a relationship, bereavement, financial problems, or career problems. It is important to make a full assessment of the personal, social, medical, and occupational circumstances of the pilots who presents with an alcohol or drug problem. Evaluation of pilots referred for an alcohol or drug problem is not always easy, as denial of the problem is quite common in spite of obvious signs such as smelling of alcohol, unkempt appearance, hand tremors, and even a prior drunk-driving conviction. The pilot referred for a drug problem may not have such obvious signs, although the pilot may demonstrate some sedation, agitation or change in pupil size. Often the more obvious signs of alcohol or drug misuse are noted by others prior to the referral for evaluation, and the pilot may present to the aeromedical examiner without any obvious signs. Urine analysis, and the more reliable method of hair analysis, is used in the workplace to screen for alcohol and drug misuse. Consumption of alcohol sufficient to cause liver damage should be considered an indication that alcohol dependence may be present.

There are two main issues that emerge from the research into the effects of alcohol upon air crew performance. The first is that even where there appears to be a low blood alcohol concentration (BAC), performance of pilots can still be adversely affected. The second is that due to the negative effects of alcohol after consumption, often characterized as post alcohol impairment (PAI), air crew should not fly until their BAC has returned to zero, and stayed at zero for some time. It has been observed that most pilots understand the importance of the first point, but underestimate the consequences of the second (Cook, 1997).

Attention Deficit Disorder (ADD) and Attention Deficit Hyperactivity Disorder (ADHD)

Attention Deficit Disorder (ADD) is a disorder associated with impaired concentration and attention, easy distractibility, and difficulty staying on task. Attention Deficit Hyperactivity Disorder (ADHD) also includes the hyperactivity component. The range of severity of symptoms varies widely. These disorders are considered disqualifying by the FAA, and require an FAA decision after review of all clinical information, including specific medications used, before a medical certificate can be issued. Although typically diagnosed in childhood, some experienced pilots may seek evaluation and treatment or may self-diagnose, particularly after learning that a child or sibling has been diagnosed with the disorder. A case example would be an experienced commercial airline pilot with a history of excellent college grades who decides to take one of his child's Ritalin tablets. Due to the stimulant effect, the pilot may feel more focussed and believe that (s)he must therefore have some form of ADD. In fact, anyone taking a stimulant may have temporary improvement in attention. ADD can present in milder forms without the need for medication. Although adult ADD is present in the population, the diagnosis of adult ADD in a pilot should only be made by experienced mental health personnel with knowledge of the aviation environment.

Organic Mental Disorders Causing Cognitive Defect

Many disorders may produce dementia syndromes or produce cognitive defects that would be disqualifying for flying, even in the absence of psychotic symptoms. Disorders such as Alzheimer's disease, Huntingdon's Disease, and Wilson's Disease can all initially present with mild cognitive difficulties and psychiatric symptoms. Metabolic disorders (such as thyroid and parathyroid disturbance, vitamin deficiencies, anoxia, and hepatic failure), chronic inflammatory conditions (collagen-vascular disorders), head trauma, neoplasms, and central nervous system infections may also cause cognitive changes. Of particular interest in the aviation environment would be toxic conditions such as exposure to industrial agents, pollutants, solvents and inhalants, and heavy metals. Other toxins which may produce cognitive defects include alcohol and illicit drugs, anticholinergic compounds, corticosteroids, antihypertensive agents, and cardiac agents. These disorders are considered disqualifying by the FAA, and even those that might resolve with treatment would require an FAA decision before any medical certificate could be issued.

Personality Disorders

Personality is distinguished from moods and emotional states that are transient. Personality *traits* become a personality *disorder* when they are found to be inflexible, maladaptive and cause significant impairment in social and occupational functioning (American Psychiatric Association, 2000). The diagnosis of personality disorder is made when an adult shows long-standing and pervasive impairments in their ability to work and co-operate with others (Colinger, Syrakic, Bayon, & Prezybeck, 1997). This negative impact on others is crucial to the diagnosis of personality disorder (Tyrer, Casey, & Ferguson, 1993). Thus, a personality disorder could have a serious impact on a pilot's functioning and is a clinically indispensable concept.

Serious personality disorders are likely to be detected at an early point in a pilot's career (probably during training) as (s)he is likely to act out, leading to disciplinary problems, or demonstrates an inability to cope with the stress of flight training due to a excessive need for control or dependency. Narcissistic, borderline and antisocial personality disorders appear to be more easily recognized by clinicians. Those with less serious personality disorders are unlikely to be removed at an early stage. These individuals typically lack flexibility and often have interpersonal problems. These problems may manifest when the pilot is required to assume greater responsibility or during crew coordination situations. Obsessive compulsive personality disorder may not be fully recognized during flight training unless quite severe. Personality traits that would likely interfere with aviation safety include impulsivity, risk-taking behavior, poor anger control, dogmatism, low sociability, and low team orientation (Berg, Moore, Retzlaff, & King, 2002).

None the less, there are serious obstacles to the accurate detection and assessment of personality disorder. While emotional disorders such as depression can be diagnosed in a single consultation with a mental health clinician, multiple sources of evidence are usually required for a reliable personality disorder assessment. Information from family, peers, and supervisors is very helpful, as the pilot being evaluated generally has little to no personal awareness that the personality traits are *maladaptive*. The presence of Axis I psychiatric disorders, such as depression or PTSD, make the diagnosis of a personality disorder more difficult and some clinicians will not make an Axis II diagnosis in the presence of a current Axis I diagnosis. Psychological testing can provide useful information in evaluating a pilot's personality problems, but the accuracy of psychological tests is not sufficient for establishing a diagnosis of personality disorder. In addition, the personality test employed needs to include a measure of defensiveness, as pilot applicants for air carriers are frequently defensive on psychological tests (Butcher, Morfitt, Rouse, & Holden, 1997). Similarly, there is a high likelihood that a pilot is likely to be defensive in a single consultation and additional sources of evidence will be needed to establish whether the pilot has significant problems. It is not uncommon for pilots referred to mental health personnel who have no prior exposure to pilot personality to be diagnosed with narcissistic personality disorder or paranoid personality disorder. The typical pilot can appear fairly paranoid when (s)he perceives that their mental health (and career) is under attack, based on a history of recent personality conflicts with co-workers and supervisors. Unfortunately, the impact of such a diagnosis for a pilot can be devastating. Such diagnoses are best made after a thorough evaluation by mental health personnel who are familiar with aerospace medicine and the psychological demands of the aviation environment.

The diagnosis of a personality disorder that is severe enough to have repeatedly manifested itself by overt acts is considered disqualifying by the FAA. The term *overt acts* generally refers to 'acting out' behavior, such as impulsivity, poor social judgment, and disregard or antagonism towards authority, especially rules and regulations. Personality disorders generally require long term psychotherapy treatment, and some personality disorders such as borderline personality disorder may require treatment with psychotropic medications.

Pilot Suicide by Aircraft

The risk of suicide in pilots diagnosed with psychiatric disorders must be emphasized, as the risk exists for pilots just as it does for the general population who experience psychiatric disorders. Pilots, however, may commit suicide by aircraft, killing not only themselves, but also the people they carry onboard the aircraft. This has been demonstrated by several aircraft crashes in the last decade where the evidence points to pilot suicide being the cause. Cases of alleged pilot suicide in commercial airline operations are generally considered to be rare, although the absence of specific data makes the true extent difficult to determine, particularly if one considers all private aviation and military aviation mishaps. It has been estimated that between 0.72 percent and 2.4 percent of general aviation accidents are as a result of pilot suicide and a history of psychiatric or domestic problems have been found in the post-crash inquiries and investigations (Cullen, 1998). One quarter of 37 cases of pilot suicide in the USA between 1983 and 2003 among general aviation pilots were associated with alcohol use, while 14 percent had used other illicit substances (Bills, Grabowski, & Li, 2005). These researchers also found that pilot suicide was more likely to occur among younger pilots who were flying alone and there was a tendency for crashes to occur away from the airport.

Pilot suicide has also been linked to certain commercial airliner crashes. In the Egypt Air Flight 990 crash of 1999, the relief first officer was recorded to have said "I rely on God" prior to disengaging the autopilot. Subsequently he was recorded as making the same statement a total of 11 times during the plane's impending crash without any obvious surprise or anxiety in his voice. The National Transportation Safety Board (NTSB) determined that the probable cause of the Egypt Air flight 990 accident was the airplane's departure from normal cruise flight and subsequent impact with the Atlantic Ocean as a result of the relief first officer's flight control inputs. The reason for the relief first officer's actions was not determined. Suicide, however, seemed likely. Pilot sabotage was suspected in the Silk Air 737 crash in December 1997 where the aircraft plummeted into a river in Indonesia, killing all 104 passengers and crew. Investigators believe that the pilot deliberately flew the plane into the ground. The former military pilot in command had a history of gambling and financial problems and had taken out a life insurance policy the day before the flight. The cause of the crash of a Royal Air Maroc commuter plane in 1994, which killed 54 people, was deemed a case of pilot suicide. In 1982, a Japan Airlines pilot was institutionalized after trying to crash the DC-8 that he was flying into Tokyo's Haneda airport, killing 24 passengers in the process. In 1998, an Air Botswana pilot informed air traffic control of his intention to fly his empty plane into the president's residence. When deterred he chose instead to crash it into the remainder of the airline's fleet at the airport at Gaborone. The pilot, believed to have been grounded after an AIDS diagnosis, died in the crash. Miraculously, no one on the ground was injured.

One can only speculate as to the possible reasons why a pilot should want to end his or her life in this way. The psychology of the self-destructive pilot has been explained in terms of the pilot's relationship with his/her plane; the aircraft is both his or her relished workplace and simultaneously the cause of frustration and stress in the pilot's life (Yanowitch, Bergin, & Yanowitch, 1973). The suicidal pilot's desire to protect his or her family and memory from shame might be bound up in the belief that any evidence of what happened would be destroyed in the ensuing crash. In some cases, the suicide crash could also be an expression of intense anger towards or revenge for problems that have their origins in the workplace, or a grudge against an employer. Failure to achieve promotion, anger regarding airline industry pay cuts, stress from military deployments, and even the stress of the threat of an employer's bankruptcy may be realistic triggers. Relationship problems, substance misuse, mood disorders, and personal problems stemming from debt are, however, more likely causes. A post-crash psychological autopsy must be carried out where the pilot's life and lifestyle are thoroughly investigated in order to gain a clearer understanding of suicidal intent prior to the event (Jones et al., 1997).

Any pilot who is diagnosed with a psychiatric disorder, or who exhibits significant psychological problems, should be temporarily or permanently grounded. Aeromedical personnel and aircrew should be aware that behavior may radically change prior to a suicide. A pilot might become more depressed and isolative, or a pilot who has appeared to have a low mood may suddenly become much brighter and calmer (signaling that the pilot may have made up his mind to commit suicide and feels better for making the decision). A vulnerable time for a pilot with serious depression is when the pilot begins to get his energy back after four to six weeks of treatment with an antidepressant. Previously the pilot may have lacked the energy to carry out a suicide plan, but during the early response phase to the medication, the pilot should be monitored carefully for suicidal ideation, intent, and plan.

Suicidal behavior is a behavior, not a diagnosis. A pilot who has demonstrated suicidal behavior should be grounded, and a mental health evaluation should be obtained. Depending on the diagnosis and prognosis, a pilot may request consideration for return to flying after treatment and resolution of symptoms, as well as a period of observation. However, the history of a suicide attempt requires an FAA decision prior to issuance of medical clearance.

Treatment

Counseling and Psychotherapies

Various types of psychotherapies and behavioral techniques have been useful in the treatment of psychiatric disorders. Anxiety, mood, and adjustment disorders are particularly responsive to psychotherapy. Cognitive-behavioral therapy (CBT) is a commonly used type of therapy for depression and anxiety, and CBT targets abnormal coping strategies and negative cognitions. Action-oriented pilots often report feeling more comfortable with CBT because CBT focuses on learning skills and utilizes workbooks and homework. Other types of psychotherapies include interpersonal psychotherapy, supportive psychotherapy, and psychodynamic psychotherapy. Behavioral techniques may include applied relaxation training, biofeedback, breathing retraining, and graded exposure (imaginal, virtual reality, and/or in vivo).

Psychotropic Medications

In those individual cases in which medication is required, it is advisable for the pilot with the psychiatric disorder to be evaluated and treated by a psychiatrist. Many pilots are quite hesitant to use psychotropic medications, believing that using an antidepressant, for example, may ground them permanently. However, in those cases in which the symptoms are moderate to severe, beginning an antidepressant or anxiolytic earlier, rather than later, may result in a faster recovery time. The following groups of medication are ones more commonly used in the treatment of pilots with common psychiatric disorders.

Selective Serotonin Reuptake Inhibitors (SSRIs) and Serotonin Norepinephrine Reuptake Inhibitors (SNRIs) This widely used class of drugs includes among others: Celexa (citalopram hydrobormide); Cymbalta (duloxetine HCL); Lexapro (escitalopram oxalate); Effexor (venlafaxine); Prozac, Sarafem (fluoxetine); Paxil, Apotex, Pevexa (paroxetine); Zoloft (sertraline HCL); and Luvox (fluvoxamine maleate). These medications are typically used to treat mood and anxiety disorders, and all are unacceptable for medical certification at the current time. Typical side effects, depending on the specific agent, may include: drowsiness, insomnia, anticholinergic effects, agitation, anxiety, tremor, gastrointestinal effect, fatigue, and sexual side effects. These medications are generally prescribed for at least 6–12 months in the treatment of depression, and require a tapering period of one to two weeks prior to discontinuation to minimize side effects. Wellbutrin (bupropion

HCL) is a non-SRI antidepressant with a risk for seizure at higher doses, and is also unacceptable for medical certification.

Although the FAA is studying the feasibility of granting medical certification to individuals who have been stable on SSRIs for treatment of depression without suicidal thoughts, aeromedical personnel must be aware that all SSRIs and SNRIs are unacceptable for medical clearance at the current time. Aeromedical personnel should be aware that pilots often decide to stop the antidepressant medications after three months if they have had a good response, hoping to return to flying sooner. However, the chance of recurrence of a depressive disorder is higher if the medication is not continued for a minimum of six months.

Tricyclic Antidepressants (TCAs)

The tricyclic antidepressants, used in the treatment of mood and anxiety disorders, as well as neurologic pain disorders, are an older group of medications with the potential for significant side effects, including significant overdose potential in suicidal patients. This class includes such medications as Elavil (amitriptylene); Norpramin (desipramine); Tofranil (imipramine); and Vivactil (protriptylene). Although the efficacy of the TCAs has been demonstrated to be equal to the SSRIs and SNRIs, these antidepressants are less frequently used now due to their side effect profile, and are disqualifying for flight when used to treat psychiatric disorders.

Benzodiazepines and Anxiolytics The benzodiazepines are typically used to treat anxiety disorders and some sleep disorders. The shorter-acting agents such as Xanax (alprazolam), Restoril (temazepam) and Ativan (lorazepam), as well as the longer-acting agents such as Klonopin (clonazepam), Valium (diazepam), and Librium (chlordiazepoxide), may produce sedation, delayed reaction times, impaired cognition. Individual response to benzodiazepines varies widely. Buspar (buspirone) is a non-benzodiazepine anxiolytic which may require three to four weeks of use before response is seen. Use of any of these agents requires grounding of the pilot, and further grounding may be required based on the diagnosis.

Alternative Treatments

St. John's Wort (hypericum) and S-Adenosylmethionine (SAMe) are two alternative treatments that have been reported to relieve depression. St. John's Wort is a herbal preparation used for many years in Europe to treat mild to moderate depression. SAMe is a component of many metabolic functions in the body, including the production of brain chemicals and antioxidants, which has also been reported to improve generalized depression, although its safety and effectiveness for long-term use has not been established. Although the FAA does not restrict the use of non-regulated nutritional supplements or herbal preparations, aeromedical examiners should always ask the pilot why (s)he is taking St. John's Wort or SAMe. If the pilot is using these alternative treatments to treat a self-diagnosed depression, then aeromedical certification must be deferred to allow for further evaluation of the depression. The primary concern is that pilots wishing to avoid being grounded for taking prescribed antidepressant medications will use one of these compounds. Given the serious nature of depression and the risk of recurrence of major depressive disorder even after treatment with counseling and antidepressant medications, a mental health evaluation is warranted prior to using alternative treatment compounds.

Conclusion

Psychiatric disorders are common in the general population and pilots are consequently not immune from them. Whilst most pilots are psychologically robust individuals, changes in domestic, social, and occupational circumstances may produce stress and other adverse reactions. Furthermore, internal biological changes can lead to a gradual decrement in performance or, in some cases, acute psychiatric disturbance. Psychiatric disorders are a threat to pilot performance and aviation safety. The presence of certain psychiatric disorders are disqualifying and will automatically prevent some of those who aspire to learn to fly from embarking on pilot training. Those who are accepted for pilot training and obtain their license are required to undergo frequent medical checks for as long as they hold a license, including assessment for the onset and presence of any disqualifying disorder. Psychiatric and psychological assessment of pilots should always be carried out in such a manner that sensitivity and respect are conveyed by the attending professional.

References

Alkov, R., Borowsky, M., & Gaynor, J. (1985). Pilot error as a symptom of inadequate stress coping. *Aviation, Space and Environmental Medicine*, 56, 244–247.

American Psychiatric Association (2000). *Diagnostic and statistical manual of mental disorders (DSM-IV-TR)*. Washington, DC: Author.

Axelson, D., Doraiswamy, P., McDonald, W., Boyko, O., Tupler, L., Patterson, L., et al. (1993). Hypercortisolemia and hippocampal changes in depression. *Psychiatry Research*, 47, 163–173.

Berg, J., Moore, J., Retzlaff, P., & King, R. (2002). Assessment of personality and crew interaction skills in successful naval aviators. *Aviation, Space and Environmental Medicine*, 73, 575–579.

Berg, J., Grieger, T., & Spira, J. (2005). Psychiatric symptoms and cognitive appraisal following the near sinking of a research submarine. *Military Medicine*, 170, 44–47.

Bills, C., Grabowski, J. & Li, G. (2005). Suicide by aircraft: A comparative analysis. *Aviation, Space and Environmental Medicine*, 76, 715–719.

Bremner, J. (2003). Functional neuroanatomy correlates of traumatic stress revisited seven years later, this time with data. *Psychopharmocology Bulletin*, 37, 6–25.

Brooks, N. & McKinley, W. (1992). Mental health consequences of the Lockerbie disaster. *Journal of Traumatic Stress*, 5, 527–543.

Butcher, J., Morfitt, R., Rouse, S., & Holden, R. (1997). Reducing MMPI-2 defensiveness: The effect of specialized instructions on retest validity in job applicant samples. *Journal of Personality Assessment*, 68, 385–401.

Colinger, C., Svrakic, D., Bayon, C., & Prezybeck, T. (1997). Personality disorders. In S. B. Guze (ed.) *Adult psychiatry* (pp. 301–317). Washington University, Mosby.

Cook, C. (1997). Aircrew alcohol and drug policies: A survey of commercial airlines. *The International Journal of Drug Policy*, 8, 153–160.

Cooper, C. & Sloan, S. (1985a). Occupational and psychological stress among commercial airline pilots. *Journal of Occupational Medicine*, 27, 570–576.

Cooper, C. & Sloan, S. (1985b). The sources of stress on the wives of commercial airline pilots. *Aviation, Space and Environmental Medicine*, 56, 317–321.

Cullen, S. (1998). Aviation suicide: A review of general aviation accidents in the UK, 1970-96. *Aviation, Space and Environmental Medicine*, 69, 696–698.

Fine, P. & Hartman, B. (1968). Psychiatric strengths and weaknesses of typical Air Force Pilots. Technical Report 68-121, USAF School of Aerospace Medicine, Brooks AFB, TX.

Fowlie, D., & Aveline, M. (1985). The emotional consequences of ejection, rescue, and rehabilitation in RAF aircrew. *British Journal of Psychiatry*, 146, 609–613.

Heinrich, T. & Grahm, G. (2003). Hypothyroidism presenting as psychosis: Myxedema madness revisited. *Journal of Clinical Psychiatry*, 5, 260–266.

Johnston, A. & Kelly, M. (1988). Post accident/incident counseling: Some exploratory findings. *Aviation, Space and Environmental Medicine*, 59, 766–769.

Jones, D., Katchen, M., Patterson, J., & Rea, M. (1997). Neuropsychiatry in aerospace medicine. In R. DeHart (ed.) *Fundamentals of aerospace medicine* (pp. 593–642). Baltimore: Williams & Wilkins Publishers.

Karlins, M., Koss, F., & McCully, L. (1989). The spousal factor in pilot stress. *Aviation, Space and Environmental Medicine*, 60, 1112–1115.

Kessler, R., Berglund, P., Demler, O., Jin, R., Koretz, D., Merikangas, K., et al. (2003). The epidemiology of major depressive disorder: Results from the National Comorbitiy Survey Replication (NCS-R). *Journal of the American Medical Association*, 289, 3095–3105.

Levy, R. (2000). Psychiatry. In R. Rayman, J. Hastings, W. Kruyer and R. Levy (eds.) *Clinical aviation medicine* (3rd ed., pp. 289–312). New York: Castle Connelly.

Little, L., Gaffney, I., Rosen, K., & Bender, M. (1990). Corporate instability is related to airline pilots' stress symptoms. *Aviation, Space and Environmental Medicine*, 61, 977–982.

Lundin, T. (1995). Transportation disasters – A review. *Journal of Traumatic Stress*, 8, 381–389.

Raschmann J., Patterson, J. & Schofield, G. (1990). A retrospective study of marital discord in pilots: The USAFSAM experience. *Aviation, Space and Environmental Medicine*, 61, 1145–1148.

Reinhardt, R. (1970). The outstanding jet pilot. *American Journal of Psychiatry, 127,* 48–52.

Sloan, S. & Cooper, C. (1986). Stress coping strategies in commercial airline pilots. *Journal of Occupational Medicine, 23,* 49–52.

Tyrer, P., Casey, P., & Ferguson, B. (1993). Personality disorders in perspective. In P. Tyrer & G. Stein (eds.) *Personality Disorder Reviewed* (pp. 1–16). Gaskell: Royal College of Psychiatrist, London.

Ustun, T., Ayuso-Mateos J., Chatterji, S., Mathers, C., & Murry, C. (2004). Global burden of depressive disorders in the year 2000. *British Journal of Psychiatry, 184,* 386–392.

Ursano, R. (1980). Stress and adaptation: The interaction of the pilot personality and disease. *Aviation, Space and Environmental Medicine, 51,* 1245–1249.

Yanowitch, R., Bergin, J., & Yanowitch, E. (1973). Aircraft as an instrument of self-destruction. *Aerospace Medicine, 44,* 675–678.

Chapter 9

The Psychiatric Evaluation of Air Crew

Gordon J. Turnbull

Introduction

Starting psychiatry was difficult. I expect it is the same for everybody. It is such a huge subject from the outside and, when you get into it, even more formidable.

When I arrived at the Neuropsychiatric Centre, the main base of RAF psychiatry worldwide at Princess Alexandra Hospital at RAF Wroughton in Wiltshire, I already had a reasonable amount of experience as a general physician. I was hoping that I would be able to graft the new "shoot" on to the established "rootstock." However, I gradually began to realize that the main change was to be in myself, as a clinician less dependent on technological tools and more reliant on my own powers of observation, rationalization, and interpretation. I will be eternally grateful to my two [tor]mentors who applied themselves to "breaking me in" to the new field and my professional and personal development. They helped me to make sense of the new discipline, showing me how to look at things in a broader, more multifaceted way and to build on hard-won clinical insights from my years as a physician.

My mentors each had their own way of teaching. One had a more practical approach, possessing an eyebrow that Freud himself would have been proud of, which he could raise to remarkable heights on his forehead that defied the laws of anatomy if my formulation went astray when I was presenting to him. This led to a very steep learning curve. The second had a much more didactic approach and introduced me, gently, to a vast new range of ideas and theories. Without doubt the strangest advice received was that effective psychiatrists had to be flexible in approach and have "several faces" to suit the job in hand.

Looking back, the "eyebrow" taught me to rely on intuition, on what I *felt* about the information that I was gathering together and that there was something lurking in myself that was important to the processing of that information. This taught me just how active the *listening* process is. My new patients almost always knew where their own problems lay and they would tell me when they were ready to. On reflection, perhaps that is where the word "patient" comes from. Then I began to understand that being like a *chameleon* was not ingenuous but it was a very good way to encourage the *accommodation* necessary to promote safety and security. I also realized that this accommodation was quite different from avoiding confrontation. Confrontation closes communication down but being flexible permitted an opening up of channels of communication. Obtaining a good history provided the solid foundation on which everything else would be built and, without any doubt, represents the only possible starting point for a psychiatric evaluation.

This is never more true than when evaluating the mental state of air crew. Pilots and other aviators need to feel safe during a psychiatric evaluation because it always represents a potential threat to livelihood and career. Air crew hope never to meet a psychiatrist and are always going to be defensive when they do. However, winning their confidence will often lead to the development of strong therapeutic alliances because pilots and other aircrew are conscientious individuals who need to know that they can depend on themselves as much as others can depend on them.

Although the focus of my comments on the psychiatric evaluation of air crew is pointed toward pilots, to some degree my comments can also be applied to others who are involved in the operations of the air transportation industry. For example, I would not be at all surprised if what I have to say can be related to the evaluation of air traffic controllers; however, in all fairness to my colleagues and their patients, I shall not overly generalize my comments to cover all persons related to flight. Whatever your curiosity, I would ask that you approach each additional group with the same respect as I urge you to have with pilots.

Establishing Safety with Pilots

Establishing ground rules at the first meeting is essential and crucially important to the quality of the psychiatric evaluation. To reiterate, most pilots would prefer never to see a psychiatrist, most of them never do and none believes that it will ever happen to them. Their fears have been reinforced by caricatures often picked up from films. The "cartoon" psychiatrist is meant to remain solemnly silent while the patient "spills the beans" on the couch and, even worse, the patient never gets to know what the psychiatrist thinks or will say in his report or to whom the opinion will be relayed.

Any such preconceptions must be dispelled as quickly as possible, especially with pilots. The psychiatrist has to take the lead; taking the pilot into the consulting room from the waiting room, or at least meeting them at the door helps. Shaking hands is even better and can provide useful indicators of just how nervous the patient is, with sweating palms, tremor etc. Those in any doubt about the reassurative value of the handshake should try deliberately avoiding the gesture once or twice to see for themselves the problems that can arise as a result.

An invitation to sit down in a comfortable chair is a good idea. Conducting the business of a psychiatric evaluation across a desk, even a wide one, is very confrontational and intimidating. In one's own domain armchairs are much better. If there is no choice but to sit at a desk because the psychiatrist is a visitor to a medical department and the room is organized for a physician or a surgeon then it is usually possible to sit alongside the desk which is much less challenging.

Commitment to confidentiality must be made at the outset. It is critically important to do this straight away. The question of dispersion of information will be in the minds of both parties in any case and it is a matter of bringing the subject up. If the patient has to introduce the topic first then the evaluation is almost guaranteed to be heading into trouble. The patient must see the psychiatrist as a person in whom it is safe to confide often very personal issues and will be encouraged considerably to do so if he can see that confidentiality is an important matter for the psychiatrist too: as it must be if he brought it up first. The psychiatrist's simple gesture of introducing the topic of confidentiality means to the patient that the purpose of the evaluation is to listen to what the patient has to say and is not a mere matter of reportage. It implies that there will be an opportunity to discuss important matters later on, that there will be time to digest information that surfaces and that there will be agreement about what will eventually go into the report. Pilots are typically conscientious and responsible individuals who will not want to conceal important matters from appearing in the report but they will seek reassurance that what is written is the truth. It is often helpful to ask the pilot if he knows why the referral has been made and to show him/her the referral documentation.

The psychiatrist should reassure the pilot that there are no preconceptions and that his mind is not already made up about the outcome of the evaluation. "Well, it's very obvious that you are depressed," would be a disastrous "opener" and would lead nowhere.

Personal experience has clearly demonstrated that making pilot's feel safe and that common sense and respect are essentials during psychiatric evaluation might tempt one to say that there is nothing exclusive about that; that these are pre-requisites for all medical consultations, whatever the speciality. That must be correct but there *is* something different about face-to-face meetings with people who fly aircraft and are accustomed to being in control. Pilots are naturally precise, meticulous, and organized individuals and special attention must be paid to the interview technique and to these factors. Pilots have been trained to perfect their natural ability to focus and have a highly-developed capacity to disconnect from their emotions so that they are not distracted from cognitive processes when dealing with emergencies. We need our pilots to have the capacity to *deny* or *suppress* their emotions in stressful situations in the cockpit so that they can remain cool-headed and deal with emergencies. However, it is precisely the same quality that can prove to be an insurmountable barrier to communication if the set-up in a psychiatric consultation is not right. It has been said that a good pilot is introverted while flying and can conveniently convert into an extravert on the ground.

When a pilot has a degree of control in a psychiatric consultation he will feel much more comfortable, just as if he has "the stick" in his hand. If the pilot lacks that sense of safety then the consultation will just not work.

Practical Problems

Psychiatrists are often under pressure to produce formulations and practical recommendations in a relatively short space of time. In my view, very few properly conducted psychiatric evaluations (interviews) can be carried out in less than one hour. In an ideal world it would be best not to organize consultations back-to-back. In an ideal world, two hours would be usefully allocated to the first assessment because of the amount of work that has to be attended to.

Most psychiatric evaluations will lead to reviews. It is not simply *preferable* that the same psychiatrist carries out the reviews, it is really vitally important. If that proves to be a practical impossibility for any reason then a lot of the work has to be repeated and, however diligently it is undertaken, the outcome will probably not have the same quality.

Sometimes it will simply prove to be impossible to arrive at useful conclusions about a particular individual as quickly as the referring agent would like. For example, assessment of the suitability of a personality for flying training might take more than a single consultation to complete. The development of uncharacteristic behaviors in an established pilot would be another example of the need to take a longer view than could be gained in a single encounter. Sometimes this can prove to be inconvenient if the pilot has traveled a long distance to see the psychiatrist or if a time scale for making a decision about fitness to fly has already been fixed, such as the starting date for a training course etc. These are situations in which good negotiating skills on the part of the psychiatrist become pre-eminent. Sensible and responsible plans can almost always be made.

The worlds of civilian and military aviation spin in parallel orbits not too far apart and they both depend on maintaining fairly narrow limits on acceptable behavior because the paramount factor is always flight safety. For this reason, it is totally predictable that when an individual steps over the line of what is generally regarded to be normal, acceptable behavior in a flying environment that this will be noticed very quickly and will provoke anxiety and concern. Aviation psychiatrists must understand this and need to take seriously the concerns of worried aviation occupational health professionals and managers and never be dismissive. Aviation psychiatrists have to make the time to communicate the findings of an evaluation to the referring agent.

How does the aviation psychiatrist learn to do this? Is there anything already established to guide psychiatric evaluations and to steer negotiations with referral agencies? Trial and error might lead to good practice in the end but the world of aviation simply has no room for attrition. *Transactional Analysis* offers a great deal.

Transactional Analysis (TA) (Berne, 1964/1996) is one of the most accessible and user-friendly theories in modern psychology. From the reference to Berne's first printing (1964) and more recent printing (1996) of *Games People Play*, one is instantly struck by the notion that TA is old, but not old-fashioned. It has wide applications in clinical, therapeutic, and organizational situations. It promotes good communication, is very useful as a management tool and can be used to harmonize relationships. It is, therefore, simply something that no aviation psychiatrist can operate effectively, or survive, without.

Transactional Analysis: Starting the Evaluation

TA provides a theory for the formation and functioning of personality, a model for communication and a study of repetitive patterns of behavior. It was founded by Eric Berne in the 1950s and is best known for its division of the personality into three components: *child*, *adult*, and *parent*. TA merits a closer look because it provides a practical and readily understandable way to conduct a psychiatric evaluation.

In the early 20th century, Sigmund Freud followed a long tradition of thinkers and philosophers who regarded human personality as having component parts in the subconscious. He described the *id*, the *ego*, and the *superego*. Freud's *id* represented the original, raw energy of the individual at the beginning of life. The *ego* was conceived as a personal executor equipped with psychological defence mechanisms which facilitate interactions with other individuals. The *superego* was the conscience factor initially derived from parental and cultural norms and regulatory mechanisms. Sometimes these three components are in a state of tranquil harmony but they could also be at war with each other, leading to conflicting thoughts, motivations and behaviors.

Broadly-speaking, Berne's *child* was the equivalent of Freud's *id*, his *adult* as the *ego* and his *parent* as the *superego*. Whichever theory is put forward, they all focus on the concept that each one of us has parts of our personality which come to the surface and affect our behavior in different circumstances.

Berne was very aware of the work performed by Dr. Wilder Penfield in the early 1950s, who showed that stimulating the temporal cortex of conscious human subjects with a weak electrical probe actually re-evoked past experiences and the feelings associated with them. The subjects replayed these events and their feelings despite not normally being able to recall them using their conventional memories.

Berne embraced Penfield's findings which suggested that the human brain acts like a tape-recorder, and while we may forget experiences, the brain still has them recorded. Additionally, the brain records feelings associated with events, and both feelings and events stay locked together. It is possible for a person to exist in two states simultaneously (because subjects replaying hidden events and feelings could talk about them objectively at the same time). Hidden experiences when replayed are vivid, and affect how we feel at the time of replaying.

Berne maintained that verbal communication, particularly face to face, is at the center of human relationships. His starting point was that when two people encounter each other, one of them will speak to the other. He called this the *Transaction Stimulus*. The reaction from the other person he called the *Transaction Response*. So, there is a transaction between two people who have three *alter ego states*, namely the *parent*, the *adult* and the *child*.

The *PARENT* is our ingrained voice of authority, absorbed conditioning, learning and attitudes from when we were young. We were conditioned by our real parents, teachers, older people, and people in authority roles in society. The PARENT is composed of a collection of recorded playbacks, both hidden and overt. It is formed by *external* influences as we develop. It is not immutable, but difficult to change.

The *ADULT* is our ability to think and reason and determine action for ourselves, based on received data. The ADULT is the means by which we keep our PARENT and CHILD under control. If we are to change our PARENT or CHILD we must do so through our ADULT.

The *CHILD* imprints *internal* reactions and feelings to external events. This is the emotional data collection within each of us and when the Child is in control it dominates reason. It can be changed too, but it is not easy.

PARENT is our *taught* concept of life.

ADULT is our *thought* concept of life.

CHILD is our *felt* concept of life.

When we communicate we are doing so from one of our own ego states. Our feelings at the time determine which one we use, and at any time something can trigger a shift from one state to another. When we respond, we are also doing this from one of the three states, and it is in the analysis of these stimuli and responses that the essence of TA lies.

At the core of TA is the rule that effective transactions must be *complementary*. That is, they must go back from the receiving ego state to the sending ego state. For example, if the stimulus is PARENT to CHILD, the response must be CHILD to PARENT, or the transaction is *crossed*, and there will be a problem between sender and receiver. If a crossed transaction occurs, there is an ineffective communication. Worse still, either or both parties will be upset. In order for the relationship to continue smoothly the parties must rescue the situation with a complementary transaction. The optimal one is ADULT to ADULT.

TA is effectively a language within a language; a language of true meaning, feeling, and motive. It can help in every situation, firstly, through being able to understand more clearly what is going on, and secondly, by virtue of this knowledge, to give choice as to which ego state to adopt, which signals to send, and where to send them. This facilitates making the most of all communications, and improves the ability to create, develop, and maintain better relationships.

There is no general rule as to the effectiveness of any ego state in any given situation in everyday life. Some people get results by being dictatorial (Parent to Child) but they will often experience crossed and unpleasant transactions with other people in Adult or Child, or by having temper tantrums (Child to Parent), but for a balanced approach to general day-to-day living, Adult to Adult is generally recommended.

For the interaction between the psychiatrist and the pilot there is only one effective transaction and that is ADULT to ADULT. TA has a highly practical value in the clinical assessment of pilots and air crew and it is to be commended to all medical professionals involved in their care and not just psychiatrists. However,

TA enables aviation psychiatrists to shift the defensive pilot from Parent to Adult and the overawed pilot from Child to Adult to enable the evaluation to proceed to a successful conclusion. Pilots are more often in Parent than they are in Child when first introduced to a psychiatrist.

TA became moderately popular with the publication of *Games People Play* (Berne, 1964/1996), but Thomas Harris's *I'm OK, You're OK* (Harris, 1976) boosted the awareness of the concept enormously. Aviation psychiatrists will find both books are extremely illuminating.

Now that the pilot patient is safely into an ADULT to ADULT transaction we will notice that p*hysically* he or she will be attentive, interested, straight-forward, with a tilted head, non-threatening *and* non-threatened. *Verbally*, he or she will be using words that convey messages such as why, what, how, who, where and when, how much, in what way, comparative expressions, reasoned statements, true, false, probably, possibly, I think, I realize, I see, I believe, in my opinion. An effective TA experience is dependent on what can be seen and felt and heard, because, only 7 percent of meaning is in the words spoken, 38 percent of meaning is paralinguistic (the *way* that words are said), and 55 percent is in facial expression.

We are now in a position to proceed with the part of the psychiatric evaluation known as the Mental State Examination.

Mental State Examination

The mental state examination (MSA) assesses general appearance, movement, and demeanor, orientation in time, place and person (known as the three spheres), moodstate, anxiety levels, form of thought, the thought stream, and the presence or absence of abnormal thoughts (delusions) and abnormal perceptions (hallucinations).

Psychoses

It is probably important to deal with the *psychoses* first because the vast majority of pilots and air crew referred are not suffering from these illnesses. They are characterized by abnormal form of thought (formal thought disorder); delusional thinking, found together with or in isolation from abnormal perceptions if they are *cognitive* in type, or involve abnormal moodstates when they are said to be *affective*. Sometimes, there is a mixture of the cognitive and the affective. *Functional psychoses* are not associated with organic brain injury or intoxication whereas *organic psychoses* are.

Psychotic illnesses can usually be readily identified. Their presentation is striking. It is said that *psychotics* believe that 2 + 2 equals 5 and are unconcerned, but that *neurotics* believe that 2 + 2 equals 4 but worry a great deal about it. The challenge is to identify the actual type of psychosis. Cognitive illnesses include the schizophrenic spectrum. There are significant changes in thought form (such as

"Knight's Move" thinking or thought loss), delusions (strange thoughts that make no sense and are unable to be challenged), and abnormal perceptions (auditory or visual hallucinations).

Affective illnesses involve changes in mood and usually also involve hallucinations together with or in isolation from delusional thinking and include unipolar mania or depression or bipolar mood disorder.

Mixtures of cognitive and affective psychosis include the spectrum of *schizoaffective states* such as schizomania, schizodepression or a mixture of schizophrenic and bipolar affective symptoms.

After identifying the type of psychosis the challenge is to organize appropriate treatment. It is extremely unusual for a pilot to be referred with a "yet-to-be-identified" psychosis. Much more common (but still, overall, very rare) is the pilot who has been receiving treatment for a psychosis and who has recovered and who needs to discuss his/her future career. This can be a very difficult encounter with a highly-motivated individual who wants nothing more than to get back into the flying role. However, the current international regulations are very clear and hold that individuals who have experienced even one episode of psychotic illness are not going to be, totally reliably, free from further episodes in the future and should not fly again. The current understanding of the natural history of such illnesses strongly supports that point of view. Psychosis is "unpredictably unpredictable." It is entirely possible for a moodstate to shift into a pathological form in a few hours or even a few minutes.

Cautionary Note: Dissociation

There is one cautionary note with regard to the diagnosis of psychotic illness and that is with regard to severe dissociative states that are actually not psychotic but can masquerade as such. Dissociation consists of a group of phenomena which have in common that the person maintains for a considerable length of time some course of action in which he appears not to be motivated by his usual self, or that his usual self does not have access to recent memories that one would normally expect him to have. Examples are sleepwalking, trances, fugues (in which the person wanders off, not knowing who or where he or she is), and loss of memory (dissociative amnesia) in which there is a gap in memory for some finite, recent period of time. Very severe dissociations occur in people who have split, dual or multiple personalities in which the personality appears to change from one mode to another. These disintegrations of an integrated identity are rare but they can occur after exposure to extreme, often life-threatening events. Derealization (when the environment seems strange or unreal) and depersonalization (out-of-body experiences) are further examples of dissociation which can occur in Acute Stress Reactions following air crashes or in posttraumatic stress disorder (PTSD) as the chronic postscript to the same. Because of the potential for extreme psychological shock in air crew who are unfortunate enough to be exposed to a major, life-threatening event such as an air crash there is a strong case for examining every air crew survivor for investigation of the traumatic impact (see Chapter 7 by Man Cheung-Chung, regarding PTSD, in this text).

Pilots typically possess strong psychological defenses and may escape the effects of psychological trauma altogether and demonstrate their resilience. The same strong defense mechanisms may conceal more immediate expression of psychological trauma and lead to the well-recognized phenomenon of *delayed-onset PTSD* which may not appear for several months or even many years. However, delayed-onset PTSD will have the same impact at the time of release as it would have had if expressed immediately. Delayed-onset PTSD tends to occur in those individuals with very robust defenses such as pilots (both military and commercial), police officers, military and emergency services personnel, and in medical and nursing professionals.

Neuroses

The neuroses are the usual conditions that present to aviation psychiatrists and are always connected with anxiety. They include the whole range of anxiety states with *generalized*, *phobic*, *panic*, and *somatic* presentations, *posttraumatic stress disorders* and *reactive* depression. These conditions are all stress-related and stress-induced. Because of the robustness of the pilot's psychological defenses they tend to present with phobic (where the anxiety is converted to another focussed form), panic (where adrenaline rushes cause episodes of physical anxiety), or somatic (in the form of a physical ailment such as irritable bowel syndrome or chronic pain).

Current insights into the etiology of these conditions emphasize their reversible biological dimension. Full recovery is entirely possible. The problems that pilots face, when they have a neurotic or stress-related condition, are not related to whether or not they can recover to fly again, but to receive appropriate treatment. Because stress reactions are so common and important in aviation psychiatry there is a special emphasis on them in this chapter.

Faced with the huge task of collating together a very large amount of information into something that is meaningful for both himself and the patient, the psychiatrist needs to develop a system which reassures him that he has reliably covered all aspects of the history. Bearing in mind that psychiatry is multidimensional; any system used needs to look at *psychological*, *social*, and *biological* dimensions.

New biological insights into the development and causation of the spectrum of stress reactions such as the spectrum of anxiety states, depression, and posttraumatic stress disorder (PTSD) make it much easier to be optimistic about the outcome of treatment because the underlying biological changes can be reversed. There should be an expectation of full recovery.

The older, more traditional psychiatric message was much more pessimistic. It tended to reinforce the pilot's worst fears; that he was inadequate, not up to the job and "once vulnerable always vulnerable." That was because the locus of the breakdown in normal functioning, the decompensation was thought to be in the structure of the personality. There was also great mistrust of psychotherapy, mainly because it was seldom thought to restore things to normal and that there would always remain a vulnerability which was anathema to a pilot's ability to control an aircraft reliably under all circumstances.

Multidimensional Psychiatry

Critics of psychiatry often regard the way that it is practised as being *reductive*. They see the psychiatrist as someone who aspires to reducing a patient's history to a psychological dimension. This could not be further from the truth because psychiatry is precisely the opposite discipline, it is actually *expansive*.

The modern psychiatrist has the task of weaving together the psychological, social, and biological strands of a history into a meaningful fabric which is the *biopsychosocial formulation*.

There is a danger in seeing the way that the mind and the body connect as merely *psycho-somatic* because this does not capture adequately the concept of a *mind-brain-body continuum* that conceptualizes the organism as a whole machine that works in harmony. Clinicians must look beyond dysfunctional behavior to the neurophysiological and autonomic nervous system dysregulation that is the source of symptoms and eventually disease. Medical science must shed the concept that a symptom not measurable by current technology is *psychological* and, therefore, invalid. Also, physicians must reject the pejorative implications of the term *somatization* and to stop any further traumatization of patients by subtly-implied rejection.

A Stress Model for Neuroses

The stress model has to embrace: 1. the sources of stress; 2. the amount of stress that is generated and; 3. the stress reaction. stressors (1) can be represented by a box, as can "stress reactions." The "stress loading" (2) has to have an open top which works as a "threshold" or limit.

The *evolution* of stress reactions moves from (1) to (2) to (3). The *resolution* of stress reactions moves in the opposite direction.

Stressors There is no practical value in simply compiling a list of stressors. The list would be too long and is much less meaningful than if the stressors are put into categories, or dimensions. There are three dimensions for practical purposes. These are the *psychological*, the *social*, and the *biological*.

The *psychological* dimension includes personal development and the three ego states identified by TA, the child, the adult and the parent influences the formation of the personality. All potentially problematic influences should be noted. This is the "living with oneself" dimension, the personal dimension.

The *social* dimension includes all interactive, relationship themes such as the family influences, marriage or partnerships, work relationships etc. which tries to identify any problem areas in interactions with others. This is the "rubbing shoulders with the rest of humanity" dimension (see also Chapter 8 by Morse & Bor, regarding the social dimension of psychiatric evaluation, in this text).

The *biological* dimension includes physical illnesses, medications, family history of illness, special stressors such as female gender (menstrual cycle), circadian rhythm disturbances, altitude stresses, heat/cold stresses, pulling "G" stresses, etc. Biological sources of stress are very important in aviation psychiatry and can be the only source of stress. I have seen cases where the only stressor that could be identified in air crew has been circadian rhythm disruption with disrupted *zeitgebers*, and the presentation has been classical depression. Each one of these dimensional stressors will generate some stress in everybody, everyday (see Chapter 6 by Waterhouse et al., regarding psychological effects of physical stressors, in this text).

Stress Loading Each individual will have some input from each of the above three dimensions of stress each day. The stress loading container has a finite capacity. It can take so much stress but no more. It helps to see the container as being filled up by "liquid" stress which really represents the amount of stress hormones that are generated to meet the current challenges. It is also important to see that the stress generated by psychological, social, and biological dimensions is all the same when it reaches the stress loading container. One source can no longer be distinguished from another. The level of stress will fluctuate, depending on circumstances. The rule is, though, that the level will rise if any single, all three, or any combination of all three dimensions is provoked. As the level rises there will be a sense of growing tension, then a "red zone" as the threshold is approached, and eventually, the level will reach the top of the container and spill over.

Stress Reactions This is a closed box, similar to the stressor box. The box itself represents individual *constitution*, the way that the individual is *primed* to react to stress over-accumulation. Psychological defence mechanisms can determine the type of stress reaction that will develop. The pilot will often become depressed when over-stressed. In addition, there appear to be genetically-predetermined ways in which stress reactions develop and an example of a genetic pre-vulnerability that leads to the development of a specific psychiatric condition would be schizophrenia and another manic-depressive illness. An example of a physical predisposition would be peptic ulcer disease.

I developed this model so that I could be sure that I had covered the necessary ground in my psychiatric evaluation, and would commend it to you. However, I made two important discoveries after I started to use it. Firstly, the identified stress reaction in box (3) often became a stressor in itself and added a fourth dimension of stressor. Furthermore, this secondary source of stress could carry on filling up container (2) even after the initial over-accumulation of stress from the original psychological, social, and biological sources had declined and would no longer be capable of filling up the container on their own. This explanation for chronicity can be very helpful to the pilot who cannot understand why his symptoms have not disappeared with waning stressor levels.

Secondly, I discovered that the model worked backwards. Forwards was *evolution* and backwards was *resolution*. Resolving factors associated with box (3) included identification of the nature of the stress reaction and appropriate biological treatments, while container (2) could be fitted with a tap which would allow draining away of the stress and might even prevent the threshold being reached and breached. This involves training in some anxiety-management method and includes a very wide range of activities. The best thing about learning how to "open the tap" is that it boosts self-esteem and boosts a sense of being in control of stress. It probably works by stimulating the production of endorphins. Bearing in mind that pilots operate in a stressful environment then this is of particular importance to them, avoiding frank illness and potential treatment with medication and all the complications that that causes. If I had my way, all pilots would utilize a stress-relieving activity as a contractual responsibility in their employment. Box (1) offered therapeutic exploration of the three dimensions of stress. For example, if there was a marital problem that was generating stress then that could be identified and treated. Circadian rhythm disruption in long-haul pilots and exhaustion in short-haul pilots are common stressors.

This stress model also permits enquiry regarding the use of alcohol, tobacco and possibly also other illicit substances which may be being used to alleviate the effects of stress.

The Neurobiology of Stress Reactions

All psychiatric evaluations dealing with stress should now include a modern explanation for stress reactions. This is essential in the evaluation of pilots. The reason for this is that an understanding of the underlying neurobiological changes in stress reactions not only appeals to the pilots' technical view of cause and effect but it also reduces their sense of helplessness, vulnerability and irreversibility, enhances their hopes for recovery and, overall, emphasizes normality and restores a sense of being in control.

These essential messages that can be imparted at the initial evaluation are varied in type and effect. In physics, a stress is a *force* that moves or distorts other objects. In biology, *adrenaline* provides the main force behind the drive to overcome problems. This is what we call the *stress response* and it is absolutely necessary if we want to "get on." Bigger challenges require *more* adrenaline. Extreme challenges require the *fight or flight* levels of adrenaline. Adrenaline is essential for normal living but these slight increases or even surges of adrenaline are only useful in the short term. If prolonged, increased adrenaline becomes a problem in itself rather than the fuel needed to solve problems.

Adrenaline is produced by the adrenal glands situated just above the kidneys. There is always some adrenaline circulating around the body. At *low* levels it is the "*get up and go*" hormone. We feel good and full of energy and enthusiasm. At *medium* levels it *sharpens* our mental and physical capacity to deal with unfamiliar challenges and permits the learning of new skills. The *"buzz"* the adrenaline gives us is enjoyable until the challenge escalates to the point of *fight or flight*. Then we begin to describe it as *"stress"* which becomes progressively more uncomfortable. *High* levels of adrenaline over longer periods of time make life very uncomfortable. Inability to sleep, feeling tense and jumpy, and irritability is both unpleasant for us and for those around us and can lead to states of anxiety and depression. However, the elastic sliding scale of adrenaline stretches and contracts to help us to cope with day-to-day living.

The system usually works very well unless we become stuck with high adrenaline levels which can happen in two situations. The first situation is when we lack the capacity to resolve the problem, and the second is when secondary stressors develop and lead to overwhelmed capacity. The brain is both the controller of the stress response and its prime target. It is equipped with bilateral sensors called *amygdala*, which resemble almonds (amygdala is Greek for almonds). The amygdala is constantly informed of what is going on in the outside world and also within the body – they act as "sentries" that guard our safety and never "go off duty." Visual, auditory, smell, taste, and touch perceptions from the outside world are continuously monitored by the amygdala for threat levels and the stress response is coordinated by sending signals to the locus coeruleus (the organ in the brain that secretes noradrenaline – the brain's version of adrenaline) and the adrenal glands to release appropriate amounts of their respective hormones to meet the challenge. Generally, the system works very well. However, sustained, high levels of stress hormones can damage certain parts of the brain (stress-induced neuron death). Of particular interest in stress reactions is damage to the *hippocampus*. The hippocampi (*"seahorse"* in Greek) are paired, curved organs embedded in both cerebral hemispheres and are the neurological seat of learning. The hippocampi convert *sensory* memory into verbally accessible memory. The hippocampi are very active parts of the brain and it is now known that this involves a turnover of nerve tissue. Up to 5,000 new neurons are generated in the hippocampus every day (neurogenesis). The discovery that the brain has the ability to produce new nerve cells in the form of *pluripotent* stem cells that can repair damaged nerve tissue was made in 2000. Exposure to elevated levels of stress hormones, and in particular cortisol and noradrenaline, has a neurotoxic effect on the hippocampus that impairs neurogenesis, causing atrophy of dendritic processes, cell dysfunction and ultimately cell death.

The attritional destruction of elevated levels of stress hormones in chronic stress reactions such as anxiety states or depression, *or*, a surge of stress hormones during exposure to trauma damages hippocampal structure and function and adversely affects the capacity to integrate new information, to form memory, to maintain normal mood, to maintain a sense of connectedness and concentration. Structural changes are most obvious in the *dentate nucleus* of the hippocampus (shaped like a tooth) which appears to act as a *fuse*, sensitive to the effects of high levels of noradrenaline and cortisol. The fuse acts like a brake in situations of unrelenting stress.

Neuronal damage to the dentate nucleus is temporary and lasts for about four weeks if stress hormone levels subside allowing the hippocampal tissue to heal. If the setback is temporary the acute stress reaction resolves but if there are complications the stress reaction moves on to a chronic phase and, as a result, the hippocampus will remain damaged. Memory processing is impaired and depressive ruminations and trauma memories, which are mainly in sensory form, remain disintegrated and fragmented, causing bewilderment and secondary stress which maintains the fault.

This turnover of nerve cells (neurons) in the hippocampus explains the onset of and recovery from stress reactions. The damage is temporary if the adrenaline levels can be brought down below the level that causes damage to the hippocampus. This can only happen if safety and control are restored quickly. Acute stress reactions, lasting for a few weeks and leading to recovery, can be seen as positively adaptive resulting in the learning of new survival skills. Since the hippocampus has a crucial role in memory processing and the development of new and innovative thinking, the memory imprint of the trauma remains unabsorbed, unassimilated and not properly understood. Depressed individuals continue to ruminate on their problems and lack insight. Psychologically traumatized individuals continue to re-experience the traumatic event together with emotional and physiological reactions. The result is that the victim remains "stuck" in the trauma, re-experiencing it over and over again, and responding to each reminder with a fresh spike of stress hormones at survival levels. This explains the three key features of PTSD: a cascade from flashbacks, to hyperarousal, to avoidance of reminding cues.

Posttraumatic Stress Disorder

This is an especially important stress reaction for air crew because of their exposure to critical incidents in the air. However, it is frequently not recognized.

The late discovery of symptoms indicating that PTSD has developed raises the possibility that the symptoms have been denied. Denial can be on the part of air crew or the examining doctor or can be a combination of both.

Davidson (1992) insisted that attending physicians and psychiatrists remain vigilant to reckon those symptoms most notably associated with trauma events in patients. Whereas depression and anxiety may or may not indicate the presence of PTSD, more pathognomonic symptoms of PTSD, such as intrusive recollections, flashbacks, and re-experiencing trauma through disturbing dreams, may present but might be diagnostically misaligned by professional psychiatrists.

There are a number of reasons why physicians will miss a diagnosis, regarding PTSD. Davidson (1992) noted that some physicians fail to ask the patient about the traumatic event. Other physicians are uncomfortable when discussing the details of the traumatic experience with their patient, so choose to pass by the retelling. Like some physicians, there are some patients who also are reluctant to talk about their horrifying or gruesome experience, which then disguises the true nature of the patient's malady. Psychiatrists might also overlook a PTSD diagnosis when the patient presents with seemingly unrelated symptoms such as headache, insomnia, tension, substance abuse, and interpersonal or professional dysfunction, further confounded by variables such as age, socioeconomic status, ethnicity, alcohol problems, and smoking history (Weathers, Keane, & Davidson, 2001).

If there has been a good beginning between the physician and patient, all of the symptoms, connected directly or indirectly to a traumatic event will be assigned their proper place. When the patient is a pilot, ready disclosure of the traumatic experience might be forestalled until there is an assurance of trust. Once this trust is established, you may find that a pilot – perhaps through exposure to higher education and technical training – is very articulate about those symptoms most wearisome to him or her.

Psychological Tests

These are especially useful in the psychiatric evaluation of pilots if used as an adjunct to clinical assessment because of their strong psychological defenses.

The General Health Questionnaire assesses psychological distress but does not identify a particular condition.

The General Health Questionnaire-28 (GHQ-28): (Goldberg & Hillier, 1979) is a 28-item self-rating scale which screens for psychological disorder in the general population. The threshold score for identifying "caseness," (i.e. above which there is an increasing likelihood that the person would be classified as suffering from significant psychological or psychiatric symptoms is 5), and the maximum possible score is 28. It focusses on breaks in normal function rather than life-long traits and, therefore, detects the appearance of new phenomena of a distressing nature. The GHQ-28 has subscales which measure: 1. somatic symptoms; 2. anxiety/insomnia; 3. social dysfunction; and 4. severe depression.

Depression is relatively common in the population of stressed pilots.

The Beck Depression Inventory (BDI): (Beck et al., 1979). The BDI measures depressive symptoms over the past seven days. It is a 21-item self-rating scale and is one of the most widely-used, validated, and reliable tests used in the assessment of depression. The Cognitive-Affective subscale measures the severity of depressive thoughts and feelings. The Somatic-Performance subscale measures the severity of the physical and social aspects of depression. A score of 0–9 is within the normal range.

10–18 indicates *mild-moderate* depression

19–29 indicates *moderate-severe* depression

30–63 indicates *extremely severe* depression.

Posttraumatic stress disorder is often denied by pilots and there are two tests that help to indicate that individuals have developed significant traumatic stress reactions. The first is a self-report screening test and the second a clinician-assisted questionnaire that identifies the core features of PTSD.

The Impact of Event Scale-Revised (IES-R). (Horowitz, Wilner, & Alvarez, 1979) is a 15-item scale designed to assess the extent to which a traumatic event has affected an individual's life over the most recent 14 days. It measures *intrusive* and *avoidance* phenomena of PTSD, that is, thoughts and images related to the trauma which intrude into the mind involuntarily, and efforts made by the individual to avoid being reminded of the trauma. This scale is commonly used as a screening tool for the presence of PTSD. However, studies have shown that it is also useful for making the diagnosis of the disorder. A cut-off score of 35 or above for making the diagnosis has been demonstrated in a British population (Neal et al., 1994). A maximum score of 75 is possible.

The CAPS (Clinician-Administered Posttraumatic stress disorder Scale) (Blake et al., 1990, revised 1998) measures PTSD symptom frequency and intensity. The instrument also assesses presence or absence of the disorder. A maximum score of 68 for symptom intensity is possible. There are scales to record currently active symptomatology and lifetime symptomatology (symptoms were present in the past but are no longer active).

Critics of the CAPS have suggested that it is too complicated, too lengthy, and that it fails to be sufficiently redundant as to frequency and intensity level redundancies. However, Weathers et al. (2001) pointed out that with proper training the instrument is not complicated, taking as little as two hours to orient the user. As to length, Weather et al. invited their critics to observe that most of the interview items were optional, allowing for more flexibility in its administration. Out of all the methods by which one could assess a patient, the CAPS has been the most investigated and therefore appears to be very reliable.

Conclusion

This chapter highlights the special conditions pertinent to the psychiatric evaluation of air crew. It offers Transactional Analysis as a means of achieving the best possible outcomes at the interview stage of the evaluation and a handy method for gathering relevant information in stress reactions, leading to discussion of available treatments.

It was suggested that there are those who see psychiatry as reductive. However, I suggested that this view does not ring true with experience. Psychiatrists, because they specialize in medicine and psychology, assess the biological, psychological, and social aspects of human behavior and performance when forming a diagnosis. Therefore, when a pilot presents with an array of symptoms, the clues toward a proper diagnosis are often those revealed in the areas of most concern to psychiatrists, not necessarily of concern to psychologists or general physicians. One can see the interplay of the biopsychosocial in the ubiquity of stress related symptoms among pilots.

Recent discoveries in the neurobiology of stress reactions have provided physicians with clues as to why patients behave in the ways they behave. Pilots, under extreme pressure to perform well while flying in inclement weather conditions, at night, while attempting to overcome the effects of disrupted circadian rhythms are often consumed by the negative side effects of prolonged performance under the effect of adrenaline. We know that neuronal damage, particularly in the hippocampi, can be caused by these prolonged states of stress. It is, therefore, of primary concern that we factor into our assessment an appreciation of those working conditions and environments that would most likely cause temporary injury to the brain, resulting in correspondingly temporary behavior changes that would most likely be arrested by the removal of the stressing event. At risk are those pilots who remain in their stressful environment for too long. Temporary neuronal damage in the hippocampi can be resolved with an intervention which stops the destruction of neurons due to stress. However, chronic exposure to damaging stress can result in permanent damage to the hippocampi. The psychiatrist must decide if the episode that triggered the assessment is of a temporary or chronic nature. You will be doing the pilot and the flying public a service if your remedy of biopsychosocial fitness, based on your assessment, coincides with the severity of the stress level and does not last any longer than is necessary to protect the health of the patient.

At the beginning of the chapter I made it clear that when evaluating the mental state of air crew you need to help your patient feel safe. I would add one more piece of advice. Not only make your patient feel safe, but make them feel that you have their long-term health interests in mind. At some point a pilot must stop flying, but this will happen long before he or she stops caring about family members, friends, spouses, and partners. The life your patient lives after his or her departure from flying should retain all its beauty and richness. It would be unconscionable to allow a pilot to drive him or herself into permanent neuronal damage because you limited your vision of your patient to just a few more years of flying service. Provide hope for your patient, by providing him or her with a long-term solution for sustained mental and physical health.

References

Beck, A., Ward, C., Mendelson, M., Mock, J., & Erbaugh, J. (1961). An inventory for measuring depression. *Archives of General Psychiatry, 4*, 561–571.

Beck, A., Rush, A., Shaw, B. F., & Emery, G. (1979). *Cognitive therapy for depression.* New York: Guildford Press.

Berne, E. (1964/1996). *Games people play: The psychology of human relationships.* New York: Ballantine Books.

Blake D., Weathers, F., Nagy, L., Kaloupek, D., Charney, D., Keane, T. (1990). *Clinician-Administered PTSD Scale (CAPS-I). Revised Clinician-Administered PTSD Scale for DSM-IV (CAPS-II).* Boston. MA: Behavioral Science Division, Boston National Center for Post-Traumatic Stress Disorder.

Davidson, J. (1992). Drug therapy of post-traumatic stress disorder. *British Journal of Psychiatry, 160*, 309–314.

Goldberg, D. P. & Hillier, V.F. (1979). A scaled version of the general health questionnaire. *Psychological Medicine, 9*, 139–145.

Harris, T.A. (1976). *I'm OK – You're OK.* New York: HarperCollins.

Horowitz, M. J., Wilner, N., & Alvarez, W. (1979). Impact of event scale: A measure of subjective stress. *Psychosomatic Medicine, 41*, 209–218.

Neal, L.A., Busuttil, W., Rollins, J., Herepath, R., Strike, P., & Turnbull, G. J. (1994). Convergent validity of measures of post-traumatic stress disorder in a mixed military and civilian population. *Journal of Traumatic Stress, 7*, 447–455.

Scaer, R.C. (2001). The neurophysiology of dissociation and chronic disease. *Applied Psychophysiology and Biofeedback, 26*(1), 73–91.

Weathers, F. W., Keane, T. M., & Davidson, J. R. T. (2001). Clinician-administered PTSD scale: A review of the first ten years of research. *Depression and Anxiety, 13*, 132–156.

Chapter 10

Psychological Assessment and Reporting of Crew Mental Health

Robert Bor

Psychologists may be called upon to formally assess air crew for their suitability to fly and to submit a report of their findings to either the crew members employer or licensing authority. Such requests place a significant burden of responsibility on the psychologist as many sensitive and complex issues are potentially at stake. These include safety and risk to the pilot, passengers, the airline and others; the possibility of the pilot losing his or her job, career and future work prospects; the pilot's personal esteem and the likely impact on family or dependents; and the broader financial risk to the airline and legal implications for insurance, among others. This chapter describes the place of psychological assessment and reporting in aviation, the essential requirements for these, and also presents an illustrative report to convey a sense of what might be considered appropriate for this undertaking.

Psychological assessments of pilots are undertaken for three main reasons. Firstly, at initial selection to determine whether the candidate has the necessary aptitude, personal resources, and skills to learn to fly. Secondly, at various points during their career advancement in order to assess whether the licensed pilot has the unique and specific temperament required to progress to more senior positions (e.g. command or a managerial role). Thirdly, and far less frequently, when there are concerns about a qualified and experienced pilot's mental state and the direct or indirect effect that this may have on his or her performance and ability to fly safely. Assessment may also be requested after an incident or accident.

This chapter is mostly concerned with the third scenario which, by definition, is exceptional and infrequent, and therefore never conducted as a matter of routine within an airline. Psychologists who have specialized in clinical, organizational (industrial) and less frequently in counseling psychology are most likely to be called upon to undertake general pilot psychological assessments. These are covered in the first two categories; however, those which involve mental state and psychopathology assessments, as is a fundamental requirement in the third category, will almost certainly be carried out by specialist clinical psychologists. It is imperative that the psychologist's report of a pilot's mental health is of the highest caliber. It must be in a format and of a quality that can be submitted as evidence in a court of law. The reasons are threefold: firstly, the pilot will normally have access to his or her report and may question some of the findings; secondly, given the possible deleterious effects on the pilot's career, his or her legal counsel may mount a challenge to the findings and the psychologist's report is likely to be closely scrutinized by other experts; and thirdly, the psychologist may be required to defend his or her report in a court of law.

Not only, therefore, must the assessment be comprehensive and robust, but it must also be conducted by a professional whose experience and specialism is conducting mental health assessments. Many different aspects of the assessment and report may come under scrutiny by lawyers and other psychologists. One that invariably diminishes the veracity of the report is the qualifications and expertise of the psychologist who undertook the assessment. The following is a short excerpt from a verbatim transcript between the advocate and psychologist in court from one such case that highlights the potential problem:

Advocate: "Can you please tell the court, Dr. Harvey, what your professional qualifications are?"

Psychologist: "I am a licensed Occupational Psychologist with a doctorate from X university."

Advocate: "Can you also please tell the court something about your experience of conducting mental state examinations and assessments of airline pilots as an occupational psychologist?"

Psychologist: "I have completed more than fifteen reports; I was trained to carry out psychological assessments in my university course. On a day-to-day basis, I consult to Airline G where I am responsible for selecting new pilots and conducting appraisal interviews. I also contribute to the assessments where first officers are being considered for a promotion to captain."

Advocate: "So you are not specifically qualified to carry out *clinical* assessments using *clinical* instruments?"

Psychologist: "Well, I do have experience in these and ..."

Advocate: (interjecting) "Then why are you not licensed as a *clinical* psychologist?"

Psychologist: "Ummm...."

In this case, the challenge to the conclusions of the report started by undermining the qualifications of the psychologist which, in turn, diminished the strength of the evidence presented. Due to the breakdown in his relationship with the airline, the pilot left his job but also received a sizeable financial payout as settlement. However, the fact that questions had been raised about his psychological condition, though not conclusively proved, made it difficult for him to find another job as a commercial pilot. This excerpt illustrates that it is not sufficient to claim expertise in using a particular psychometric test when reporting on a pilot. The psychologist must also demonstrate that he or she has the training, qualifications, and experience to interpret such tests and present a balanced account of the pilot's performance in the context both of their background as well as their mental state at assessment.

The quality of psychological reports on air crew has never been the subject of a systematic study and therefore it is difficult to gain insight into the extent of their use and their usefulness. However, there is anecdotal evidence that some are fundamentally faulty and may unfairly jeopardize a pilot's career. The shortcomings are varied and include: incomprehensibility, lacking in validity, being based on faulty assumptions or interpretations, inappropriate use or choice tests or over-reliance on testing, lacking in rigour and plainly biased. It is not surprising, therefore, that a sizeable proportion of pilots are skeptical about the place of psychological insights in aviation and show great antipathy towards both psychologists and the usefulness and fairness of psychometric testing. This may also be due to the fact that testing almost always benefits a third party such as the employer or licensing authority, leaving the pilot to draw the conclusion that he or she is a dispensable commodity. In some cases, little effort is made to assist the pilot or devise a program of treatment or rehabilitation, where this may be indicated. Some pilots argue that psychologists have only a very limited understanding of what their job really entails. This is regrettable as, used correctly, and taking into account certain limitations, psychological insights and methods have served aviation well in many different contexts. Psychology may not be a *precise* science, but this does not diminish its potential for value and usefulness.

Psychologists themselves face significant challenges in their assessment of pilots. The person requesting assessment may place unrealistic or unattainable demands on the assessment process and outcome. Determining psychological fitness to fly is complex and the results of testing may prove inconclusive. Apart from clear and incontrovertible cases (e.g. a pilot sustaining a severe and irreversible head injury following a fall whilst enjoying recreational rock climbing), the psychologist more usually has to make recommendations in his or her report based on probability, degree and risk. The degree of complexity increases even further where the pilot's pre-morbid state has to be inferred and where the pilot has many years of experience in their job and logged a considerable number of hours of flying time. The ability to predict a person's behavior in given situations nearly always poses a significant challenge. However, the need to protect passengers, crew, those on the ground and the airline from the actions of a pilot suffering from a disqualifying mental disorder may necessitate applying more stringent cut-offs when interpreting tests since the risks to third parties outweigh the level of risk that might be acceptable where only the individual may be affected.

Disqualifying Conditions

There can be few other professionals that require such frequent and rigorous health checks as a condition of certification to work as must be endured by pilots. These are checks are carried out as frequently as every six months for commercial pilots. The medical standards that have to be met by pilots are laid out by aviation authorities, and are broadly similar (see also Chapter 8 on psychiatric issues and syndromes by Morse & Bor). The US Federal Aviation Administration (FAA) and the European Joint Aviation Regulation (JAR) medical standards stipulate almost identical criteria although the former are arguably more clearly defined. These standards are considered the benchmark for determining pilots' medical fitness. The psychological requirements for medical certification state that pilots must not be suffering from any mental disorders, neurological conditions, or be dependent upon alcohol or recreational drugs. A past history of any of these can also serve as grounds for exclusion (see Reinhart, 1997 for a more thorough list of exclusions).

Pilots may not act in command of an aircraft or as part of a flight crew if they have a medical or psychological problem that would make them unable to meet the requirements for their current medical certificate. The use of any prescribed or non-prescribed drug that may interfere with a pilot's mental and physical faculties and threaten safety also invalidates medical certification. A recent history or sudden onset of a medical or psychological problem is therefore covered through certification. In view of the stringent requirements for psychological fitness, a history or diagnosis of any of the following mental disorders or states automatically leads to initial denial of medical certification and consequently spells the end of any hope of acquiring a pilot's license. A medical certificate will also usually be denied where a qualified pilot develops any of these disorders:

a. psychosis
b. affective disorders, including bipolar disorder [although there has been a move in exceptional cases to change the regulations pertaining to the use of selective serotonin re-uptake inhibitors]
c. personality disorders (especially where there has been evidence of overt acts of violence, a pattern of interpersonal problems or any other acting out behavior)
d. substance dependence (alcohol, sedatives, hypnotics, anti-depressants, recreational and illicit drugs, and inhalers)
e. neurosis
f. self-destructive acts
g. disturbance or loss of consciousness
h. transient loss of control of nervous system functioning without satisfactory explanation of the cause
i. epilepsy or convulsive disorders
j. progressive disease of the nervous system.

Medical and psychological assessment of pilots is by interview, physical examination and appropriate tests where indicated. Collateral information may be requested from the pilot's employer, line manager, occupational health department or personal family physician or general practitioner. Random checks (e.g. for excessive alcohol or illicit drug use) can also be made at any time. Candidates for assessment are also required to complete a questionnaire that authorizes the FAA or JAR to investigate further for alcohol offences, and provide examiners with information about visits to psychologists, clinical social workers, or related professionals for substance abuse or psychiatric evaluation or treatment within the preceding three years. This specifically excludes consultations for help with stress, relationships, and other situational problems. The penalties for failing to disclose information or being untruthful include hefty fines and custodial sentences. The pilot's insurance may also be jeopardized as a consequence of falsification of his or her health records. Even if a pilot were to pass an initial medical examination based on the absence of symptoms of either psychological or medical impairment, the onset of a new problem can automatically disqualify a pilot who is already licensed. Where the pilot believes that an unfair decision has been reached over medical certification, there is an appeals procedure. Evidence can then be submitted from other sources testifying to the pilot's abilities and health, in an attempt to persuade the licensing authority to reverse their decision (Bor, Field, & Scragg, 2002).

The most common causes of deleterious changes to a licensed pilot's mental health are unforeseen organic syndromes (e.g. late onset major depressive disorder), mental disorders due to a general medical condition and those problems that may be triggered by lifestyle and associated changes. These include alcoholism, drug misuse and anxiety (e.g. posttraumatic stress disorder, generalized anxiety disorder, and panic disorder), among others.

Psychological examination of a pilot is normally requested by an authorized medical examiner, employer, licensing authority or legal counsel. There may be simultaneous referral to a psychiatrist or other specialist. Psychological assessment is conducted in order to determine whether the individual currently suffers from any of the excluding psychological problems. The psychologist must, therefore, describe and report the *presence* of psychopathology or clinical syndromes. A further requirement is to offer an opinion as to the extent or severity of the problem and the likely impact that it will have on the pilot's performance. Prediction in this context is almost always challenging and in this section of the report, the psychologist should seek to draw on findings from published research to substantiate his or her opinion. It does not diminish the usefulness and quality of the report to state that it is not possible to speculate as to the likely consequences of having the problem if, indeed, this is the case. It is most important to be truthful and fair and to give a balanced account of probabilities. For this reason, it may be both appropriate and necessary to point out several possible outcomes without stating with certainty which one is most likely to occur.

Whilst the list of psychological problems and therefore exclusions is the discretion of the licensing authority, the definition of "caseness" and specific criteria for diagnosis is made in relation to the American Psychiatric Association's classification system for mental illness, the *Diagnostic and Statistical Manual of Mental Disorders* (DSM-IV, 1994). This nosologic system defines clinical syndromes and lists the unique symptoms that must be present in order to diagnose an individual as suffering from a particular condition. Not every clinical syndrome listed in DSM-IV automatically disqualifies a pilot from passing their medical test and consequently from obtaining or remaining licensed. In other words, it is possible to experience certain adverse symptoms or even suffer from some psychological problems without automatic disqualification. However, it may be that medical treatment for a specific problem disqualifies a pilot. For example, a male pilot who undergoes gender reassignment surgery after extensive and long-term medical and psychiatric assessment need not necessarily face disqualification on the basis of this procedure. It may be, however, that the pilot's reliance on prescribed hormone and related therapies may subsequently disqualify her after surgery.

Purpose of Reports

Psychologists are trained to communicate with and to other professionals about and with their patients in various ways. An essential part of the communication is report writing. Reports facilitate the sharing of information, mapping, and formulation of clinical problems and describing possible psychotherapeutic interventions. Psychologists should aim to produce accurate, clear, credible, useful, and persuasive reports (Benn & Brady, 1998). It is an unfortunate fact that many reports are criticized for not being sufficiently useful. They may be deficient in certain or many respects. This may be due to the fact that some of those who undertake such assessments have learned more from experience than formal training. The more frequently cited deficiencies of reports include vagueness, excessive speculation, failure to include data from which inferences are drawn, excessive use of jargon and unbalanced (i.e. overly negative or positive) opinions being expressed.

The UK Data Protection Act (1998) and Access to Health Records Act (1990) enable people to have access to their health records and psychological reports are no exception. Any information written about the individual, including rough notes or the results of psychological tests, is included within the scope of this. The only exception to access is where there is a risk of serious harm to the individual or to others. The individual may also challenge the accuracy of information held about them although the psychologist can stand by his or her opinions or recorded facts as long as a note is made of the disputed sections. It, therefore, almost goes without saying that any pilot who is assessed can request access to the report and therefore it should always be written with this outcome in mind. This is distinct from who *owns* the assessment. The report is the property of the person or agency that requested it. Psychological reports should always be used in their entirety and never in edited sections.

The purpose of a report of a clinical assessment of a pilot is guided by the need to answer five main questions:

1. What psychological problems, if any, does the pilot currently experience?
2. Could these psychological problems prevent the pilot from passing their medical examination and therefore from obtaining a license?
3. Are these problems that could impair the pilot's performance and, therefore, have a direct or indirect impact on safety?
4. Are the problems likely to be transient or more permanent?
5. Are these treatable problems and, if so, is psychotherapeutic intervention or treatment likely to be effective?

Reports are typically structured in a format that allows the reader to follow the process and outcome of the assessment and the judgments made. This enables other specialist psychologists as well as anyone who is not a psychologist (e.g. authorized medical examiner, the pilot him- or herself, or lawyer) to make sense of: a. the problem that has been assessed and the context in which it occurs; b. how it has been assessed; c. what has been found through the assessment; d. the psychologist's opinion as to the likely implications of the problems and; e. the psychologist's conclusions and recommendations.

The first section of the report contains biographical details about the pilot and also includes important contextual information. Included in this section are the pilot's name, date of birth, the place and date of assessment, the date of the report, reason for referral and the name of the referrer. Some psychologists also summarize their qualifications and experience in this section. If intended for court, this first section will also include a short CV or resume of the psychologist writing the report. The second section outlines the different sources that the psychologist has drawn on to undertake the assessment. This might include psychological tests, previous tests carried out on the pilot, their company health or flight records and related information, such as information gleaned from other sources (e.g. interview with the pilot's line manager). The third section typically describes the background to the problem and the pilot's account of this. A summary of what the pilot tells the psychologist should be included in this section. The results of the clinical and mental state assessment of the pilot are described in the fourth section. The person writing the report should keep in mind that technical information should be presented in a format that is easy to understand. It is normal practice to include references to the tests used. The fifth section is optional and may summarize the findings from any additional relevant information that has been gathered (e.g. interview of a line manager). The psychologist's formulation and opinion comprises the sixth section. The psychologist includes an "educated guess" as to the causes of the problem in this section. The limitations of the assessment should be included here. Only claims which can be substantiated should be included in the report. Finally, the author makes his or her recommendations in the seventh section. This may include recommendations for further assessment, psychological treatment, and crew licensing. Some authors attach a brief summary of the report at the end and also include a legal declaration that the report reflects the psychologist's best professional judgment and has not been unduly influenced by the person who has requested or paid for the report.

All statements in the report about the individual should be credible and persuasive. The conclusions reached should be consistent with the data presented. Opinions that are based on the professional's personal thoughts or subjective reactions to the pilot have no place in a psychologist's report. Care should be taken to present a succinct, readable, and balanced report comprising several sub-headings, that is free from jargon, typographical errors, wordiness or unsubstantiated findings. It is also important to keep in mind the person for whom the report is written. Specialist knowledge of the reader should not be assumed. Terms that may be more familiar to psychologists (e.g. *transference* or *neurotic*) but less well understood by others should be avoided.

It is most important for the report writer to bear in mind his duties not only to those commissioning the report, but also to the pilot and third parties to whom the report will have relevance. If, for example, the report is written in the light of legal proceedings, the writer must feel confident that the report is accurate, based on well established scientific reasoning and that it will stand up to cross examination. Above all, the writer must be satisfied that the report is true and believed to be true. To compromise on this point could lead to the report writer having to explain himself in a criminal court. Psychologists and other experts who are able to establish themselves as unbiased and balanced whilst able to address all relevant issues based on scientific reasoning will make a good impression on the reader and the fear of being contradicted can therefore recede. A good reputation should surely follow.

Conclusion

Those who have specialist training and registration as clinical psychologists should preferably write psychological assessment reports of pilots that address mental health concerns. Occupational and counseling psychologists also undertake psychological assessments of pilots. The findings of the assessment should always be presented in a balanced and useful way. The report should follow a clear structure and be comprehensible to a non-specialist audience. Whilst the report is the property of the person requesting it, the author should keep in mind that the pilot and others might view it and have comments to make about its accuracy. For this reason, the report should always be of a medico-legal standard.

Appendix

The following is a sample report that is entirely fictitious and included for illustrative purposes only. Some sections are briefer than is normally the case due to space considerations.

Sample Report

Touchdown Airlines Psychologist's Report

Name:	First Officer John Smith
Date of birth:	31 May 1969
On the instructions of:	Dr. Stephen Harris, Authorised Medical Examiner
Location:	Touchdown Airlines Head Office, One Mile Island, United Kingdom
Date of assessment:	15 January 2006
Date of report:	20 January 2006
Reason for referral:	F.O. Smith was referred for assessment of his psychological state and to determine whether he is mentally fit to return to full flying duties having been off work for more than 21 days due to marriage difficulties.
Psychologist and author:	Dr. Peter Johnson

Basis of the report:

This report is based on my:

a. clinical interview of F.O. Smith,
b. interpretation of a psychometric test completed by F.O. Smith,
c. understanding and interpretation of background information supplied to me by Dr. Stephen Harris, Aero medical Examiner at Touchdown Airlines, and
d. understanding and interpretation of information supplied to me during a meeting with Captain Jane Elford, Chief Pilot, Boeing Fleet at Touchdown Airlines.

Background history as F.O. Smith presented it to me:

a. Personal and piloting background:
F.O. Smith is a 36-year-old man who grew up in Canada. After completing his schooling, he was accepted into Bank Airlines pilot *ab initio* pilot scheme, where he obtained his commercial pilot's license and for whom he flew for ten years. He later joined Practical Airlines where he also flew commercially, mainly operating Boeing 737 and 767 types. He has been employed at Touchdown Airlines for the past eight years and he currently operates Boeing 737 Classic and NG types.

b. Family history and personal background:
F.O. Smith's parents and two younger brothers all live in Canada. He married his wife, Juliet, ten years ago. They have two daughters, aged eight and six years and an eighteen month old son. He told me that he has a good relationship with his children.

c. The present problem:
He told me that his wife has recently questioned him about his relationships with female crew and she has expressed concerns that he has been having an extra marital relationship with one particular flight attendant. Things came to a head two months ago when his wife threatened to leave him to go back to her parents in Scotland. F.O. Smith told me that he has been experiencing problems in his marriage for several months. He said: "I have had personal issues with my wife; we have not been seeing eye-to-eye since the beginning of this year."

F.O. Smith explained to me that his wife comes from a very close and emotionally "claustrophobic" family background. He told me that his in-laws telephone his wife three to four times a day and give her advice as to how she should manage her life both in terms of major decisions as well as in seemingly insignificant details such as where she should shop for the cheapest milk. He explained that he has sometimes felt displaced by his in-laws and has argued with his wife about her overly close attachment to her parents.

I asked him to explain more about the nature of the marriage problems. He told me that it was difficult to pinpoint the precise cause of the problems, but they amounted to a breakdown of trust. He explained that there had been several situations in which trust had presented as a problem; for example, his wife's concerns about his relationship with a colleague at work. F.O. Smith also pointed out to me that he felt that his wife confiding in her parents about issues that he felt were personal to them as a couple made him feel wary of discussing some things openly with her as he could not trust her not to convey the issue to his in-laws.

When the problem came to a head two months ago, his wife threatened to return to Scotland; she told him that she needed the physical space to reflect on the future of their relationship. Recognizing the emotional repercussions to himself of this decision, F.O. Smith took himself off line and presented himself to one of the specialist doctors in the airline's Aero medical service and he also arranged for counseling through the airline's Employee Assistance Programme.

He told me that while he is able to normally maintain separation between his working life and his domestic/personal life, he said: "It is important to take the responsibility onto myself and to remove myself from a situation if I think that there is any danger." He told me that in his work context, he is able to take charge in challenging and difficult situations and apply the lessons taught in CRM. He recognizes though that he cannot manage problems in the same way in his personal relationships. He explained that work is an antidote to his personal difficulties as it is somewhere where he can feel good about himself and where his mind is taken off his domestic problems.

I asked F.O. Smith to explain how he felt at the time that his wife threatened to leave. He told me that he was shocked that his wife should want to leave him and was very distressed that she should want to return to Scotland, taking the three children with her. He also felt resentful and lacking in personal control when "the rumor network in Touchdown [Airlines]" became involved in his personal situation. He said that the rumors about him made him feel very angry, as he did not feel that he had control of aspects of his personal life.

I understand from F.O. Smith that his own parents in Canada are aware of his marriage problems and they have encouraged him to "sort things out" within the relationship.

I asked him what he was hoping to achieve through counseling and psychological support. He said: "I am prepared to face my demons. I need to know what is causing this [the problems in the marriage relationship]." He explained that he felt that he was more motivated to overcoming the problems at this time than he perceives his wife to be. Nonetheless, he expressed an interest in couples' therapy.

In order to determine whether he has a realistic sense of the possible outcomes, I asked F.O. Smith what he thinks may happen to him if his wife decides to leave him. He replied in a calm tone: "If she leaves me, I accept this; she is entitled to move on. It would be traumatic at first, but I would heal in time."

F.O. Smith told me that he was surprised that his personal situation had "got out of control" and he felt both confused and resentful that his taking the positive step of voluntarily and proactively coming off line had resulted in the need for a specialist report. Whilst he understood that his exceeding the 21-day threshold for sick leave necessitated a report in the light of the automatic suspension of license pending medical clearance, he emphasized that this was a very stressful time for him. He told me that with the threat of the loss of his marriage relationship and his license, the situation was beginning to feel like a witch-hunt. He acknowledged that some of his recent behavior might be an excessive reaction to these threats, which in turn, might have increased people's concerns about him.

I discussed with him several hypothetical outcomes to the assessment and recommendations. These ranged, on the one hand, from loss of license to a recommendation that he be deemed fit to return to work, as well as other outcomes. He said that he understood and accepted these possible outcomes.

Assessment of F.O. Smith:

F.O. Smith is a 36-year-old male pilot employed by Touchdown Airlines. He arrived on time for the assessment. He was neatly, though casually, dressed. He appeared clean, well kempt and physically healthy.

In interview, he was pleasant, friendly and fully cooperative. He answered all questions I put to him in an open and honest way. He maintained good eye contact. He gave clear answers to both open and closed questions. His affect was at all times appropriate. A good rapport was established and, overall, he seemed to be at ease in the situation. He displayed good insight into his difficulties although evidence of his insight required some prompting from my side at times.

He was fluent in English, this being his first language although he was brought up in a French-Canadian community and he is bilingual.

There was no suggestion of any major deficit when he was asked to recall things from both his short- and long-term memory. Verbally, he was fluent and his pace of speech measured suggesting no major cognitive problems.

There was no evidence of delusions or hallucinations, and he was fully oriented in time and place. There were no psychotic features or suicidal ideation.

In order to emphasize some points, he gesticulated in a confident manner. There was no evidence of any involuntary behavior, inappropriate gestures or responses, or thought disorder.

Overall, his manner was pleasant. It was clear from his background, educational achievements, demeanor and communication skills that he is probably of at least average intelligence.

He told me that his mood is usually "jovial" and "happy-go-lucky" though he has felt more depressed since he has been unable to resolve his marriage problems. He also has a tendency to become "moody" after a night flight.

F.O. Smith told me that his appetite and weight are both "fine" and "stable."

For recreation, he enjoys fishing, golf, aerobics, and aerobatics. He owns his own light aircraft.

He told me that he drinks alcohol, although in moderation and almost only socially. He denies using any recreational drugs.

I explored with him how he manages stress and conflict both in the workplace as well as in his personal life. He seemed to give a balanced account of some of his personal shortcomings. I noted his inability to respond flexibly to certain problems. His need to be in control is consonant with this. At times, he can be impulsive and demanding, and he may lack insight into some of his own behavior. Where his needs are not met, he responds in certain situations by becoming more demanding and dominant. He then loses the ability to be diplomatic and empathic to others. Some of this pattern can be quite child-like. He may pull on others for a response and is unremitting in this until he gets a reaction; after which, he then backs off. One consequence is to alienate himself from those around him, as he appears to lack insight into the effect his behavior has on others and especially into how his relationship with his wife has deteriorated.

When he feels frustrated, this may be marked by a rapid change in his temper. It appears that he does not always adequately manage his personal stress. Where this is the case, he has a tendency to become defensive and argumentative. He may place self-interest above all else when he feels threatened in a situation. When challenged or threatened, he either acts aggressively or tactically in order to achieve his goals.

I confined my *psychometric assessment* to his personality.

I asked F.O. Smith to complete the Personality Assessment Inventory (Morey, 1991), which he did. I personally undertook the scoring and interpretation of his test. He was fully cooperative and the results of each of this measure were usable. A brief description of the instrument together with the results in his case is presented below.

Personality Assessment Inventory:
The PAI is a rigorously constructed psychometric instrument designed to provide clinicians with a reliable and valid measure of personality and psychopathology. The PAI contains 344 items and respondents are required to answer whether each item is "totally false," "slightly true," "mainly true," or "very true." The items of the test are scored onto twenty-two non-overlapping scales, some of which help to determine whether the respondent is consistent, careful, and honest in relation to the test items.

Validity of Test Results:
The PAI provides a number of validity indices that are designed to provide an assessment of factors that could distort the results of testing. For this protocol, the number of uncompleted items is within acceptable limits. Also evaluated is the extent to which the respondent attended appropriately and responded consistently to the content of test items. F.O. Smith's scores suggest that he did attend appropriately to item content and responded in a consistent fashion to similar items. The degree to which response styles may have affected or distorted the report of symptomatology on the inventory is also assessed. The scores for these indicators fall in the normal range, suggesting that he answered in a reasonably forthright manner and did not attempt to present an unrealistic or inaccurate impression that was either more negative or more positive than the clinical picture would warrant.

Clinical Features:
The PAI clinical profile is entirely within normal limits. There are no indications of significant psychopathology in the areas that are tapped by the individual clinical scales. According to F.O. Smith's self-report, he describes no significant problems in the following areas: unusual thoughts or peculiar experiences; antisocial behavior; problems with empathy; undue suspiciousness or hostility; extreme moodiness and impulsivity; unhappiness and depression; unusually elevated mood or heightened activity; marked anxiety; problematic behaviors used to manage anxiety; difficulties with health or physical functioning. Also, he reports NO significant problems with alcohol or drug abuse or dependence. F.O. Smith, however, acknowledges that he

sometimes loses his temper and is prone to lose control when he is emotionally stressed.

Self-Concept:
F.O. Smith's self-concept appears to involve a generally stable and positive self-evaluation. He is normally a confident and optimistic person who approaches life with a clear sense of purpose and distinct convictions. These characteristics are valuable in that they allow him to be resilient and adaptive in the face of most stressors. He describes being reasonably self-satisfied, with a well-articulated sense of who he is and what his goals are.

Interpersonal:
F.O. Smith's interpersonal style seems best characterized as friendly and extraverted. He will usually present a cheerful and positive picture in the presence of others. He is able to communicate his interest in others in an open and straightforward manner. He usually prefers activities that bring him into contact with others, rather than solitary pursuits, and he is probably quick to offer help to those in need of it. He sees himself as a person with many friends and as one who is comfortable in most social situations.

Background of F.O. Smith as presented to me by his fleet Captain:
I learned from the Chief Pilot Boeing Fleet that F.O. Smith's flying skills are excellent. All of his line checks have been to a high standard. There have been no reports of risk-taking or actions that would pose a danger. There have been no reported interpersonal problems in the workplace.

Opinion:
1. F.O. Smith is not currently suffering from any psychological condition that would prevent him from resuming flying duties. There is specifically *no evidence that he suffers from any definable mood or personality disorder.*
2. His marriage relationship has deteriorated over the past few months and this has given rise to *significant personal stress.* It was a positive decision and also a measure and sign of his prudence that he elected to take himself off line whilst undergoing counseling for his marriage problem.
3. The present crisis in his life was precipitated by his wife's threat to leave him. His reaction to this demonstrated that he is prone to behave impulsively at times with the aim of trying to stabilize or control events. This may also signal an urge to control his own mood, as he fears personal loss and being exposed to unpleasant feelings (e.g. low mood or depression). This tendency to act defensively and to try to control events is not a definable psychological problem *per se*, but a sign that he has *traits of personality problems.* The most likely of these are a tendency towards paranoia and narcissism, particularly when he is under stress. In my opinion, his personal problems have exacerbated several personality traits, which in turn, have compounded his difficulties with other people. This pattern

escalated significantly in the past week. However, there was clear evidence to me that under professional guidance and by confronting him with how he mismanaged his problem, this pattern all but dissipated.
4. When he is under stress, his logic and reasoning tends to become one-sided. He has a tendency to become self-referential, child-like in his behavior and display a lack of empathy for others. His attributions to events and situations can be impaired or biased by his determination to view problems only from his perspective. His solutions to certain problems exacerbate his personal and interpersonal difficulties. He also has a tendency to act out his feelings (e.g. anger, feeling out of control; lacking in trust) when he feels stressed, rather than finding solutions to the problems he faces. His reactions to events may, at times, be inappropriate or disproportionate. He has a tendency to compound his difficulties by misjudging problems and solutions. An example of this was his reaction to his misjudging the clerical problem that arose over his exceeding the 21-day limit for being sick after which he would be grounded. There are times when he experiences intense anger and this can break through when he is under pressure.

It must be stressed that there is no evidence that any of these traits, tendencies or behaviors occur in the workplace. On the contrary, it appears that these traits are confined to his domestic situation and personal relationships. In my opinion, he is currently suffering from an *adjustment reaction* (as distinct from an adjustment disorder) to an adverse life event. The fact is that this event has not passed and a solution to his marriage problem is not imminently at hand. This must imply that he continues to be prone to the self-defeating cycles of behavior relating to stress, as outlined above.
5. In my opinion, given his pre-morbid personality and psychological state, his track record as a pilot, as well as the opinion of a senior colleague whose opinions I was able to solicit and from my own clinical assessment of F.O. Smith, *most of his psychological difficulties are transient* and not that dissimilar to those seen in some other pilots. *They reflect personal shortcomings rather than gross or enduring psychopathology.* I would expect his problematic behavior to dissipate once the source of his stress has been resolved. There is evidence, however, that some of his personality traits and behavioral patterns cause ongoing difficulties.

Recommendations:
In the light of the above findings, my recommendations are as follows:

1. To give F.O. Smith the opportunity to bring resolution to his marriage problems through his own efforts as well as with the help of a marriage therapist. It would be prudent to take him off line whilst he undertakes this.
2. To encourage him to undergo psychological counseling for a minimum of ten sessions with a trained, sympathetic specialist therapist who can help him to gain insight, both into his behavior when under stress as well as into his personality traits, and to help him to acquire more adaptive skills for dealing with anger.

There is no evidence that long term psychological treatment would produce more favorable results in his case (Bor et al., 2004).
3. To enable him to maintain currency of his pilot's license through simulator sessions; and to maintain professional involvement in the airline.
4. To review and reassess his situation in not less than three months with a view to determining the extent to which he has resolved his marital difficulties and demonstrated that he can apply alternate responses to challenging situations. This could be in the form of a report from his therapist and/or further psychological assessment.

Declaration of the author:

1. I understand that my duty in writing this report is to provide an accurate professional account of my work with this individual. I understand that this duty overrides any obligation to the person from whom I have received instructions or by whom I am paid.
2. I confirm that I have complied with that duty in writing this report.
3. I believe that the facts I have stated in this report are true and that the opinions I have expressed are correct.

Dr Peter Johnson PhD
Consultant Clinical Psychologist

References

American Psychiatric Association (1994). *Diagnostic and statistical manual of mental disorders* (4th ed.). Washington DC: American Psychiatric Association.
Benn, A. & Brady, C. (1994). Forensic report writing. In M. McCurran & J. Hodge (eds.) *The assessment of criminal behaviors of clients in secure settings* (pp.127–145). London: Jessica Kingsley Publishers.
Bor, R., Field, G., & Scragg, P. (2002). The mental health of pilots: an overview. *Counselling Psychology Quarterly*, 15, 3, 293–256.
Bor, R., Gill, S., Miller, R., & Parrott, C. (2004). *Doing therapy briefly*. Basingstoke: Palgrave Macmillan.
Data Protection Act (1998). *Public and General Acts of 1998*. Chapter 29: 963–969.
Morey, L. (1991). *Personal Assessment Inventory Professional Manual*. Odessa, FL: Psychological Assessment Resources.
National Health Service Management Executive. (1991). *Health Service Guidelines: Access to Health Records Act (1990)* HSG (91) 6. London: Department of Health.
Reinhart, R. (1997). *FAA Medical certification: guidelines for pilots*. Ames: Iowa State University Press.

Chapter 11
Psychological Factors in Cockpit Crew Selection

Robert A. Roe and Pieter H. Hermans[1]

Introduction

At present it is impossible to become a professional air pilot without passing through some kind of selection process. At the beginning of powered flight, anyone with an interest in flying an airplane could do so without having to bother about extensive training programs or being burdened by regulations concerning one's health and aptitude. Nowadays the business of aviation is heavily regulated and those who want to become a pilot must minimally undergo an assessment of physical and mental fitness, receive training by a certified flight instructor, and gather some 40 or 50 flight hours to get a first license. In order to become a professional commercial or military pilot one must pass several additional assessment hurdles. The costs and risks involved in aviation, for the individual pilot, the employing organization, and society are so high that serious selection has become a necessity. However, not all selection processes are equally rigorous and effective. Although the use of formal assessments and decision methods has become more widespread during the last decade, especially among the larger airlines and armed forces of developed countries, there is still room for improvement.

The aim of this chapter is to inform readers about the state-of-the-art in pilot selection and to highlight issues that are critical when looking at opportunities for improvement. We assume that most readers will be experts in human resources management (HRM) within the field of aviation. Nevertheless, we believe that the chapter will also offer some useful information to others with an interest in pilot selection, including flight instructors, medical doctors, and psychologists. Those with expertise in the design of pilot selection systems may be find some useful references to the recent literature.

The main questions to be addressed in this chapter are the following: What purposes does selection serve? What can one expect from selection? How does one develop and run an effective selection program? What role does mental health play

---

1 The authors gratefully acknowledge the comments of Eugene Burke on an earlier version of the text.

in selection? In answering these questions we will draw upon the literature on air pilot selection as well as on practical experiences from the domains of HRM and aviation. We will begin with a brief discussion of the main forms and functions of selection. Next, we will consider pilot competences and look at the role they should play in developing a selection system. After reviewing the main methods and instruments currently in use, we discuss the architecture of modern selection systems and the methods of systems design. Subsequently, we will present an example of an application in the design of an existing airline selection system, and discuss some typical issues considered in the design process. The way in which selection can contribute to pilots' mental health will be the subject of a separate section. We will conclude with some recommendations regarding air pilot selection in the future.

Forms and Functions of Pilot Selection

There are three major forms of selection which aim at different purposes. The first is selection for pilot training, usually referred to as *ab initio* selection. Its primary function is to establish the chance of success in completing the training program and to inform the candidate and the school in order to avoid high costs of failure. The emphasis is on psychological factors required for effective learning, adaptation to the school environment and the specific world of aviation, and the psychomotor and cognitive skills implied in flying. Selection for training usually happens once in a pilot's career.

The second form is *selection for employment* by an airline or another commercial or military operator. Here, all candidates have a license to fly a particular type of aircraft and at least some degree of operational experience. The function of the selection is to identify those who are expected to perform best according to standards set by the organization. These standards reach beyond handling the aircraft and typically include aspects such as following procedures, interacting with other crew members, and fulfilling leadership and representation roles. The emphasis is on responsible behavior; that is, acting safely at all times, while balancing the sometimes conflicting interests of the company, the crew, the passengers, and the environment. This kind of selection may occur during a pilot's career a number of times; that is, every time he or she applies for a job with another operator.

The third form, *clinical selection*, is of a different nature. Its function is to identify signs of psychopathology among candidates as well as employed pilots, and to prevent behaviors (both during training and in operation) that might endanger human life and corporate property. It is often part of a periodical medical check-up, such as required by Joint Aviation Requirements on Flight Crew Licensing in commercial aviation or similar military standards. Although the content of these three types of selection is different, they have in common that candidates are subjected to a predictive assessment. This means that the assessment focuses at forecasts of future behaviors (criteria), which are derived from currently displayed characteristics and behaviors (predictors) on the basis of both statistical evidence and expert judgment.

Performance and Competence in Pilot Selection

The ultimate criteria in pilot selection relate to the performance demonstrated by candidates at some time beyond the moment of the psychological examination. Performance is known to depend on many personal and situational factors (Roe, 1999), some of which cannot be measured beforehand. For example, when assessing candidates' suitability in a pilot selection program it is hard to account for variable environmental factors, such as crew composition, specific properties of aircraft and airspace, and system malfunction, and for changeable personal factors, such as motivation, fatigue, or (mental) illness. In recent years it has become customary to concentrate on competences rather than on performance per se. Competence can be defined as "a learned capacity to adequately perform a given task, role or mission" (Roe, 2002, p. 206), within a certain array of conditions. That is, competence can be conceived as a proxy of performance, assuming that external and internal (i.e. personal) conditions remain within a certain range. In addition to competences, there are a number of other concepts which also play a role in pilot selection (as well as in the processes of recruitment and training) preceding it. The model in Figure 11.1 specifies the relevant concepts and shows how they relate to the concept of competence (Roe, 2002).

Two main characteristics define the nature of competences, i.e. the fact: 1. that they are task and context specific, and 2. that they integrate knowledge, skills and attitudes. An example of a competence is effectively carrying out a landing or coping with engine failure. Sub-competences are similar to competences but are more limited in scope. They refer to activities that are part of the performance of tasks or roles but lack an independent purpose. Certain sub-competences, such as communicating, taking decisions, or reading dials, can be part of multiple competences and can hence be called *basic competences*. Competences (and sub-competences) are typically acquired in a process of learning-by-doing, that is, by performing the actual task in a simulated or real-life setting. In contrast, knowledge, skills, and attitudes are often learned in scholastic settings over a longer period, and are much less tied to a particular work situation. While competences, knowledge, skills, and attitudes are all learned, other personal characteristics are supposed to be more stable. Most of these can be referred to as dispositions, which points at the fact that they are largely innate. Although psychologists tend to classify these dispositions in many different ways, we distinguish between three main classes only: intellectual and psychomotor abilities, personality traits, and other characteristics (which include physical constitution, values, interests, biographical characteristics etc.). These characteristics are important for two reasons: they help to predict success and failure in training, and add to the prediction of actual performance. We will come back to this aspect later.

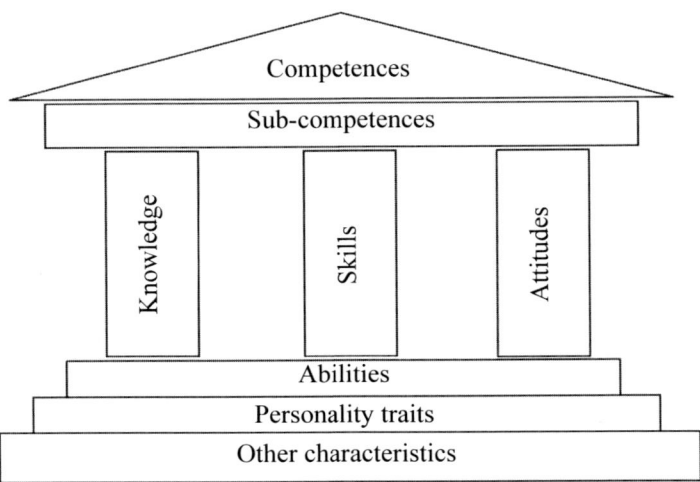

Figure 11.1 Competence Architecture Model (Roe, 2002)

With the help of this model we can clarify the differences and similarities between the three types of selection that were introduced above. In *ab initio selection*, candidates have no experience with flying and usually very little relevant knowledge and skills. Actually, the training at the flight school aims at providing them with the necessary knowledge and skills, as well as relevant attitudes (e.g. concerning risk and safety). The ultimate aim of the training is to let the students acquire a set of sub-competences required for a beginning pilot. When selecting candidates, the assessment focusses largely on the dispositions, to a limited degree on general elements of knowledge, skill, and attitude that are supposed to be useful in further training. As a consequence the instruments used for assessment will be ability and personality tests, supplemented with grades, biodata, and ratings. In addition, simulation-based instruments will be used to estimate success in attaining flying competence. While these attributes and instruments are the predictors, the competences to be acquired at the end of the training are the criteria.

In the *selection of experienced pilots*, the situation is different. Most candidates can be expected to master at least some of the competences required for flying as second-officer or first-officer, and perhaps as captain, as well as the associated knowledge, skills, and attitudes. However, their exact qualifications may differ as a result of differences in work experience and training background. Some candidates will have prior experience as commercial air pilot, while others may come from the fields of general aviation or military aviation. Therefore, the selection will focus on the degree to which candidates meet the standards following from the position or career trajectory recruited for. The requirements may well differ between operators. Depending on their business profile and the type of operations carried out, they may put an emphasis on particular competences and pose higher than average demands. For instance, they may emphasize competences related to crew interaction, leadership, and representation. The assessment will focus on competences that are already present. But since there is always a need for further training – be it conversion training or non-technical training – some information on knowledge, skills, attitudes, and abilities and personality, will be useful to predict the likely success in such training and the efforts that should be put into it. Many operators are inclined to look beyond competences and prefer to identify performance risks, resulting from attitudes or dispositions. Thus, the criteria for selection cover both competences and performance, whereas the predictors may cover the full range of human attributes from competences to dispositions. The assessment battery used for selection will comprise a series of assessment center techniques, used for establishing non-technical competences, psychological tests, accomplishment records, and possibly other devices.

What was said about identifying performance risks also applies to *clinical selection*. Here, the aim is to identify signs of psychopathology, mainly among employed pilots. The focus is primarily on personality traits that predispose the person towards psychopathy, delusions, depression, schizophrenia, etc. In rare cases such traits may be present without having produced a clinical record. These traits may not result in deviant behavior under normal conditions, but do so when the person faces extreme demands at work or in his or her personal life. The assessment may include deviant attitudes and behaviors, including addiction (e.g. heavy drinking). It may also cover temporary states that make the person unfit to fly. Thus, the major criteria for selection relate to the pilot's performance, while the predictors are personality traits and indicators of behaviors and attitudes. These predictors are measured by means of personality tests, clinical interviews, and behavior observations. Some of these predictors may be included as add-ons in systems for *ab initio* selection and employment selection.

The most important competences involved in pilot selection relate to safe and economic flying, effective teamwork and leadership, the representation of the organization, and more generally the promotion of the organization's business or defense interests. The precise description of the competences differs across profiling systems. Moreover, the number of competences in each category depends on the granularity of the analysis. Examples of typical competences and sub-competences can be found in Table 11.1 (based on Roe, 2003), which gives an extract from a competence profile for civil pilots. It also gives examples of knowledge, skills, attitudes, abilities, personality, and other characteristics.

Predictors in Pilot Selection

An important general distinction in personnel selection (see Wernimont & Campbell, 1968) is between *sign* instruments which measure personal dispositions (abilities, personality traits, values etc.) that are conditional for the development of competences and for effective performance, and *sample* instruments that directly assess competences and specific components thereof (sub-competences, knowledge,

Table 11.1 Extract from a Civil Pilot Competence Profile (Roe, 2003)

COMPETENCES	KNOWLEDGE
1. Flight preparation • Analyzing and interpreting the flight plan • Analyzing and interpreting meteorological information • Determining alternative routes and airports 2. Flight execution • Execution the flight in a safe, efficient and comfortable manner • Recognizing and diagnosing technical problems during the flight • Solving irregularities and problems during the flight 3. Flight completion • Reporting particularities of the aircraft • Reporting particularities of the flight • Reporting events and incidents 4. Providing leadership • Leading a two-person cockpit crew • Leading ground personnel • Changing leadership styles (consulting/ negotiating) 5. Representing the airline and the public authority • Following legal regulations and rules of conduct • Acting as host • Acting as representative of the airline	• Knowledge of aircraft systems • Knowledge of flight procedures • Knowledge or corporate procedures & current instructions SKILLS • Coding information • Working with checklist • Negotiating ATTITUDES • Openness to criticism • Integrity • Service attitude MENTAL ABILITIES • Analytical ability • Spatial ability • Perceptual speed & precision PSYCHOMOTOR & INFORMATION PROCESSING • Selective Attention • Visual pattern recognition • Time-sharing
SUB-COMPETENCES • Reading maps • Entering data in the flight computer • Maintaining and updating a complex mental model (situational awareness) • Giving clear instructions • Monitoring own actions and performance • Changing roles (e.g. leading - following, diverging roles)	PERSONALITY • Initiative • Emotional stability • Self confidence OTHER CHARACTERISTICS • Strategic thinking • Environmental awareness ("larger picture") • Practical style

skills, attitudes). Typical sign instruments are mental ability tests, personality and motivation questionnaires, and biographical inventories. Sample instruments are job samples (or work simulations), assessment center exercises, and situational interviews. The recommended use of sign and sample instruments depends on the type of selection. In *ab initio* and employment selection best results are obtained by combining both measurement techniques. As will be discussed later, the choice of specific predictor instruments depends on psychometric properties, operational qualities (e.g. group-wise or computer-based administration) and costs.

Sign Instruments

There are two major categories of sign instruments. The first category comprises tests, questionnaires and other devices for the measurement of *maximum performance*, i.e. tests for mental ability, psychomotor skills, cognitive processing. The second category consists of *typical performance* instruments, i.e. personality questionnaires, motivation and value-inventories, projective tests (using pictures, ink blots and tachistoscopic slides), and clinical questionnaires. Generally speaking, maximum performance tests have produced modest to good validities in pilot selection; among them tests of cognitive and psychomotor functions have obtained the highest validities (Hunter & Burke, 1994; Martinussen, 1996).

Predictive validities for typical performance tests tend to be much lower. In many studies researchers have found low levels of validity for personality tests (Dolgin & Gibb, 1989; Hunter, 1989; Hunter & Burke, 1994; Martinussen, 1996; Siem, 1992). In the literature there is no agreement about the reasons for this finding. Since personality questionnaires tend to capitalize on the self-description of the candidate and are open to social desirability, the resulting measurements may be poor. However, validity outcomes may also have been low as a consequence of using inadequate prediction models. When personality variables are used indiscriminately – just to find out whether they do or do not relate to job criteria – many validities will turn out to be zero or low. And this obviously suppresses average validity levels (Tett, Jackson, & Rothstein, 1991). Also, the effect of personality traits may be indirect in the sense that they influence the process by which competences are learned but less so the utilization of the competences in everyday work settings (Jessup & Jessup, 1971). This would explain why there tends to be a wide range of personality traits in any occupation. For example, introverts can become good salesmen, and good pilots can be intellectually curious persons but also pragmatic doers with little interest for theory. In this context it is worth noting that researchers have found different personality profiles among air pilots (Fitzgibbons, Davis, & Schutte, 2004), even though pilots differ from the general population in certain respects (Butcher, 1994). Given the fact that pilot behavior is highly controlled by routines and that crew members are expected to adjust to each other, differences in personality might indeed not be a powerful factor.

In spite of these arguments, many experts consider personality to be an indispensable ingredient of a pilot selection system, particularly because personality seems related to crew interaction, behavior under stressful conditions, and accidents (Hough & Ones, 2002). In addition, it may be argued that small contributions to the prediction of success (Hormann & Maschke, 1996) may be meaningful because they can have large implications. Therefore, efforts should be made to use more objective measures, to select only scales that are known to be relevant for particular competences or performance dimensions, and to tailor prediction models in such a way that only relevant scores range (extreme or optimal ranges) are taken into account. Evidence from research studies shows that a focussed approach to personality measurement, selecting only specific attributes (e.g. neuroticism) can lead to acceptable validities (Bartram & Baxter, 1996; Chidester, Helmreich, Gregorich, & Geis, 1991). Researchers have recently aimed at the development of personality tests that are based on the maximum performance principle (James, 2001; Rubio Franco, 2003). That is, personality traits (e.g. risk taking) are derived from the way in which subjects perform certain types of tasks.

Sample Instruments

Unlike sign instruments, which can also be applied for selection in other domains, sample instruments are specifically constructed for one domain, in this case, for flying. There is a wide range of sample instruments, with different measurement qualities. A distinction can be made between two main categories of sample instruments, i.e. *high-fidelity* job samples which measure pilot's competences by assessing performance in realistic settings that strongly resemble the pilot's actual work situation (e.g. type-simulator), and *low-fidelity* job samples which measure sub-competences such as communication, decision-making, or coding. Assessment centers and situational interviews can also be considered as sample instruments, in the sense that they confront candidates with certain types of situations and aim at assessing responses in their behavior repertoire. All sample instruments require candidates to perform real-life tasks or parts thereof. Examples of tasks are carrying out a landing in a flight-simulator, giving instructions to a live co-pilot (a role player), addressing live passengers through a public address system, organizing an operational task, or answering to a complaint brought in live by a situational interviewer playing the role of an angry passenger, confused cabin-assistant, or impatient duty officer (all high fidelity). Other examples are: estimating the speed and heading of moving objects on a screen, solving problems and taking decisions in a game, communicating and collaborating with others in a group task (low fidelity).

Most sample instruments are valid predictors and they are well accepted by candidates. Predictive validity tends to be higher than for ability tests, but not all job samples perform equally well. Predictive validities obtained with high fidelity samples, i.e. flight simulators, are relative high. However, these job samples are expensive and time consuming, both in administration and scoring; they also require the presence of trained observers (assessors). Low-fidelity job samples tend to give lower but still acceptable validities. This is true for samples with psychomotor tasks (Hunter & Burke, 1994; Martinussen, 1996) and for assessment centers (Damitz, Manzey, Kleinmann, & Severin, 2003). The costs are lower than for simulators but can still be substantial. In recent years many computer-based instruments for assessing information processing and psychomotor functions have been generated that are low in cost and operationally similar to tests e.g. CogScreen (Kay, 1995; Westerman, Draby, Maruff, & Collie, 2001) and PILAPT (Burke, Kitching, & Valsler, 1997).

Other Instruments

Some other instruments applied in pilot selection cannot be easily categorized as sign or sample. Among them are various types of interviews, including the unstructured and structured interview, the clinical interview, and biographical inventories. Their predictive validity varies depending on the specific content, the way of interpreting the answers, and the expertise of those using them. Unstructured interviews have little value for predictive purposes, but they may serve the purpose of exchanging information with the candidate. Structured interviews can cover traits and competences, and produce ratings that can serve as predictors. In general, they appear to have moderate predictive validities (McDaniel, Whetzel, Schmidt, & Maurer, 1994). There is also some evidence of construct validity (Salgado & Moscoso, 2002). There is not much proof of the validity of interviews in pilot selection. A study by Walters, Miller and Ree (1993) found a structured interview to be valid but to lack incremental validity compared to a test battery. Biographical inventories have been quite popular in earlier decades and enjoyed good predictive validities. More recent studies (Hunter & Burke, 1994; Martinussen, 1996) show a decline of validity. This might be attributed to societal changes which have modified people's pattern of social participation, their exposure to education, various life events, leisure activities, as well as their careers, all of which can have eroded their significance for the prediction of future behavior.

Predictive Validities

Summary information about the predictive validity of various tests and other predictors can be found in two meta-analyses that were published in the 1990s by Hunter and Burke (1994) and Martinussen (1996). The first study covered 468 correlations taken from 68 validity studies conducted in the 1940–90 period and published in the US, Canada, and Britain. Most of these pertained to *ab initio* selection for military training, but the results seem applicable to civil aviation as well. The second study covered 173 correlations from 66 samples reported in 50 studies during 1919–94. Again the emphasis is on military selection for military but there is a greater variety of countries, including non English speaking countries (Scandinavia, Spain, Germany, Israel, and Turkey). Also, there appears to be less emphasis on *ab initio* selection. Both studies included fixed-wing and rotary-wing training. Below we summarize the main findings from these two studies, which were not corrected for restriction of range and criterion unreliability because of lack of information on these artifacts. The left part of Table 11.2 gives mean sample-weighted validities along with Lower and Upper 95 percent confidence interval from Hunter and Burke's (1994) study. The right part of the table gives the sample-weighted validities along with the 90 percent Lower Credibility Value from the Martinussen (1996) study.

Although the mean rs seem to be low, the evidence shows that most types of predictors possess some validity (greater than zero). It must be emphasized that the given values underestimate the predictive validities, due to the dichotomous nature of

Table 11.2 Meta-analysis from Hunter and Burke (1994) and Martinussen (1996)

Study 1: Hunter & Burke (1994)				Study 2: Martinussen (1996)		
	r-mean	L95	U95		r-mean	90%CV
General ability	0.13	-0.05*	0.30	Intelligence tests	0.13	0.03
Verbal ability	0.12	-0.09*	0.33	Cognitive test	0.22	0.07
Quantitative ability	0.11	0.01	0.21			
Spatial ability	0.19	0.05	0.32			
Mechanical	0.29	0.11	0.48			
General information	0.25	0.06	0.44	Aviation information	0.22	0.14
Aviation information	0.22	0.06	0.38			
Gross dexterity	0.32	0.15	0.49	Psychomotor/ information processing	0.20	0.10
Fine dexterity	0.10	-0.09*	0.29			
Perceptual Speed	0.20	0.05	0.35			
Reaction time	0.28	0.16	0.39			
Job sample	0.34	0.19	0.55			
Personality	0.10	-0.16*	0.37	Personality	0.13	0.00
Biodata inventory	0.27	0.07	0.47	Biographical inventories	0.21	0.00
Education	0.06	-0.16	0.27	Academics	0.15	0.11
Age	-0.10	-0.25*	0.05	Combined index	0.31	0.19
				Training experience	0.25	0.07

non-generalizable results

the criterion (Pass-Fail in most cases), restriction of range and criterion unreliability. This implies that corrections for these artifacts will produce substantially higher values (see also Burke, Hobson, & Linksy, 1997). When predictors are combined into composites (corrected) validities of .50-.60 can be reached.

Systems Architecture

A pilot selection system is more than a set of predictive instruments. Valid tests, assessment exercises, and other instruments represent only part of the system. Equally important are the procedures for administering these instruments and handling information, and the people using the instruments and applying the procedures. It is essential to distinguish instruments from procedures, as good instruments may produce bad results if they are used in an ineffective way.

> For example, there may be too much redundancy in the test battery, weights may be chosen inadequately, cutoffs may be set too high or vary over time, the prediction may target the wrong type of criteria, the composition of the battery or the manner of administration

may be at odds with candidates' expectations and lead to unwanted withdrawals, users may be doubtful about the adequacy of the selection procedure and therefore disregard its outcomes, and so on. (Roe, 2005, p. 75)

People represent another crucial component of a selection system. If good instruments are put in the hands of untrained staff they can lead to poor results as well. Selection can only be effective if it is carried out by qualified people performing well-defined roles. All this is to say that selection systems have multiple components. They consist of computers and other equipment for testing and administration (hardware), tests and other assessment instruments linked together by some administrative procedure (software), and people responsible for carrying out the test examination, conducting interviews and assessments, giving judgments, and taking decisions (humanware).

Systems for *ab initio* selection and employment selection in aviation have a similar architecture as other selection systems (Roe, 2005), but at the same time, there are certain distinguishing features. Firstly, there are typically multiple criteria reflecting the different demands posed in training and employment. Secondly, candidates are usually confronted with several groups of predictors that are administered sequentially, e.g. ability tests, knowledge tests, personality tests, low-fidelity job samples, assessment centers, interview(s), and high fidelity job samples. Thirdly, the information generated by these predictors is typically combined by means of composite scores. A composite is a weighted or unweighted sum of a number of test scores or ratings. Combination of information by clinical judgment happens as well. Fourthly, the examination and decision making process are carried out by psychologists and support staff on the one hand, and subject matter experts (SMEs) on the other hand. The role of SMEs, mostly experienced instructors and captains, is typically giving ratings in an assessment center or high fidelity sample (e.g. Link trainer; see below). Sometimes they also participate as members of a selection board that takes decisions about the admission of candidates.

Clinical selection does not stand on its own in most cases. It can either be incorporated in a selection procedure or be a separate part of it. The first happens in many systems for ab initio or employment selection, the second occurs in systems that select people for special duties, particularly in the military. Clinical selection can also be part of a periodical psycho-medical check-up of pilots. Although psychopathology is known to occur among cadets and air crew – Butcher (2002) mentions personality disorders, substance abuse, bipolar disorders and psychosis – it is difficult to predict and identify. There are two major approaches. One approach is to test candidates for personality traits that predispose to (particular types of) psychopathology. This typically happens by means of objectively scored clinical personality tests (e.g. MMPI, ALAPS, see below). A second approach is to assess subjects' current mental health by means of observations in an assessment setting or a clinical interview carried out by a psychologist.

In spite of the prevalence of a general architecture there is a clear need for *tailoring* selection systems to the unique situation in which they are supposed to operate. Although the pilot's job seems rather standard, there are significant variations in the settings in which pilots are recruited, educated, and employed. There may be differences with regard to the populations from which candidates are recruited and selected (e.g. in type and level of education, language, and values, norms, and expectations) as well as fluctuations in demand and supply on the labor market. In addition, there are differences between fixed and rotary wing aircraft, between fleets of aircraft, between civil and military operations, cargo and passenger transport, line and charter flights, etc. As a consequence there are legitimate differences in selection requirements and in ways of handling candidates which show up in the selection system. The time that every flight school or airline had to start developing a selection system from scratch lies well behind us, but there are still many choices to be made in developing a well-functioning system. In the next section we will describe the major steps in the design of a selection system and discuss the major choices.

Systems Design

The need for systems design is clearest when a new operator enters the field and intends to build up a new pilot workforce. In such cases one has to develop a comprehensive selection system right from the beginning. Most airlines and military operators have well-developed selection systems in place, however. For them the need for design emerges from the necessity to review and improve the existing system, usually in response to changing conditions. This is typically the case when there the number of candidates shows a substantial increase (or decrease), when costs reductions call for greater speed and efficiency in recruitment and selection, when new investments in testing equipment are to be made, when new technologies are introduced, when changing views on safety, operations, or business strategy lead to new requirements on pilots or cadets, when organizations opt for large-scale restructuring, outsourcing or merger. In some of these situations it will be sufficient to modify or replace certain components of the selection system, in others it will be commendable to launch a project aiming at redesign of the existing system or the design of a wholly new system, along the lines described in the previous paragraph.

There are three important points that have to kept in mind when the decision to design a new selection system is taken. Firstly, as was pointed out above, a selection system can only be expected to function well if it fits into the context. Copying a selection system is rarely an option, even though the validities of certain predictors may generalize. An Italian price-fighter airline cannot simply adopt the selection system of a large international airline. Nor can the selection system of the Czech Air Force simply be transferred to the Norwegian Coast Guard. There are operational, organizational, educational, labor market, cultural and other factors that must be taken into account, and this will give each system certain unique features. Secondly, there is usually more than one way to build an effective selection system. This has to do with the complexity of the system and the fact that multiple ways of selecting and arranging its components can give equally good results. Thirdly, the best solution cannot be found deductively. The logic of design is an inductive one, that is, one

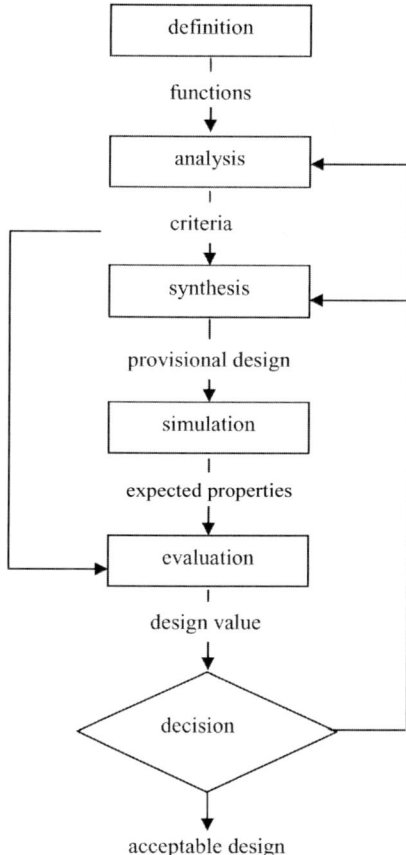

Figure 11.2 Basic Design Model (after Roozenburg & Eekels, 1991; Roe, 1998)

will need to work in a cyclical way, going through a number of iterations before a good result is found. Fourthly, the design process can become very complex when there are multiple components and hence many ways to proceed. For this reason it is segmented into architectural design, which deals with the overall structure of the system, and detailed design which deals with separate components (Roe, 2005). Here we will only focus on architectural design.

The general principles of selection systems design have been outlined by Roe in a number of publications (Roe, 1989, 1998, 2005). There are two tools that designers may use to facilitate their work. The first is the Design Cycle Model (DCM) (Roozenburg & Eekels, 1998) which is depicted in Figure 11.2. The DCM is helpful in showing the logic the design. It comprises a number of stages, some of which will have to be passed through repeatedly till a satisfactory result has been obtained. These stages are:

1. Definition of the functions the selection system should fulfill, such as collecting information, making predictions performance, measuring performance, taking decisions, informing people.
2. Analysis of requirements, constraints and specifications, operationalizing and weighting them, and formulating a Program of Requirements (POR).
3. Synthesis, or making a provisional design of the system that has the potential to fulfill the desired functions, stays within the limits of the constraints, and meets the specifications.
4. Simulation, or testing the operational, predictive, and economic properties of the system by means of models or field tests.
5. Evaluation of the system against the POR in order to assess its acceptability.
6. Decision making resulting in revision of the (provisional) design or accepting it for operational use.

The second tool is the POR, mentioned in the second stage. It is established with the help of various stakeholders, parties that have an interest in the process and outcomes of selection. The POR specifies the intended use of the selection system (e.g. decisions about entry or about career options) and typically contains requirements regarding quota, the competence level of selected candidates, information given to commissioners and candidates, reporting formats, speed of processing etc. It will also list constraints in terms of costs, time, portability, legal issues etc. What makes the POR most useful is that it helps to clarify the expectations of various parties in advance and to identify and resolve conflicting demands.

The core of the design process of course lies in synthesis and simulation, the activities that actually shape the system and its various components. With regard to the components that aim at prediction and decision-making – the most critical functions of the system – designers will take the following steps (for more detail see: Roe, 2005).

Defining Criteria

This implies choosing the number of criteria and their specific nature. Due to the complexity of the pilot's job multiple criteria are needed. These criteria are typically competences and performance dimensions relating to safe and economic flying, effective human interaction and leadership, and the promotion of organizational interests (see above).

Choosing Predictors

Depending on the human attributes required to satisfy the criteria a set of potential predictors has to be selected. In order to attain sufficient predictive validity there are normally multiple predictors to cover each criterion. This process is guided by knowledge of the psychological literature, especially by data on predictive validity of various instruments and the attributes measured by them. Predictors will typically be sign and sample based.

Setting Stages and Batches

Given the fact that pilot selection involves several criteria and a substantial number of instruments the procedures tend to become complex and costly. This is especially true in *ab initio* selection since selection ratio's are low (sometimes only 5 percent or less) and the numbers of candidates high. Substantial reductions of complexity and cost may be achieved by dividing the selection process into stages and processing the candidates in batches. The exact way of doing this depends on logistic and economic considerations. Special precautions are needed to make sure that the way in which information is handled is in accordance with the aimed for prediction model.

Deciding on Compensation

The next issue is to decide whether the predictors should be kept separate or rather be combined. Selecting candidates on the basis of separate predictors (multiple hurdles model) increases the number of wrong decisions (false positives and false negatives) and leads to a drastic reduction of the number of candidates. The preferred approach is to combine information in composite scores, which implies that good scores on one test can compensate poor scores on another test. In pilot selection procedures one would normally use two types of composites, i.e. primary composites to combine scores on tests within the same category (e.g. ability tests, psychomotor tests) and secondary composites to combine these primary composites. Often a single secondary (overall) composite is used to predict success in training or operation. Compensation can also be applied to the criteria, which means that some overall index of competence or performance is created against which predictor composites can be validated.

Determining Weights

Adopting composites implies that one must assign weights to different predictors (possibly also criteria). The practice of using empirically established weights based on regression analysis has become less popular because of capitalization of chance in the procedure. Rational weights set by the designer offer an easy to use alternative that gives good results. A simple weighting scheme (e.g. weights 1, 1.5, 2, 2.5 and 3.5) can produce composite validities comparable to those found by complex statistical optimization procedures (Wherry, 1975). An advantage of rational weighting is that it allows the designer to take into account the results from meta-analyses and other published research data.

Setting Cutoffs

There are two issues here: firstly, whether to opt for absolute (fixed) cutoff scores, relative (variable) cutoff scores, and secondly, what cutoff level to choose. Fixed cutoffs can lead to too high or too low numbers of selected candidates. Variable cutoffs will give the right number of candidates but their level of qualification may differ. Therefore a compromise may be sought in a quasi-absolute cutoff that is only used for a certain period of time and revised afterwards.

An assumption behind the description of the latter steps is that information about candidates will be processed in a mechanical rather than a clinical way. This is indeed the case in most current selection systems for *ab initio* and employment selection, and it is in agreement with research findings on mechanical and clinical methods. With very few exceptions mechanical methods were found to give better or equal results compared to clinical methods (Grove & Meehl, 1996; Grove, Zald, Lebow, Snitz, & Nelson, 2000; Sawyer, 1966). Clinical methods do have their place in selection, particularly in clinical selection where the use of objective tests interpretation rules does not suffice and the expert judgment of psychologists is needed. Some current systems for pilot selection use clinical judgment in the final stages of the selection process when a great deal of diverse information has to be integrated into a single decision. This is usually seen as the role of a selection board. A risk inherent in this practice is that members of the selection board rely more heavily on their own overall judgment of the candidate and ignore test results (Carretta, 2000), which may lead to a drop in predictive validity.

Assessment Methods

Since pilot selection has a long history there is much accumulated knowledge about which methods do or do not work. In this section we will discuss the current state-of-affairs with regard to the use of various assessment tools. It should be kept in mind that designers will choose predictors in accordance with the requirements in the POR, that is, consider operational aspects, costs and acceptance by candidates in addition to the traditional requirements of reliability and validity.

Ability Tests

Instruments for testing abilities are widely available. There are many single tests but also several batteries covering a variety of verbal, numerical, figural, spatial and other abilities. Examples of the latter are: Ability Screening On-Line and other SHL test batteries, the Multidimensional Aptitude battery (MAB), the Differential Aptitude Test (DAT), the General Aptitude Test battery (GAT-B), and the Wechsler Adult Intelligence Scale (WAIS). There are also multiple ability batteries used within particular organizations, such as the Air Force Officer Qualification Test (AFOQT) of the US Air Force. Ability tests are often part of compound batteries (e.g. PILAPT and MICROPAT; see below) as well. Although paper-and-pencil versions are still in use, a growing number of test batteries has been computerized and can nowadays be administered through the internet. Ability tests are a standard component of pilot selection system, even though the validities are not always high. These tests serve to warrant a standard level of mental ability as a base for effective learning and job performance. The costs of ability tests are relatively low and they are reasonably well accepted by candidates.

Personality Tests

The most frequently used personality tests are multi-trait inventories, such as the Occupational Personality Questionnaire, the Eysenck Personality Inventory, the Personality Research Form, the California Psychological Inventory, and Edward's Personal Preference Scale. Projective tests, such as the Thematic Appercection Test (TAT), Defense Mechanism Tests (DMT), and the Rorschach Inkblot Test are no longer in use due to negative findings concerning validity (e.g. Hermans, Jacobs, & Van Niekerk, 1987). There is no consensus as to the way of using personality tests. In the context of *ab initio* selection and selection for employment one might adopt a prediction model that focusses on critical scores ranges (e.g. extreme scores) or unusual score profiles. Hermans and Mulder (1999) have proposed to consider extreme scores on various scales of a personality test as signs of possible psychopathology and to use a count of these signs (Unified Personality Indicator, UPI) as a negatively weighted predictor of training success. Another approach is to include selected personality scales in the prediction model. When multiple criteria are used (e.g. flight performance, crew leadership and safety), different sets of personality scales can be used for each of them. Tests may also be used in an indirect way, i.e. as tools for self-description that allow recruiters to evaluate candidates on their degree of self-understanding. Hofstee (1983) has suggested inviting candidates to pick the most desirable answers when filling out personality questionnaires at selection time. This might enable the recruiter to find out whether the candidate understands the advantages and disadvantages of certain dispositions, such as, for instance, high levels of thrill seeking or authoritarianism. In clinical selection one would use instruments such as the Minnesota Multiphasic Personality Inventory (MMPI, see Butcher, 1994) or the Armstrong Laboratory Aviator Personality Survey (ALAPS, see Berg, Moore, Retzlaff, & King, 2002) in combination with clinical interviews. Occasionally projective tests might also be used in this context. Except for clinical use, which requires the involvement of experts, the costs of personality testing are low. Their acceptance by candidates is not very high, however.

LF Job Samples/Simulations

There are two major types of tools in this category. The first type covers cognitive and psychomotor tasks. Candidates must respond to visual and auditory signals with a variety of motor operations, using computer-based equipment. Typical instruments in this category address attention span, reaction time, tracking, time-sharing and situational awareness. An instrument for the measurement of situational awareness is the WOMBAT, which was not only found to improve the prediction by abilities but also to be associated with very high levels of pilot performance (O'Hare, 1997). Nowadays there are several batteries available that include a variety of such tasks. The CogScreen battery and the Vienna Test System are examples used in civil selection, the Basic Attributes Test (BAT) (Carretta, 1989) is an example from military selection, and the Pilot Aptitude Tester (PILAPT) (Burke, Kitching et al., 1997) has been used in military as well as civil settings. The second type of tools covers more complex tasks to be performed in a social setting. Their aim is to find out how competent candidates are in the realm of interaction (crew resource management and teamwork), time sharing (anticipation, priority-setting and decision making under information overload and time sharing), leadership and coaching, and servicing customers. For KLM Royal Dutch Airlines (Hermans, Vlug, & Vogelaar, 1994), a job sample was developed in the form of a two-crew command & control computer game, in which live-interaction and system monitoring were combined. The game has been in use for about ten years for both *ab initio* and employment selection (Kleewein, 1996). The combined use was possible because the game tapped into aircrew skills such as decision making, systems monitoring, teamwork and coordination, without requiring technical flying skills or knowledge of jargon. The costs of these simulations are low to moderate, their acceptability is generally rather high.

HF Job Samples/Simulations

Originally many selection systems made use of a Link Trainer, an aircraft-like device that could move round three axes, named after its designer Edwards Link. Although originally designed for instruction purposes the trainer was widely adopted as a selection tool. Nowadays several other simulator devices are used. Popular among them is the FRASCA, which simulates a light single-engine aircraft. After instruction candidates have to fly several legs while various parameters of their flight performance are scored by a computer. Similar systems are the Automated Pilot Selection System (APSS) developed by the Royal Netherlands Air Force (Tier, 2004) and the Canadian Automated Pilot Selection System (CAPSS) developed by the Canadian Armed Forces. This type of high fidelity simulator is costly to use but possesses substantial validity in the prediction of flight training success for *ab initio* candidates (Woychesin, 2002). The acceptability is normally very high.

Integrated Test Batteries

Although it is possible to build a selection system from modules that include instruments as were just described, there are also packaged batteries that combine a number of sign and sample based instruments. A good example is the Microcomputerized Personnel Aptitude Tester (MICROPAT) developed by Bartram (1987; 2002). This battery comprises landing, scheduling, and navigational calculation tasks next to tracking, mental arithmetic, and spatial orientation. Research has shown its validity for *ab initio* selection in civil aviation (Bartram & Baxter, 1996). There are several test batteries for the field of military aviation. Examples are the Computerized Test Battery of the NATO Air Crew selection Working Group (Hermans & Mulder, 1998), and the Aviation Selection Test battery (ASTB) which has been developed for applicant screening in the US Navy and Marine Corps. Kantor and Carretta (1988) have developed a computerized test battery for pilots in the US Air Force, comprising two psychomotor tests and tests of perceptual speed, decision-making speed, and memory function. These test batteries are generally cost-effective and enjoy moderate to good acceptability.

Interviews

There are several techniques for building useful structured interviews. Criterion-referenced interviews focus on competences and ask questions about the candidate's experience with and conduct in specific job-related situations (Hermans & Mulder, 1998). Situational interviews have the same purpose, but ask candidates what they would do in particular situations. Both techniques use follow-up questions to get specific enough information from the candidate. In addition one can use the STAR technique to elicit precise answers (S=situation, T=Task, A=action, R=result). Interviews may also address questions that are meant to reveal traits. For instance, candidates are asked to provide live accounts of their behavior in critical work-situations, relevant for a specific competence ("tell me about one particular instance in which you had to correct someone, what exactly did you say?"). The situations presented in situational interviews are usually dilemmas. The candidate's answers may be hypothetical or intentional ("if that was the case, I would certainly try to make it understood that ..."). To overcome this, the interviewer may involve the candidate in a short episode of here-and-now role playing, acting as a live crew member or customer for a while ("let me be the co-pilot, say, Jim just shouted at the cabin-attendant"). The pros and cons of structured and situational interviewing are described in Latham and Saari (1984). Hermans and Mulder (1998) provide a step by step guide to competence-based and situational interviewing for recruiters of air crew, with worksheets and checklists. Interviews tend to be costly because they are conducted individually by qualified personnel. Their acceptability is moderate to good.

Methods of Delivery

Although paper-and-pencil tests are still in use, computerized tests are currently most prevalent. For both modes of delivery good validities have been found. Although the use of personal computers – either stand-alone or in a network environment – remains popular, the administration of tests through the internet is developing rapidly. Internet-testing has great advantages in terms of cost reduction, processing speed and data-gathering. There is less need for testing rooms, for travel to testing sites by candidates and for administration staff. Also, there is better control over the quality of the data and it is easier to build up data-bases for research. However, there are also problems (see Bartram, 2001, for an overview). Firstly, the test items, when made public, may quickly become known among candidates and lose their power. Secondly, candidates will be able to make use of stand-ins, that is, (smarter) friends or relatives may make the tests for them. One way out is the so called *kiosk model*. Temp-agencies or satellite offices of airlines can instantiate walk-in locations where candidates can make the tests, on line but under supervision of personnel, and after having proven their identity. Current internet applications cover ability tests, cognitive tests, personality tests, low-fidelity job samples, biographical questionnaires and interviews – i.e. instruments used in early selection stages. In a recent field experiment by Chapman and Rowe, videoconferencing tools were used in an attempt to determine how candidates and interviewers respond to web-based interviewing (Chapman & Rowe, 2001). Further developments of multi-media technology may expand the scope of internet-based assessment and open the way to the use of web-based simulators.

Statistical Issues

Although the principles of design are straightforward their application to the design of pilot selection systems is not easy (Carretta & Ree, 2003; Damos, 1996). In this section we will discuss some key issues related to validation research and the interpretation of its outcomes. While designers will rarely plan to actually conduct a validation study while embarking on the (re)design of a selection system, they will always look at the results of studies conducted by others and therefore be confronted with the complexities of validation research. A first difficulty relates to *sampling size* and lies in the fact that the numbers of candidates whose test scores are to be validated are often are not large enough to reach the sample sizes needed for sufficiently powerful analysis (Burke, Hobson et al., 1997; Hunter & Burke, 1994). This difficulty can be overcome by stretching the time-interval, that is, by putting together candidates who have been tested over a longer time-period. However, this normally creates another difficulty, i.e. that candidates have been assessed (and selected) with non-identical instruments which increases the heterogeneity of the sample and reduces the relevance of the resulting validity. A second difficulty is that validity coefficients can only be calculated for those candidates who have surpassed certain threshold scores, which implies that samples are censored (Carretta & Ree, 2000). This problem, known as *restriction of range* (Thorndike, 1949) has a variety of unwanted consequences. It reduces the value of the validity coefficient (sometimes even gives it a negative sign) in a degree that is dependent on the predictor's role in the selection. As a result the rank order of validities may change and a predictor that was more valid may appear to be less valid. Moreover, restriction of range has an impact on intercorrelations and factor structures, as well as on reliabilities. There are a number of techniques to correct for the effects of restriction of range on validities, but the corrected values are on average slightly biased and are less accurate, particularly in small samples. When true validities are low this poses a risk of overcorrection (Roe, 2004).

The interpretation of published validity estimates is seriously hampered as a result of these effects, especially since many studies fail to give information about restriction of range. This is especially troublesome when results from a diversity of studies have to be interpreted. The application of *meta-analysis* (also: validity generalization, Schmidt & Hunter, 1977, 2003) has done much to change this situation by integrating the results from multiple studies while correcting for the effects of sample size, restriction of range and criterion unreliability. Although meta-analysis has been helpful, its application is not without problems. Since there is no in-built check on the comparability of the original studies there is a risk that unwarranted conclusions are drawn. In the case of pilot selection one must take into account that aviation has undergone dramatic changes which make earlier and later validities incomparable. Various studies have observed a decline of predictive validities over time (e.g. Bartram & Dale, 1982, Hunter & Burke, 1994). "Because the studies were collected over a time period from the Wright Brothers to modern jet fighters, perhaps one would expect changes in test validity" (Martinussen, 1996, p.14). Bartram (1987) is more specific in noting that the use of advanced computer technology has changed the nature of the pilot's job from manual to supervisory control, which has apparent implications for the validities of various abilities. Apart from changes in aviation, there may also have been changes in the supply of candidates in terms of numbers applying, level of education, spontaneous self-selection, and the reliance on tests in taking selection decisions. It is also important to pay attention to the precise nature of the criteria (Carretta, 1992). Validities against a pass-fail criterion in basic training may not generalize to validities against supervisory ratings.

In interpreting validity data from the field of aviation, one must also be aware of *population differences*. An important factor is the heterogeneity of the general working population and hence of applicants in the field of aviation. Many published validity estimates are based on data from more homogeneous populations that may not generalize well to current conditions. Selection from heterogeneous populations requires a check on predictive bias in test batteries that was not needed in the past (e.g. Carretta, 1997; Weeks & Zelenski, 1998). In addition, there are national and cultural factors to consider. Most validity data derive from studies conducted in the United States and other English speaking countries. Although research in other domains has revealed similarities between the US and Europe (Salgado, 1997; Salgado et al., 2003), research in pilot selection has found some differences in the validity of job samples (Hunter & Burke, 1994). Further research will have to establish whether validities in pilot selection are generalizable and to what degree. A result of all this, is that a straightforward interpretation of validity coefficients is not possible and that expert judgment is required to draw the right conclusions.

A final point is that interpretations of validity data may depend on *theoretical positions* taken. For instance, some researchers posit that prediction in personnel selection should start from general mental ability or g – which corresponds to a common factor – and should be complemented with group or specific factors whenever incrementally valid. They assign great importance to the predictive power of the first unrotated factor found in factor analysis. This approach has been advocated by Carretta and Ree (2000) who analyzed validity data from a set of ability and psychomotor tests in terms of improvements to the validity of g. In this way neither specific abilities nor psychomotor factors seem to make a great contribution. Others argue that g cannot be measured directly but is derived from tests for verbal, numerical, figural, spatial and other abilities. To them, information on these factors gives a better view of the mental abilities required by the pilot's job. There are also different views on *incremental validity*, a concept that stems from the application of multiple regression analysis. A predictor is said to incrementally valid if it produces a significant increase in the multiple correlation. Evidence of incremental validity offers a strong argument to expand an existing battery. For instance, Caretta and Ree (1994) starting from paper-and-pencil tests found that flying experience, psychomotor skills, and attitude toward risk incremented the prediction of military training criteria. Adding computer-administered information-processing measures did raise validity. The question is what meaning should be assigned to instruments that have low or zero incremental validity (see e.g. Walters, Miller & Ree, 1993 for structured interviews in military pilot trainees). Following the line of interpretation mentioned above, these instruments might be interpreted to "measure nothing but g" and they might consequentially be dropped. One can also take another stance and argue that a lack of incremental validity (in combination with acceptable validity of the predictor itself), points at redundancy in the battery that can improve the robustness of the prediction. This could be seen as an advantage, provided that the costs of the additional predictor are acceptable. Another argument against assigning much weight to lack of incremental validity is that there is some risk of capitalization on chance in the regression procedure, which can lead to an unduly rejection of the predictor.

A Practical Example

In this section we describe a selection system that has been developed for and used by KLM Royal Dutch Airlines. The development of the system took place over the period 1994–99 and included a number of revisions. It replaced the older selection system, which had been largely clinical. The development process was guided by a protocol that reflects most of the principles mentioned in this chapter. An overview of the resulting system is given in Table 11.3. The development began with systematic work profiling based on critical incident interviews with subject-matter experts. This produced five target areas that provided criteria for the selection. The criteria are Work Attitude, Information Management, Leadership, Stress Management, and Cooperation. They were operationalized by means of Behaviorally Anchored Rating

Table 11.3 Selection System for KLM Royal Dutch Airlines (1994–1999)

Target areas (Criteria)	Work Attitude, Information Management, Leadership, Stress Management, Cooperation
Instruments (Predictors)	*Mental Ability Tests*
	Tests for Logical and Numerical Reasoning, Technical Comprehension, Spatial Orientation, Fault Finding
	Occupational Personality Questionnaire
	Counter-indicative signs from various scales
	Psychomotor Coordination Task
	Unstable target tracking task with memory load
	Command & Control Job Sample
	Job sample assessing crew resource management & emergency handling
	Structured Situational Interviews
	Competence based interviews by psychologists & trained chief pilots
	Clinical Personality Test
	MMPI-2 with limits
	Mental State Interview
	Clinical interview for applicants who exceed MMPI-2 limits by psychologist / psychiatrist
Administration and Decision Making (Procedure)	*Applicants for ab initio training*
	Stage I: Mainly based on Mental Ability composite, marginal selection on UPI composite
	Stage II: Based on Command & Control Job Sample and Situational Interviews
	Stage III: Based on MMPI-2 and Mental State Interview
	Applicants for licensed pilot positions
	Stage I: Selection on Mental Ability, Command & Control Job Sample and Situational Interviews
	Stage II: Based on MMPI-2 and Mental State Interview

Scales. Next, a shortlist of predictor instruments was made using the results of a literature survey and the best predictors were included in the selection system. Five tests for mental ability were chosen from the SHL Occupational Test Series, i.e. Logical Reasoning, Numerical Reasoning, Technical Comprehension, Spatial Orientation, and Fault Finding. The scores on these tests were combined into a Mental Ability Composite. SHL's Occupational Personality Inventory was used to identify a set of counter-indicative signs, i.e. high scores on thrill seeking, dominance, and antisocial tendencies, and low scores on decisiveness, emotional stability, team-orientation and conscientiousness. From these sign scores a regression-based composite, the Unified Personality Indicator (Mulder & Hermans, 1999) was calculated.

Additional instruments were a Psychomotor Coordination Task, an unstable tracking task with memory load (AGARD, 1989), and a computer-based Command & Control Job Sample (*Future City*, Hermans et al., 1994) mimicking a two-crew operation that calls for crew resource management and emergency handling. At a later moment a Structured Situational Interview was developed, to be administered by psychologists and trained chief pilots. The interviewing method is available in the form of a stepwise guide (Hermans & Mulder, 1998). Moreover, two clinical instruments were included for use at the final stage of selection, i.e. The Minnesota Multiphasic Personality Questionnaire-2 and a Mental State Interview to be administered by a clinical psychologist or psychiatrist.

In the system a distinction is made between selection of *ab initio* candidates and selection of licensed pilots. The instruments are essentially the same, but the way they are arranged and used in the procedure differ. The *ab initio* procedure is segmented in three stages, and the licensed pilot procedure in two stages.

The above description of the system focuses on its content in terms of predictors and criteria and its overall structure. There are many other facets that are to be considered in the context of systems design. One of them is shaping the expectations of applicants and giving them the opportunity to provide feedback on their experiences. For most applicants there is some mysteriousness around selection and psychological testing. They tend to see selection as a hurdle and those involved in it as temporary enemies, or at least as a possible source of mishaps and troubles for which no manuals or standard operating procedures are available. It is important that candidates develop realistic ideas about the nature and purpose of the selection procedure because this will enhance the validity of the outcomes. Moreover, allowing candidates to give feedback on their experiences, e.g. on perceived arrogance or indifference on the part of the selecting staff, will contribute to more open and businesslike communication, something airlines will benefit from in the shorter or longer run. Table 11.4 shortlists typical expectations (and experiences) along with recommended actions.

How to Account For Mental Health in Pilot Selection

In the examples of selection systems and in the general approach that was outlined in before, it almost seems as if mental health is not really an issue anymore in modern aircrew selection. Throughout the selection process the emphasis seems to be on identifying competence in air crew and mental health aspects seems to receive hardly any attention. This is puzzling to many people in the field, both novice and experienced. Since the purpose is to select competent and mentally healthy air crew, one would expect personality inventories and questionnaires, directly aiming at identifying signs of psychopathology, to appear up front in the system. One would think that such instruments should be applied first with considerable sternness and with appropriate defensive measures against lying, faking and socially desirable answering tendencies. What could and should come next would then be the searching

Table 11.4 **Some Typical Aircrew Applicant Expectations and Recommended Actions**

Typical applicant expectations	Recommended action
Lengthy examinations by one or more medical doctors.	Provide (on line) information on who the applicant is going to meet.
Relentless competition with other candidates.	Inform applicants about the program. Make clear if and when applicants will work on their own.
Immediate disqualification when a single mistake is made.	Provide realistic information on the way performance will be evaluated.
Tests and interviews with hidden purposes.	Provide full information on the target competences, skills and attitudes, both before and during testing.
Raised eyebrows when questions are asked.	Make clear that questions will be welcomed.
A focus on motives, traits and flaws outside the realm of work.	Explain shortly but clearly about the relationship between test content and work.
Long shelf life of test outcomes.	Provide details on shelf life and distribution of results and reports.
A linear relation between preparatory effort and outcomes.	Explain what kind of preparation helps, and to what extent. Provide exercise materials.
Outcomes in terms of binary pass/fail decisions without further detail.	Provide access to the results in full.[2]
Long delays and difficulty when trying to contact the company.	Provide details on timing, and be available.[3]

for the most competent aircrew amongst those who pass, i.e. the mentally healthy. In a cookbook for air crew selection, the novice reader would probably expect this kind of recipe, which reflects a "safety first, so let's apply inventories first" philosophy. However, following this kind of recipe would actually result in washing out a large group of mentally healthy candidate air crew in the early stages of selection. In other words there would be a large number of *false negatives*. One might argue that such false negatives are entirely acceptable because of the absolute primate of safety. But, perhaps surprisingly to those favoring this approach, the strict and early use of inventories would fail to ensure that those who pass will actually be (and remain) the mentally healthy candidates. Many cases of pathology would remain undetected and thus represent *false positives*. This means that the safety first inventory approach would reduce the number of candidates without bringing greater safety in the air.

---

2 In many European countries, aviation psychologists will follow the code of *prior perusal & veto*. Applicants will be the first party to receive the outcome. Usually, there is no room for negotiation on the content, but applicants do have the possibility to have their files closed, that is, to withhold consent for any further distribution of the results.

3 Air France's website is a place where applicants are informed and questions welcomed. SHL provides test examples and exercises on this website.

The root cause for this counterintuitive but established fact lies in the moderate to poor predictive validity of inventory data (Dolgin & Gibb, 1989; Hunter & Burke, 1994). As was explained earlier in this chapter many explanations have been suggested for the low predictive validities of inventory data. Some authors argue that inventories are valid and useful in therapy, coaching and career counseling, but not really in selection, when few candidates are willing to reveal bad traits and behaviors (Hofstee, 1983). Low validities of inventory data may also be explained by the fact that mental health problems often develop over time, not only as a result of psychopathology that crops up, but also due to other factors, such as excessive work demands or working conditions.

The bottom line is that the early and stern use of the outcomes of inventory data is to be advised against. Also later on in the selection procedure, their use will not add substantially to the filtering out of aircrew on the basis of insufficient competence. One might make a case for fully abandoning personality inventories and inventories aimed at identifying early signs of psychopathology. The argument could be that, apart from low validity, there is no real need to search for underlying pathology when selection directly targets behaviors and competences (e.g. teamwork, alertness, judgment under time pressure) that are inherently incompatible with pathology (e.g. antisocial personality, confused states and indecisiveness). There are two main arguments against fully abandoning inventory data. Firstly, and especially in aviation, the primate of safety obliges companies to make the best possible use of available instruments, even weak ones, when it comes to maximizing safety (Hoermann & Maschke, 1996). When clinical instruments are weak from a perspective of prediction, the implication should rather be that there is need to develop stronger ones (Butcher, 2002). Secondly, even though competence based selection will help to reject candidates with mental health problems, one may not assume that this effect is sufficiently strong. Clinical symptoms have been found to occur in intelligent and socially competent persons. Thirdly, one would do well to search for signs of psychopathology at a stage that is close in time to the moment of performance, that is, late in the selection process and during actual employment. Probably the most advisable approach is to make prudent use of inventory data, in combination with modern, competence oriented methods in the early stages of selection and to screen for symptoms of psychopathology at the end of the selection process as well as in periodical check-ups. More specifically, we would suggest the following.

Pre-Selection Stage

Personality inventories should be part of the test battery but be used only to reject candidates showing traits that are incompatible with healthy functioning. This means filtering on the presence of one or more extreme scores (e.g. <10th or >90th percentile) on scales that known to be associated with pathology or poor performance (Dolgin & Gibb, 1989; Hunter & Burke, 1994). Very similar is the use of the Unified Personality Indicator (Mulder & Hermans, 1999) that adds up counter-indication-points given to candidates with very low scores on decisiveness, emotional stability, team-orientation and conscientiousness, and very high scores on thrill seeking, dominance and anti-social tendencies. Only candidates passing a cutoff on this scale will be removed from the candidate pool at this early stage. The cutoff on the UPI should be low (e.g. 20th percentile) compared to the cutoff on an mental ability composite (e.g. 60th percentile).

Mid-Selection Stage

The use of sample type instruments (e.g. job samples and situational interviews) will normally remove many of the unsuited candidates since the behavioral repertoire they tap into requires minimal levels of mentally healthy functioning (e.g. judgment under pressure, time sharing, divided attention, crew communication, servicing demanding customers and resolving conflicts). Beyond this, one might employ specific personality scales to predict particular criteria. For instance, self reported traits such as agreeableness and conscientiousness might be used to predict service and representation competence. When there are also sample instruments to predict these criteria[4] – for instance, a role play with an angry customer – these should be allowed to outweigh the inventory data, especially when experienced applicants are sought.

Final Selection and Post-Selection Stage

Candidates who have passed the previous selection stages may be subjected to a clinical examination, using instruments such as the MMPI-2 (Butcher, 1994) or the ALAPS (Berg et al., 2002), and an interview by a trained clinical psychologist or psychiatrist. These techniques may also be applied during periodical check-ups of candidates after they have entered employment.

Recommendations on Pilot Selection

In this chapter, we have outlined the principles and methods of cockpit crew selection and discussed how effective selection systems can be developed. We have tried to recapitulate the state-of-the-art and given some practical examples in order for our readers to understand which steps must be taken and which decisions are to be made in designing (or redesigning) a selection system. Although few of our readers will be system designers, most of them will be somehow involved in pilot selection and share responsibility for the way in which selection is carried out. To summarize in practical terms the major points that we have tried to make, we conclude this chapter with a list of "dos and don'ts" in systems design (see Table 11.5).

---

4   Hermans & Mulder call this "hybrid mapping."

Table 11.5 Do's and Don'ts in the Design of Aircrew Selection Systems

Do	Don't
Develop a program of requirements (POR) first.	Take an exclusively psychometric look at selection that ignores the political, commercial and practical of selection.
Apply work profiling. Shortlist target (sub)competences, skills & attitudes.	Exceed the magical number of 7 when short listing target competences for assessment.
Draw up a prediction model and map the links between instruments and competences.	Omit mapping potential mental health risks to (subscales of) selection instruments.
Use weighted composites rather than multiple cut-offs.	Use complex decision rules overloaded with cut-offs and no-go.
Apply structure to interviews.	Trust impressions developed in unstructured interviews.
Have recruiters follow training in competence based interviewing.	Allow board-interviews by untrained recruiters to rehash selection.
Make use of sample type selection instruments when available and feasible.	Use projective tests for selection purposes.
Do apply personality inventories, but use marginal cut-offs on combined scales.	Apply large weights to the outcome of inventories in the early stages of selection.
Consider porting selection instruments to the internet. Consider the "kiosk model".	Disregard limitations of unattended testing.
Consider cost, predictive validity and face validity when porting tests to the internet.	Disregard attractiveness and transparency when designing an on-line recruitment pages.
Consider online situational interviewing using conferencing technology.	Have unstructured and informal interviewing take place on the internet.
Apply utility analysis to guide decisions on the content and stages of selection and to calculate return on investment.	Expect executives to be compelled by numbers that are not linked to strategic goals.
Evaluate the system with regard to all requirements in the POR.	Leave out the impact of accidents, bad service, and inefficiency in operations.

Although many international airlines and military operators have good pilot selection systems in place and make efforts to keep these systems up to date, it may not be assumed that all selection systems currently in use reflect the state-of-the-art. On the contrary, there are also cases in which operators base their selection decisions on applicants' flying experience and common sense. In view of the critical role of the human factor in the cockpit and the high stakes involved in aviation, we think that operators should ask themselves whether their selection practices do meet contemporary standards and whether there is room for improvement. More specifically, we feel that they should engage in a periodical evaluation of their selection systems and consider re-design whenever doubts about the systems' adequacy emerge. The Program of Requirements (POR), while primarily meant to provide input to the design of a new selection system, is also an excellent tool for evaluating existing systems. Therefore, establishing a POR is a worthwhile investment for any civilian or military operator. Making stakeholders' views, expectations and requirements explicit, will raise the awareness of the importance of selection and will facilitate quality assurance regarding the selection system. The actual process of designing or redesigning a pilot selection system may not a simple task that can be carried out routinely, but the efforts to do so – when needed – will certainly contribute to better performance and greater safety in aviation.

References

AGARD (1989). *Human performance assessment methods* (No. 308). Neuilly-sur-Seine: NATO Advisory Group for Aerospace Research and Development.
Bartram, D. (1987). The development of an automated testing system for pilot selection: The MICROPAT project. *Applied Psychology: An International Review, 36*(3-4), 279–298.
Bartram, D. (2001). Internet recruitment and selection: Kissing frogs to find princes. *International Journal of Selection and Assessment, 8*, 261–274.
Bartram, D. (2002). The MICROPAT Pilot Selection Battery: Application of generative techniques for item-based and task-based tests. In Irvine, Sidney H. (ed.); Kyllonen, Patrick C. (ed.). (2002). *Item generation for test development.* (pp. 317–337). Mahwah, NJ: Lawrence Erlbaum Associates, Publishers.
Bartram, D. & Baxter, P. (1996). Validation of the Cathay Pacific Airways pilot selection program. *International Journal of Aviation Psychology, 6*, 2, 149–169.
Berg, J. S., Moore, J. L., Retzlaff, P. D., & King, R. E. (2002). Assessment of personality and crew interaction skills in successful naval aviators. *Aviation, Space and Environmental Medicine, 73*, 6, 575–579.
Burke, E., Hobson, C., & Linksy, C. (1997). Large sample validations of three general predictors of pilot training success. *International Journal of Aviation Psychology, 7*, 3, 225–234.
Burke, E., Kitching, A., & Valsler, C. (1997). The Pilot Aptitude Tester (PILAPT). On the development and validation of a new computer-based test battery for selecting pilots. In *Proceedings of the 9th IInternational Symposium on Aviation Psychology, Columbus, OH; 27 April -1 May 1997* (pp. 1286–1291).
Butcher, J. N. (1994). Psychological assessment of airline pilot applicants with the MMPI-2. *Journal of Personality Assessment, 62*, 1, 31–44.
Butcher, J. N. (2002). Assessing pilots with "the wrong stuff": A call for research on emotional health factors in commercial aviators. *International Journal of Selection and Assessment, 10*, 1–2, 168–184.
Carretta, T. R. (1989). USAF pilot selection and classification systems. *Aviation, Space and Environmental Medicine, 60*, 1, 46–49.
Carretta, T. R. (1992). Short-term retest reliability of an experimental U.S. Air Force pilot candidate selection test battery. *International Journal of Aviation Psychology, 2*, 3, 161–173.

Carretta, T. R. (1997). Group differences on US Air Force pilot selection tests. *International Journal of Selection and Assessment, 5*, 2, 115–127.

Carretta, T. R. (2000). US air force pilot selection and training methods. *Aviation, Space and Environmental Medicine, 71*, 950–956.

Carretta, T. R. & Ree, M. J. (1994). Pilot-candidate selection method: Sources of validity. *International Journal of Aviation Psychology, 4*, 2, 103–117.

Carretta, T. R. & Ree, M. J. (2000). General and specific cognitive and psychomotor abilities in personnel selection: The prediction of training and job performance. *International Journal of Selection and Assessment, 8*, 4, 227–236.

Carretta, T. R.. & Ree, M. J. (2003). Pilot selection methods. In M. A. Vidulich & P. S. Tsang (eds.), *Principles and practice of aviation psychology*. (pp.357–396). Mahwah, NJ: Lawrence Erlbaum Associates.

Chapman, D. S. & Rowe, P. M. (2001). The impact of videoconference technology, interview structure, and interviewer gender on interviewer evaluations in the employment interview: A field experiment. *Journal of Occupational and Organizational Psychology, 74*, 3, 279–298.

Chidester, T. R., Helmreich, R. L., Gregorich, S. E., & Geis, C. E. (1991). Pilot personality and crew coordination: Implications for training and selection. *International Journal of Aviation Psychology, 1*, 1, 25–44.

Damitz, M., Manzey, D., Kleinmann, M., & Severin, K. (2003). Assessment center for pilot selection: Construct and criterion validity and the impact of assessor type. *Applied Psychology: An International Review, 52*, 2, 193–212.

Damos, D. L. (1996). Pilot selection batteries: Shortcomings and perspectives. *International Journal of Aviation Psychology, 6*, 2, 199–209.

Dolgin, D. L. & Gibb, G. D. (1989). Personality assessment in aviator selection. In. R. S. Jensen (ed.), *Aviation psychology*. (pp.288–320). Brookfield, VT: Gower Publishing Co.

Fitzgibbons, A., Davis, D., & Schutte, P. C. (2004). *Pilot personality profile using the NEO PI-R* (No. NASA/TM-204-213237). Hampton, VI: NASA.

Grove, W. M. & Meehl, P. E. (1996). Comparative efficiency of informal (subjective, impressionistic) and formal (mechanical, algorithmic) prediction procedures: The clinical-statistical controversy. *Psychology, Public Policy, and Law, 2*, 2, 293–323.

Grove, W. M., Zald, D. H., Lebow, B. S., Snitz, B. E., & Nelson, C. (2000). Clinical Versus Mechanical Prediction: A Meta-Analysis. *Psychological Assessment, 12*, 1, 19–30.

Hermans, P. H., Jacobs, M., & Van Niekerk, P. (1987). *The Defense Mechanism Test and Success In Flying Training*. Paper presented by F. Rameckers at the Conference of the Western European Association for Aviation Psychology WEAAP 1987, Vienna. Later published in: Farmer, E. (ed.) (1991) *Proceedings of the XVIII WEAAP Conference*. Aldershot: Avebury Aviation.

Hermans, P. H. & Mulder, H. W. (1998). Job analysis and the selection interview. In K. M. Goeters (ed.), *Aviation Psychology: A Science and a Profession* (pp. 81–95). Aldershot: Ashgate.

Hermans, P. H., Vlug, T., & Vogelaar, V. (1994). *Future City: Job sample for Command & Control, Software & Instruction Manual.* Utrecht, Schiphol: PsychoTechniek, KLM Royal Dutch Airlines.

Hoermann, H. J. & Maschke, P. (1996). On the relation between personality and job performance of airline pilots. *International Journal of Aviation Psychology, 6,* 2, 171–178.

Hofstee, W. K. B. (1983). *Selectie: begrip, theorie, procedures & ethiek. Met bijzondere aandacht voor de problemen van personeelsselectie en de positie van de sollicitant.* Utrecht: Het Specturm.

Hormann, H. J. & Maschke, P. (1996). On the relation between personality and job performance of airline pilots. *International Journal of Aviation Psychology, 6,* 171–178.

Hough, L. M. & Ones, D. S. (2002). The structure, measurement, validity, and use of personality variables in industrial, work, and organizational psychology. In (2002) D. S. Ones & N. Anderson (eds.) *Handbook of industrial, work and organizational psychology, Volume 1: Personnel psychology* (pp.233–277). London: Sage Publications Ltd.

Hunter, D. R. (1989). Aviator selection. In M. F. Wiskoff & G. N. Rampton (eds.), *Military personnel measurement* (pp. 129–167). New York: Praeger.

Hunter, D. R. & Burke, E. F. (1994). Predicting aircraft pilot-training success: A meta-analysis of published research. *International Journal of Aviation Psychology, 4,* 4, 297–313.

Jessup, G., & Jessup, H. (1971). Validity of the Eysenck Personality Inventory in pilot selection. *Occupational Psychology.*

Kantor, J. E. & Carretta, T. R. (1988). Aircrew selection systems. *Aviation, Space and Environmental Medicine, 59,* 11, Sect 2, 32–38.

Kay, G. G. (1995). *CogScreen Aeromedical Edition Professional Manual.* Odessa FL: Psychological Assessment Resources.

Latham, G. P. & Saari, L. M. (1984). Do People Do What They Say? Further Studies on the Situational Interview, *Journal of Applied Psychology* (Vol. 69, pp. 569): American Psychological Association.

Martinussen, M. (1996). Psychological measures as predictors of pilot performance: A meta-analysis. *International Journal of Aviation Psychology, 6,* 1, 1–20.

McDaniel, M. A., Whetzel, D. L., Schmidt, F. L., & Maurer, S. D. (1994). The Validity of Employment Interviews: A Comprehensive Review and Meta-Analysis., *Journal of Applied Psychology* (Vol. 79, pp. 599–616): American Psychological Association.

Mulder, H. W. & Hermans, P. H. (1999). *A Unified Personality (Counter-) Indicator and its Usage for Pilot Selection. Internal memo.* Utrecht / Rotterdam: SHL Netherlands / Zeelenberg Advisory Group.

O'Hare, D. (1997). Cognitive ability determinants of elite pilot performance. *Human Factors, 39,* 4, 540–552.

Roe, R. A. (1989). Designing selectiion procedures. In P. Herriot (ed.), *Assessment and selection in organizations* (pp. 127–142). Chichester: John Wiley & Sons.

Roe, R. A. (1998). Personnel selection: Principles, models and techniques. In P. J. D. Drenth, H. Thierry & C. J. De Wolff (eds.), *Handbook of work and organizational psychology - Vol. 3* (pp. 5–32). Hove: Psychology Press.

Roe, R. A. (1999). Work performance: A multiple regulation perspective. In C. L. Cooper & I. T. Robertson (eds.), *International Review of Industrial and Organizational Psychology* (Vol. 14, pp. 231–335). Chichester: John Wiley & Sons.

Roe, R. A. (2002). Competenties - Een sleutel tot integratie in theorie en praktijk van de A&O-psychologie. *Gedrag & Organisatie, 15*, 4, 203–224.

Roe, R. A. (2004). *Are we under- or overcorrecting for restriction of range? Results from a Monte Carlo study of seven correction formulas.* Maastricht: Universiteit Maastricht, Faculty of Economics and Business Administration.

Roe, R. A. (2005). The design of selection systems: Context, principles, issues. In A. Evers, O. Smit & N. Anderson (eds.), *Handbook of personnel selection*. Oxford: Blackwell.

Roozenburg, N. F. M. & Eekels, J. (1998). *Productontwerpen, structuur en methoden* (2nd edn.). Utrecht: LEMMA.

Salgado, J. F. (1997). The five factor model of personality and job performance in the European Community. *Journal of Applied Psychology, 82*, 30–43.

Salgado, J. F., Anderson, N., Moscoso, S., Bertua, C., de Fruyt, F., & Rolland, J.-P. (2003). A Meta-Analytic Study of General Mental Ability Validity for Different Occupations in the European Community. *Journal of Applied Psychology, 88*, 6, 1068–1081.

Salgado, J. F. & Moscoso, S. (2002). Comprehensive meta-analysis of the construct validity of the employment interview. *European Journal of Work and Organizational Psychology, 11*, 3, 299–324.

Sawyer, J. (1966). Measurement and prediction, clinical and statistical. *Psychological Bulletin, 66*, 3, 178–200.

Schmidt, F. L. & Hunter, J. (2003). History, development, evolution, and impact of validity generalization and meta-analysis methods, 1975-2001. In K. R. Murphy (ed.), *Validity generalization: A critical review* (pp. 31–65). Mahwah, NJ: Lawrence Erlbaum Associates.

Schmidt, F. L. & Hunter, J. E. (1977). Development of a general solution to the problem of validity generalization. *Journal of Applied Psychology, 62*, 5, 529–540.

Siem, F. M. (1992). Predictive validity of an automated personality inventory for Air Force pilot selection. *International Journal of Aviation Psychology, 2*, 261–270.

Tett, R. P., Jackson, D. N., & Rothstein, M. (1991). Personality measures as predictors of job performance: A meta-analytic review. *Personnel Psychology, 44*, 703–742.

Thorndike, R. L. (1949). *Personnel selection*. New York: Wiley.

Tier, R. (2004). Aviation and occupational psychology in the Royal Netherlands Air Force. In *Proceedings of the 26th Conference of the European Association for Aviation Psychology, Seimbra, Portugal, 3-7 October 2004.* (pp. 56–59).

Walters, L. C., Miller, M. R., & Ree, M. J. (1993). Structured interviews for pilot selection: No incremental validity. *International Journal of Aviation Psychology*, *3*, 1, 25–38.

Weeks, J. L. & Zelenski, W. E. (1998). *Entry to USAF undergraduate flying training* (No. AFRL-HE-AZ-TR-1988-0077). Mesa AZ: US Air Force Research Laboratory.

Wernimont, R. F. & Campbell, J. (1968). Signs, samples and criteria. *Journal of Applied Psychology*, *52*, 372–376).

Westerman, R., Draby, D. G., Maruff, P., & Collie, A. (2001). Computer-assisted cognitive function assessment of pilots. *ADF Health*, *2*, 29–36.

Wherry, R. J. (1975). Underprediction from overfitting: 45 years of shrinkage. *Personnel Psychology*, *28*, 1, 1–18.

Woychesin, D. E. (2002). Validation of the Canadian Automated Pilot Selection System (CAPSS) against primary flying training results. *Canadian Journal of Behavioural Science*, *34*, 2, 84–91.

Chapter 12

Psychological Aspects of Selection of Flight Attendants

Ferenc Albert

Introduction

> We hire great attitudes and we'll teach them any functionality they need. (Herb Kelleher, Southwest Airlines in Farkas et al., 1997, p. 73)

When flight 092, a British Midland B737, suffered a fan blade failure at the top of a climb which damaged one of the engines, the pilots quickly analyzed the situation and found that they had to shut down the starboard engine. The flight data and cockpit voice recorder revealed that the co-pilot was hesitant, but even though he implied that it was an incorrect decision, he soon accommodated to his captain's decision. The pilots also informed the passengers that they had to shut down the starboard engine and they intended to return to the airport for a single engine landing. Since the pilots could no longer smell any smoke, they were convinced that the situation was under control.

In the cabin, the passengers and the cabin crew listened carefully to an announcement from the flight deck about the situation. One member of the crew, a young and inexperienced flight attendant, was anxiously watching the engines when he made a quite remarkable observation: the pilots had mistakenly performed a precautionary engine shutdown of the wrong engine. However, since he expected the flight deck crew to be fully competent and experienced to handle this critical incident, he strictly followed the *sterile cockpit* concept he recently had been trained to do and refrained from interfering with the work on the flight deck.

In the subsequent accident investigation report (Air Accidents Investigation Branch, 1990), *pilot error* was discussed while *cabin crew error* was hardly mentioned. There are a number of situations where the outcome of a flight incident or an accident could have developed quite differently, depending on the decisions and actions of the cabin crew. In 1985, another B737 suffered an uncontained failure of the left engine which tore a fuel line and ignited setting off a massive fire. Although the take-off was successfully aborted, 55 passengers and crew members could not escape the aircraft before being overcome by smoke and fumes. One contributing factor to the high number of fatalities was the confusion in the evacuation led by the cabin crew. Nevertheless, there are also countless other cases where the cabin crew made a more positive difference and saved many lives. The problem is however, that while flight academies and airlines around the world train flight attendant candidates to conduct themselves professionally in the cabin, they are seldom offered courses in understanding the role of personality and decision making, judgment, independence, self-esteem and other traits and skills, similar to Crew Resource Management training undertaken by the pilots. However, given their crucial role in the safe operation of a flight, a careful selection of flight attendants is of the same vital importance as the selection of flight deck crew. Expressed in another way, we should train flight attendants for skills and select them for attitude.

Brief History

In the early days of commercial aviation, co-pilots had the responsibility for the cabin before the first "hostesses" were employed and had the responsibility for passing around boxes with lunches and serving coffee (Musbach & Davis, 1980). They were also expected to inform passengers not to throw cigarette butts out of the window, particularly over populated areas. Among the first profile descriptions of the ideal stewardess was the following: "blue eyed with brown hair, poised and self-possessed, slender, 5 feet 3 inches tall, weighs 115 pounds, is 23 years old, actively engaged in some participant sport, an expert swimmer, a high-school graduate, with business training and attractive" (Beaty, 1976, in Edwards & Edwards, 1990, p. 59). Young, nice-looking and charming females were employed but had to leave the job as soon as they got married or became pregnant. They were responsible for serving customers, who were usually wealthy enough to afford an airline ticket. This has, of course, contributed to the air of glamour we can still observe, although the professional status of the flight attendant has not really been accorded to those who perform this task.

A number of regulations now specify what cabin crew should accomplish and have been added to the original requirements. The focus on duties on board has changed from serving to safeguarding the passengers by providing leadership in case of emergency – although the popular image has not caught up with this. Larger aircraft, wider socio-economic groups of passengers, increasing need of profitability and additional business factors have changed the character, status and the power of attraction of the job as well. A luxurious environment with exclusive privileges has been substituted with a mass market scenario with decreasing benefits and increasing duties and stress. Although the glamor has faded in many airlines, the competition in the airline industry still keeps the glamorous image alive. The "Singapore Girl" is one classic example of an image aimed at directing customers' expectations to charming pleasure, unlimited service and warm care. Some passengers have difficulties swapping their perception of flight attendants from this *caring* service-minded image into an authority to respect and follow instructions in an emergency. It is also understandable if many flight attendants themselves still emphasize the service-oriented parts of the job and place less focus on and professional pride in the extremely demanding safeguarding aspects.

An additional factor contributing to this stereotypic image is the fact that cabin crews seldom use the competencies that mainly justify their presence on board. On board fire, decompression and emergency landings are not daily events – fortunately. On the other hand we can observe an increasing need of other types of interference on board. Management of unruly passengers and air rage requires a trustworthy authoritative and firm figure, partly in contradiction to the caring service role normally associated with flight attendants. Pursers or cabin managers also have added responsibilities as leaders for the whole cabin crew and they need even more determination and firmness than their more junior colleagues.

From a psychological perspective this is a contradiction not so easily solved. Airlines attract customers with promises of high levels of service delivered by pleasant flight attendants. At the same time, these people are expected to perform assertively in certain situations, radiating authority.

Selection of Flight Attendants

Each airline has its own specific model when selecting flight attendants. Typical criteria include suitable age, height, weight, sight, hearing and other physical factors related to circumstances and limitations in the cabin environment. Educational background is typically focused on communication and language skills. An ability to swim is also necessary. Professional experience from the service sector or nursing are often considered as an advantage. Together, these qualities form the basis of the capacities and skills needed as a flight attendant.

Most skills and capacities are also trainable. However, the true test of proficiency is undertaken when these skills are assessed under actual or simulated emergency conditions. To identify who is therefore best suited in a selection process is complicated. Besides the skills and capacities, there is also a personality dimension to include in the assessment. The task is therefore to determine to what extent the person's personality helps or hinders in utilizing their skills under pressure. Between our skills and our effective competence (see Figure 12.1) is a personality filter which is either able to reduce the output of our skills or able to enhance them.

A flight attendant with accurate basic skills and capacities, who does not possess the self-esteem to trust his/her correct observations, must be considered to be less effective or competent. A reliable selection process should strive to assess the effective competence of the applicants and not only their capacities and skills. This does not automatically imply that airlines should look for a certain personality, but rather dynamically evaluate the influence each candidate's personality traits have on their capacities and skills. If presented with a well-educated applicant with brilliant language skills and experience from both care and service, we should also take into consideration his/her introvert personality and decide how this combination will affect the job in the cabin.

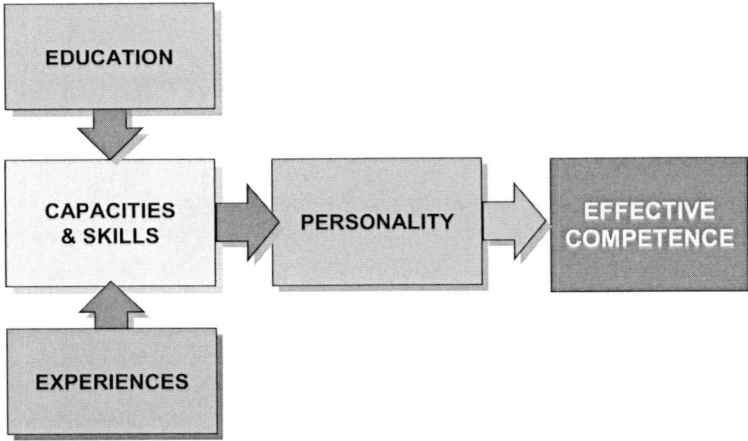

Figure 12.1 Effective Competence Model

There is a large range of factors and attributes airlines emphasise in their selection of flight attendants. Examples of typical requirement dimensions, covering education, experience, skills, personality characteristics and social competence, are: value of education, value of experience, verbal ability, language skills, judgment, personal appearance and health, personal maturity, panic tolerance, self-esteem, independence, self-knowledge, empathy, responsibility, social flexibility, co-operative ability, and value of motivation. The following paragraphs address these dimensions more closely.

Value of Education

Theoretical and practical education should be evaluated both in relation to the probability to succeed in the training and with regard to expectations passengers have about the flight attendant as a representative of a respected airline. A possible future career in the airline administration should also be taken into consideration when the value of education is assessed.

Value of Experience

This refers not only to experience gained from one's working life but to general life experiences as well. An extended trip around the world is a valuable experience in the perspective of an international, multi-cultural working environment. The assessment must be qualitative and take the value and not only the number of experiences into consideration. Experience from other airlines is not automatically an advantage, since these applicants might have learnt non-compatible procedures and might possess non-desirable attitudes from another airline company culture.

Communication Ability

One of a flight attendant's main requirements is effective communication. Both in the service part of the job and when it comes to safeguarding, a crucial success factor is sensitive listening and an ability to explain things in a concise, uncomplicated and persuasive way. Active listening is a prerequisite to understand passengers' needs, and clear oral communication from a flight attendant offers the assurance many passengers look for.

Language Skills

English is the official language used in aviation. A reasonably good command of English is therefore usually a compulsory requirement. To offer excellent service in the cabin by speaking additional languages is, of course, a competitive advantage for an airline. In safety demonstrations and in events of emergency, English is not always sufficient. Several accidents where direct communication between flight attendants and passengers was prevented by language problems bear witness to this problem (Edwards & Edwards, 1990).

Judgment

Judgment refers to the ability to evaluate the total situation adequately in order to make proper priorities and correct decisions. Sound judgment requires intellectual capacity but also experience and intuition covering both operative and social situations. To select able bodies, to decide when it is necessary to overrule sterile cockpit conditions, to handle unruly passengers and to give a personalized service are just a few of the situations in the cabin demanding a sound judgment. It is tempting to ask applicants to solve a hypothetical problem based on events in the cabin but these are often *if* questions producing theoretical responses. To ask applicants in an interview about problems they have really solved gives a more reliable indication of the applicant's judgment.

Personal Appearance and Health

Although this is a subjective dimension, mostly assessed by superficial observations, it gives an indication of passengers' "first impression" and is sometimes associated with the corporate image of the carrier. A pleasing youthful and attractive appearance characterizes what some airlines are looking for, but the definitions of these terms fluctuate not only due to culture but also personal preferences. A negative selection is probably most used here to screen out those whose physical appearance does not conform to the corporate image. Of course, those who suffer from certain disabilities e.g. hearing or eyesight problems, neurological problems or motor problems would not normally be suited to the task.

Personal Maturity

This dimension refers to attitudes and behavior adequately adjusted to age. Personal maturity is reflected in lifestyle, relations, attitudes to work, attitudes to others. Airlines usually select from candidates who are approximately 25 years of age, at the same time they wish the candidates demonstrate a maturity of 40+. One should be more cautious in relation to those young individuals who still have tight bonds to parents and other significant parental figures and who behave dependently and generally immaturely. Even if youth and a certain level of *innocence* is attractive on board, a reasonable maturity is necessary both in meeting passengers in service situations and in emergencies. For example to handle a passenger with oversized hand luggage demands some assertiveness and leadership traits.

Flight attendants are also exposed to many different temptations. On stops far from their usual social environment and control, with access to alcohol and drugs, there may be the risk of promiscuous sex and other self-destructive behaviors. A mature person with a well-developed super-ego and judgment may be better equipped to handle some of these challenges and temptations.

Panic Tolerance

While pilots focus on flying the aircraft and communicating with other professionals when an incident occurs, flight attendants have to manage both themselves and the passengers. Both on flight deck and in the cabin, panic reactions would be disastrous. In a selection process, it is a serious challenge to make a valid prediction of each applicant's tendency to be seized by panic in a given situation. Panic reaction is not a single-dimensional, clearly defined factor, so the assessor has to capture the whole dynamic complexity behind this phenomenon. When Scandinavian Airlines (SAS) faced a series of cases with panic-stricken flight attendants after aborted take-offs, the flight operative department consulted the author in order to improve the screening of candidates with a potential disposition to panic. Paper-and-pencil tests and computer-aided tests were developed, but we came to the conclusion that test data should always be supplemented with psychological background data for valid interpretations. Ego-strengths, emotional stability, absence of severe traumatic experiences, a flexible defence organization and a strong drive are factors we considered as important for a higher panic tolerance (Albert, 1991). These factors are useful when exploring background experiences and how they influence our behavior in the present. A person with some traumatic experiences in the past who has not been able to develop mature defences might, for instance, react with avoidance, denial or repression when critical situations emerge. By collecting data from different sources (tests, interviews) and different themes (background, present life, here-and-now) and comparing them systematically, a valid assessment of the panic tolerance – and other variables – is possible to attain.

In a selection process, it has been found that women who have recently become mothers, may have an increased though temporary tendency to become more anxious

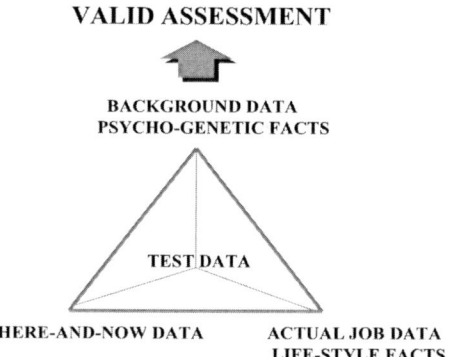

Figure 12.2 Valid Assessment based on Different Data Sources

onboard. Among flight attendants in this situation, we have also come across several cases of fear of flying. When these reactions do not have other cooperating sources or influences, however, they will soon disappear.

It should be noted, that well-trained mechanical operations, drilled repeatedly, are better executed even under very strenuous circumstances. Thorough mental preparation increases the probability of responding appropriately in an emergency, but at the end of the day we all fall back on the strength of our psychological defences. A sound and well-balanced personality has more surplus energy to efficiently handle anxiety and fright.

Some airlines measure stress tolerance among applicants seeking to become flight attendants since many of the working operations in the cabin are performed under time-pressure or simultaneously with other tasks. Keeping performance at an acceptable level, while the workload increases, is a relevant dimension in a cabin crew selection. Nevertheless, stress tolerance should not be confused with panic tolerance. Stress is related to time, multi-tasking and a pressure to achieve. Panic is an emotional reaction of fright, anxiety and loss of control.

Self-esteem

Self-esteem refers to the degree of trust in the individual's own aptitudes. A flight attendant with realistic self-esteem handles praise and criticism relevantly, asserts her/himself naturally without being overly authoritarian or boastful and is not reluctant to re-evaluate their opinions. A *strong self-esteem* is not the same as a realistic self-esteem. Someone with *too high* a self-esteem might behave in an adventurous or risky way and may also be regarded as pompous. A realistic and well-founded self-image is a good base for optimum behavior in the cabin in relation to passengers, flight deck and colleagues. It is also valuable when it comes to acting assertively in an emergency situation where authoritative behavior is desirable and necessary.

Independence

The capacity to be able to act autonomously based on own intentions and responsibilities and without being irrelevantly influenced by others, is necessary for all flight attendants. Too many pleasant and humble young flight attendant applicants hold the idea that humbleness and servitude are important for a successful job in the cabin. To correctly assess the applicant's independence requires an examination of behavior in context, but also of the person's background to check if previous family and work conditions have offered opportunities to develop a reasonable independence. Independence should not be confused with sovereignty and to avoid this risk, one should rather talk about co-ordinated independence or an ability to function both with a crew as well as independently.

Self-knowledge

This factor refers to an ability to understand, survey and handle one's own ideas, feelings and actions in a professional and self-reflective manner. A self-reflecting flight attendant, whose behavior is based on genuine drives and not based on stereotypic role-expectations, will probably develop appropriate service characteristics and would still be able to demonstrate assertiveness in an emergency.

Empathy

To listen with a "third ear" and to try to understand people by putting yourself in their place and visualize their reactions and needs, are irreplaceable skills for a flight attendant. Conflicts, passenger complaints, fear of flying are just a few of the problem situations on board requiring the flight attendant to listen actively with empathy. Many passengers, who behave aggressively, may have consumed too much alcohol and some have an underlying fear of flying. If the flight attendant is able to understand and interpret what may be behind these undesirable behaviors, she or he is better prepared to meet the genuine needs in the person and may be able to improve the passengers' comfort. When passengers are anxious or frightened, different emotions are provoked and by empathy, flight attendants can learn to become a "container" of all this feelings. By understanding instead of counter-acting, the flight attendant might also escape a good portion of frustration.

Responsibility

A responsible flight attendant not only recognizes and accepts the duties of the job, but continuously scans the cabin for potential challenges and problems. Responsibility is also to the airline and requires a measure of loyalty and conviction. Many candidates describe themselves as responsible, but on closer examination flaws are usually exposed. Therefore the assessor has to delve deeper both in background, present work and social relations to find out the appropriate level of responsibility. An active concern for the time, the issues discussed, the manner of response and behavior in the interview are all important indications of appropriate responsibility. In one interview for example, a candidate, who had described himself as responsible, laid his sports car keys on the table and the interviewer asked him how fast he could drive his car. The reply revealed that he had exceeded all the speed limits on his journey and he proudly added that he had driven in the bus lane all the way to the interview to ensure he was not late. Whilst his motivation could not be faulted, his sense of responsibility could.

Social Flexibility

Social flexibility refers to the ability to interact and work efficiently in a multi-cultural environment. A flight attendant continuously works in a multi/cultural environment and encounters people from different social backgrounds, religions and sexual preferences, among others. They have to relate to the pilots, ground staff, colleagues and passengers, all of whom may be different to themselves in terms of their demographics. Applicants with broad social experiences and from different environments and cultures, are possibly more likely to be socially flexible and to tackle diversity on board.

An issue to consider in a selection process is if the airline should strive for diversity among the flight attendants themselves. Some airlines deliberately choose a cultural mix on board to meet passenger needs in the best way, while other airlines prefer to demonstrate their cultural individuality by forming homogenous crews. Also concerning educational level some airlines have a tendency to choose young university graduates, while the customers have a broader social distribution.

Cooperative Ability

One usual misinterpretation of the concept of cooperation is to place it on a par with adaptability. A surly attitude is not the most desired one in a team, but an exaggerated adaptability also does not engender a dynamic productivity. A moderate level of conflict within a team is far more productive than pure harmony (Stott & Walker, 1992). Genuine cooperation is built upon collaboration, compromises, but also assertiveness. On the flight deck there have been many examples from disasters when a co-pilot has not been sufficiently assertive. The same circumstances are also valid in the cabin. Candidates, who avoid conflict often have a pleasant appearance, but they are not compatible with responsible work in the cabin – or on flight deck.

Value of Motivation

Many applicants to the airline business are highly motivated, but the reasons behind may be muddled. In a selection process, the assessors should look for genuine motives which are based on a realistic perception of both the positive and more trying aspects of the profession as a flight attendant. Among cabin crew applicants, there are sometimes failed pilot candidates – not always the best motivation to become a flight attendant. We should not forget either, that the subject of motivation is probably the most prepared and all applicants have many well thought-out *correct* and rehearsed motives in the interview. The assessor is advised to use all her/his ability and creativity to challenge these expected answers to explore the genuine motives. An accepted and trained flight attendant, who leaves the job after s/he found out that the conditions on board were too challenging, is an unnecessary and expensive investment.

Selection of Pursers or Cabin Managers

There are sometimes puzzling assumptions in many airlines concerning promotion of flight attendants to pursers or cabin managers. This assumption says that flight attendants with a lengthy work record will normally develop into capable managers with sufficient leadership skills. Therefore in many airlines, promotion to a leading position in the cabin is based on seniority, disregarding all references to leadership potential. Unless severe drug or alcohol abuse or a criminal record exists the experienced flight attendant is regarded as a suitable leader.

The task of the purser or cabin manager is to distribute and lead the work on board in the cabin. Candidates for promotion to these positions should be examined in relation to leadership qualifications and promoted not solely based on their efficiency as flight attendants. These leadership dimensions are, however, outside the scope of this chapter.

Assessment Procedure

There is no standard procedure in use when selecting flight attendants. However, there are a variety of assessment methods which can be included and combined in a selection design. Typical methods are: ability tests, personality tests, interviews, and referral to assessment centers.

Ability Tests Ability tests include for example language, verbal, analytical, numerical tests. Regarding flight attendants, stress tests, tests of ability to handle simultaneous workload and panic resistance could exemplify the special tests. Since many airlines place greater emphasis on physical attraction and a pleasing appearance, they tend to depreciate the value of abilities in their selection of flight attendants. At Scandinavian Airlines, several ability tests have been used. Besides language tests, tests of the applicants' ability to organize and, as mentioned earlier, tests measuring stress tolerance and panic resistance have been included in the test battery in the cabin crew selection process. It is useful to carefully integrate the test results into the total assessment. An applicant with superb results in an English vocabulary test, without accompanying acceptable communication skills would probably not become the most successful flight attendant. There are also applicants who demonstrate satisfactory capabilities in verbal or numerical tests, but they soon lose their abilities when applied in a social context or under stress.

Personality Tests Most published research on personality tests and aviation focuses on pilots and very few studies on flight attendants. Some personality tests do not have the same reliability and validity as ability tests and there is a general point of view not as much research data that have verified the benefit of personality tests in relation to aviation. One reason is probably that most traditional personality tests are designed to detect psychopathology (such as the MMPI – Minnesota Multiphasic Personality Inventory) even if there are tests developed to assess "normal" personalities. The 16PF – Sixteen Personality Factor (Cattell et al., 1970) is one such test that measures dimensions such as warmth, impulsivity, group conformity, anxiety etc. From a flight attendant selection perspective these are interesting dimensions, but even if the test detects "fake good/bad" responses, the risk of distorted result remains. Regardless of whether personality tests are used in the selection process of flight attendants, the results have to be cross-validated with other sorts of data before final conclusions are drawn. For an overview of personality tests in aviation, see Dolgin and Gibbs (1989).

Interviews One of the most frequent assessment methods is the interview. The ways they are conducted vary widely from the superficial to clinical in-depth interviews. The objections raised against interviews have always centered on the subjectivity. Nevertheless, the advantage of interviews is the direct relationship between the interviewer and the candidate. All of us may accept that a flight simulator is a superior method for checking up most of a pilot's aviator abilities. In the same way we should strive for a "flight attendant simulator" and the interview meets many of the qualifications for such a device. Often we would like to know if the applicant meets the criteria for independence. In an interview we can assess whether the candidate is able to take an independent position/stand in relation to the interviewer. If we want to know the applicant's stress tolerance, we have the opportunity to test this too in the simulator by stressing the applicant. In these cases we not only talk about how the applicant behaves but we can also observe any critical conduct. In the same way we can directly observe if the candidate is able to cooperate in the interview and so on.

In a well-conducted interview with mutual trust, there are also opportunities to gather information on the background of the applicant. By comparing background data with data from present work and life patterns and an observable here-and-now behavior in the interview, the validity can be substantially improved. Another method is to let two independent interviewers conduct one interview each and afterwards discuss the findings and agree on a common rating of the applicant. With two assessors many of the subjective elements in the interview method are eliminated.

In pilot recruitment, a team of aviation psychologists and airline pilots usually constitute the selection team. Their results are then presented for the flight operative executives. In flight attendant selection, there is no natural composition of the recruitment team. Some airlines use experienced flight attendants as interviewers and some airlines offer them training in selection, but far from all airlines. Others use administrative staff or managers and even aviation psychologists to accomplish the selection. It is highly recommended to use the complementary skills of an interviewer who is well-known to the flight attendant's working conditions and another interviewer who is skilled in assessing people. Using in-house interviewers presupposes extensive interview and assessment training to avoid the most basic mistakes.

Referral to Assessment Center If certain core abilities are tested and some well-established personality tests are used and integrated together with interview data, the probability of reaching a good decision about the applicants' potentials is high. With different interviewers involved – even well-trained interviewers – we need a calibration of the candidates. Taking into consideration the fact that flight attendants work in teams, a final step is relevant to add to the selection procedure. An assessment center meets both these criteria. In an assessment center, several candidates are exposed to different work-similar tasks which they have to accomplish in different team-settings while a couple of observers are rating and recording their performances. In such a setting, the candidates are exposed to several assessors and a natural calibration takes place. The candidates also expose themselves to more work-similar circumstances and in teams with the same dynamics as on board. In an assessment center, the author has conducted for airlines, it is striking that in the interviews the candidates give away voluntary data, while in the center they also expose involuntarily data.

In a well-designed assessment center with work-similar team tasks, team-work ability, tolerance, coordinated independence, social competence, social confidence, power of initiative, dominance, natural authority, working speed, stress tolerance etc. are displayed. If the assessment center has a straightforward design without any mystical elements that can only be unravelled by personal specialists, these traits should be comparatively uncomplicated to observe and to relate to the work in the cabin. A participant who exposes a dominant manner in several exercises in the assessment center has probably the same tendency in the cabin and a person who repeatedly takes natural initiatives in a team exercise is likely to behave in the same way among colleagues on board.

Conclusion

The future changes in the work conditions of flight attendants depend on technical and business developments in aviation. Aircraft will become larger and able to fly longer both in distance and time. With the globalization of the airline industry and the open skies policy, we can expect more multi-cultural teams working in the cabin. There are also an increasing number of budget-airlines starting up which attract a wider range of passengers and offer a different service concept on board. These and other trends have an influence both on the attraction of and motivation to work as flight attendant and influence on what the airlines should look for in their selection.

Long distance flights need not only robust physical, but also psychological qualities in flight attendants. To work and rest in very close proximity to other people during 17 hours or longer might be a trying intimacy – as it is for the passengers. It is in few situations except on a long-haul flight that we may wake up with the head on some stranger's shoulder. With the increasing size of the aircraft, the work in the cabin requires more of organization, overview and consequently also leadership. With less focus on of glamour and status in the job, probably well educated young people may be more reluctant to apply for a job in the cabin. On the other hand we can expect flight attendants from other countries and nationalities to apply to Western airlines in search of higher salaries. This raises the demand of cultural awareness in the selection staff. While the status of the job is changing with the trend towards no-frill airlines, we can also expect a shift in the motivation of cabin staff. From the selection perspective, airlines have to take these factors into consideration.

The big challenge for the future selection of flight attendants is, however, still to balance the safety and service aspects of the requirements. Frontline personnel are the airline's face to the customers and from a commercial standpoint they are expected to be excellent service-givers with a sensitive attitude. On the other hand, safety is the major task of a flight attendant and an overly emotionally sensitive person is probably not suitable for calming down passengers during a decompression incident or to take the lead when an evacuation is necessary in an emergency. While the service qualities may be more pleasant and uncomplicated to assess in a selection interview, the interviewer has to make conscious efforts not to forget the over-riding safety aspects.

References

Air Accidents Investigation Branch (1990). *Report on the accident to Boeing 737-400-G OBME near Kegworth, Leicestershire on 8 January 1989.* The Department of Transport, UK.

Albert, F. (1991). *Utvärdering av paniktolerans test PTT 21.* Report in Swedish to Scandinavian Airlines (SAS).

Beaty, D. (1976). *The water jump.* London: Secker and Warburg.

Cattell, R.. Eber, H. & Tatsuoka, M. (1970). *Handbook for the sixteen personality factor questionnaire.* Champaign IL: Institute for Personality and Ability Testing.

Chute, R., Dunbar, M., Wiener, E., & Hoang, V. (1995). *Cockpit/Cabin Crew Performance: Recent Research.* Proceedings of the 48th International Flight Safety Seminar Seattle W A November 7–9, 1995.

Dolgin, D., & Gibbs, G. (1989). Personality assessment in aviator selection. In R. Jensen (Ed.), *Aviation psychology,* (pp. 288–320). Vermont: Gower.

Edwards, M. & Edwards, E. (1990). *The aircraft cabin.* Aldershot: Gower.

Farkas, C., De Backer, P., & Sheppard, A. (1997). *Maximum leadership 2000.* London: Orion Business Books.

Musbach, A. & Davis, B. (1980). *Flight attendant.* New York: Crown Publishers Inc.

Stott, K. & Walker, A. (1992). *Making management work.* Singapore: Prentice Hall.

Chapter 13

How Cabin Crew Cope with Work Stress

Carina Eriksen

Introduction

Despite the importance placed upon cabin crew's psychological wellbeing, a better understanding of the complex processes involved in the management of physical and psycho-social work related stress is clearly needed before one can begin the course of appreciating cabin crew mental health.

This chapter examines how cabin crew deal with their somewhat abnormal lifestyle, compounded by organizational structures that are often beyond the control of the individual crew member. Of special interest is the apparent interconnectedness between a range of diverse stressors, which taken together, appear to exert a far more powerful threat to psychological ill health than the isolated consequences of a single event. Decision making competencies becomes a critical part of cabin crew coping when faced with conflicting life choices, often under the pressure of physical fatigue and jet lag. These realities may represent a departure from the existing thinking in the field of air crew mental health. Traditionally, the focus has been on adverse cognitive effects associated with repeated jet lag and non-consistent shift work. Although this reflects the importance placed on crew performance, especially in terms of passenger safety on board the aircraft, the implication of emotional ill health on performance standards may be just as detrimental as any direct cognitive degradation.

Background

Work stress has become a major concern in recent years because of the potential impact on both employee well – being and performance. Although stress has been defined in different ways over the years, the generally accepted definition today is one of interaction between the situation or context and the individual. It is the psychological and physical state that results when the resources of the individual are insufficient to cope with the demands and pressure of the situation. The most common detrimental outcomes include mental and physical ill health such as anxiety, depression and heart disease (Cooper & Marshall, 1976; Michie, 2002).

In recent years, there has been an increase in the number of reports on occupational health, with a particular focus on stress management in the workplace (Cox & Griffiths, 1995; Kagan, 1995). The prevention and management of workplace stress requires both individual approaches (for example training in assertiveness, time management, problem solving) and organizational approaches, ranging from structural (staffing levels, work schedule) to psychological (control over work, social support) interventions (Michie, 2002). One reason for this is that there are many sources of stress that the individual is likely to perceive as outside his or her power to change, such as the structure, management style or culture of the organization. Thus, stress management approaches that concentrate on changing the individual without changing the sources of stress may be of limited effectiveness, and may ultimately be counterproductive by masking these sources.

Research into Work Related Sources of Stress in Commercial Aviation

Airline cabin crew are in many ways a unique occupational group in terms of their irregular work patterns, unique set of job demands and lifestyle. The working environment of the modern commercial aircraft and the conditions in which airline cabin crew operate have been largely ignored in the industrial relations literature, and organizational psychology. The apparent success stories concerning new low cost airlines such as Southwest Airlines, EasyJet and Ryanair seem to have intensified the competition within the airline industry, which in turn, have significantly increased workloads and additional stresses affecting cabin crew (Brown, Rushton, Schucher, Stevens, & Warren, 2001). Stress has also been triggered by the testing financial climate in which airlines operate, which in turn affect workloads and job security.

In a review of existing literature on the health, safety and working conditions of airline cabin crews, a number of health and safety risks, about which workers received little, if any, information was identified (Boyd & Bain, 1997). This included exposure to poor air quality in cabins leading to increased level of airborne pollutants, recurring lethargy, headaches, and a range of influenza-type complaints. In addition, greater risk of exposure from ozone formed in the upper layer of the atmosphere could generate pulmonary symptoms such as loss of concentration. Another distinctive feature of the airline industry is that both aircraft and cabin crew are required to operate around the clock in order to maximize revenue causing sleeping difficulties and fatigue (Caldwell, 1997; Samel, Wegman, & Vejvoda, 1997). Moreover, cabin crew on transmeridian flights, crossing different time zones have been found to suffer jet lag and significant disturbance in sleep quality after such journeys, regardless of directions of travel (Beh & McClaughlin, 1997; Sharma & Schrivastava, 2004).

Given the overall direction of the development discussed above, and its impact upon health and safety, further research that seeks to communicate the health risks associated with cabin crew working conditions would be beneficial, with the ultimate aim of preventing and managing work stress for this particular occupational group.

Empirical Research into the Influence of the Working Environment upon Air Crews' Performance and Health

The results of a number of studies reveal the potential influence of the occupational environment upon air crew's health and performance. Haugli, Skogstad and Hellesoy (1994) undertook a survey of a sample of crew members to establish perceived health risks amongst the air crew population. Tiredness and fatigue was seen as the strongest threat to ill health followed by digestive difficulties, dry skin, lower back pain, and cold symptoms such as sore throat, eyes, and sinus problems. These findings were replicated in a similar survey study of the Norwegian cabin crew population (Lindgren, Andersson, Dammerstrom, & Norback, 2002).

Although research on commercial aviation covers a wide range of risks, the majority of studies tend to focus on negative consequences surrounding issues of sleep deprivation and fatigue. The topic has been well researched and findings from the studies have been taken into account in the development and implementation of flight time regulations (Price & Holley, 1990). Fatigue associated with sleep loss, shift work and long duty cycles was found to cause air crew to become *sloppy, inattentive, careless, and inefficient* (Caldwell, 1997). The only cure for fatigue is adequate sleep: however, gaining sufficient amounts of sleep is often difficult because of work requirements, family demands or poor sleep habits. Although it may not be possible to avoid some of these problems, it has been suggested that air crew can improve their sleep habits and thus gain more restful sleep by using self administering relaxation therapy, and avoid certain activities or substances prior to sleep (Caldwell, 1997).

However, air crew are also prone to suffer from jet lag or disruption of the circadian rhythm caused by flying across time zones. The combined effects of shift work and long distance travel can cause a cumulative build up of sleep deprivation, and the gradual onset of reduced concentration and alertness can sometimes be difficult for the individual to pick up on (Caldwell, 1997; Price & Holley, 1990). In a series of surveys comprising nearly 500 cabin crew, it was revealed that jet lag degrades attention, short-term memory, and decision making abilities (Sharma & Schrivastava, 2004). Consequently, the entire job content of cabin crew suffers. Coping mechanisms included use of alcohol or drugs (anxiolytics) to induce sleep, and individual coping strategies included exercise at the layover hotel or forced sleep, to combat jet-lag. Cabin crew need to be educated on the various aspects of jet lag and the organization has to implement a flight schedule to help them overcome the peculiar job related disorder. There is a pressing need to develop and implement medical standards for cabin crew, along the lines of aircrew medical categories, to prevent fatigued crew from turning up at work either due to pressure from management or failure to notice signs and symptoms of fatigue (Price & Holley, 1990).

Research into Psychosocial Work Related Sources of Stress on Cabin Crew's Wellbeing

To date, studies on the effect of jet-lag and shift work patterns on aircrew have focused on work performance, and therefore the impact on safety of aircraft and passengers. However, there has been less focus the impact of the combined effects of jet lag, shift work and lifestyle on the psychological health of crew. It is not known how other factors associated with the particular lifestyle of cabin crew (such as lack of control over work schedule, sleeping patterns, domestic and social life) may interact with jet lag and shift work to adversely affect wellbeing.

In interviews with cabin crew from major UK airlines, respondents voiced their concern over the combined effects of jet lag plus the stresses of work on their health, and it was felt that this had not been thoroughly studied (Brown et al., 2001). It is recognized that although there have been research on some of the individual aspects of employment as aircraft crew, the combination of jet lag, shift work, changing time zones and lifestyle is unique to this particular occupational group. As such, a plausible recommendation may be that aircraft crew should ideally be studied in a long-term epidemiological study, where participants are followed up not just through their working life but for their lifetime (Brown et al., 2001).

Literature Review on Pilots' Mental Health

In an overview of the mental health of pilots, it was found that the consequences of disruption to personal relationship upon air crew mental health are not as yet sufficiently understood and appreciated (Bor, Field, & Scragg, 2002). It is the nature of the aviation industry that individuals are removed from their home or environment, often for long periods of time, and this may be one reason for the high percentage of failed marriages or divorce rates amongst workers. Whereas some studies demonstrate the importance of relationship support in predicting pilot performance (Rigg & Cosgrove, 1994), it may equally be that pilot competency leads to a happier domestic life. Although it has been long recognized that stable, happy personal and social relationships can reduce the effect of stress in the workplace, disruption to personal relationships may exacerbate stress leading to impaired performance at work and mental health risks (Michie, 2002). In a study that compared married and non-married flight attendants, it was revealed that those who were in a stable relationship showed higher levels of role conflict, stress, and work dissatisfaction than those who were not in a relationship (Levy et al., 1984). Taken together, these studies show that relationships may buffer crew from stress, though work stress can also affect the stability of their personal relationships.

Qualitative Investigations of the Effects of Work Stress on Cabin Crew's Mental Health

In a qualitative study of cabin crew, six female flight attendants were interviewed to identify possible work related sources of psychosocial stress and their health effects (Lauria et al., 2004). The authors identified fifteen common themes, each of which was organized into five domains. In the first domain, orientating towards *positive and negative aspects* of the job, respondents learned over time that the benefits of travel and variable schedule exacted a price in terms of stability of family and social relationships, general wellbeing and good health. This consisted of: a. sleep loss; b. constant fatigue and; c. lack of control over the work schedule. In the second domain concerning *perception of occupational risk for serious disease*, participants expressed lack of protection by employer with respect to workplace exposure. This consisted of: a. typical work-related illness (frequent colds, gastritis, back pain, and circulatory problems); b. deafness and memory loss, and; c. increased risk of cancer and reproduction difficulties. In the third domain *compatibility of work with family roles and social relationships*, participants talked about the stress of having to adapt to home life after returning from a long tour of duty. This consisted of: a. tension between the public self and the private self; b. failed marriage, and; c. guilt derived from absence at events such as marriages or birthdays. In the fourth domain *relationship with colleagues and passengers*, participants described relationships with colleagues as supportive, especially in terms of normalizing job related experiences, but of short-duration due to the composition of the crew changes with each flight. In addition, *handling* angry passengers seemed to generate unnatural ways of dealing with feelings. This consisted of: a. limited opportunities to develop a steady group of work place friends for mutual support and b. being treated badly by passengers. In the fifth domain, *experience of work related sources of stress and their effects on health*, participants indicated that mental health was a major concern, and several work related risk factors related to adverse outcomes, such as depression and anxiety, were highlighted: a. social isolation and solitude; b. fear of being an inadequate partner, mother, daughter, or friend; c. lack of detail provided by the employer with respect to occupational health risks, and; d. passenger relationship such as violence and abusive behavior.

Within each domain, participants' appeared to dedicate much discussion to the topic of mental health, stress, and balancing work and life outside of work (Lauria et al., 2004). In other words, stress and mental health were of concern to this category of working people and a key source of adverse effects on wellbeing seemed to be the difficulties involved in balancing work and personal life due to the frequent absence from home and little control over work schedules. The need for follow up qualitative research in order to put the findings of the current research into context is clearly indicated. In particular, it may be useful to explore the actions or strategies used by this population to deal with their experience of work related sources of psychosocial stress. The findings could then be incorporated into the airline's training procedures, which may serve to reduce the risk of serious mental health problems for both the existing staff as well as new recruits.

Summary of Past Literature on the Effects of Work Stress on Cabin Crew's Psychological Health

As we have seen, the majority of empirical researches into the adverse effects of continuous long haul travels have primarily focused on physical and cognitive outcome measures of fatigue and jet lag (Caldwell, 1997; Price & Holley, 1990; Sharma & Schrivastava, 2004). Whilst this has led to a greater understanding of issues surrounding air crew performance and physical health, the extent to which the emotional wellbeing of this population is threatened by, for example, accumulative sleep deprivation, is still a controversial topic. Although qualitative interviews with cabin crew have given insight into the richness of the full dynamics that constitute the cabin crew profession (Lauria et al., 2004); it is still uncertain whether the experience of psycho-social stress is compounded by the reality of excessive tiredness and jet lag. In other words, further knowledge pertaining to the simultaneous interactional effects of both physical and psycho-social health threats may be necessary in order to develop an appreciation of their combined impact on workers emotional wellbeing (Brown et al., 2001).

In addition, the way in which air crew cope with physical and mental health threats forms an important part of the management and prevention of work-related stress issues (Bor et al., 2002; Michie, 2002). To date, this topic has not been thoroughly studied, and there is currently no reference to theories or models that could explain air crew's active (or non-active) attempts to combat stress in the work place. By failing to consider the extent to which coping skills may ameliorate or exacerbate mental health, it may become difficult, if not impossible, to assess the psychological wellbeing of any crew member.

The criticism of former investigations then, and perhaps the underlying cause of the limited knowledge available on cabin crew mental health issues, seems to be the pre-occupation with symptoms of individual "threats" as opposed to the coping process involved in handling potential intimidations to air crew's wellbeing.

Psychological Theory on How Long Haul Cabin Crew Potentially Copes With Work Stress

The ongoing enlargement of the airline industry indicates that there has also been a steady increase in the employment of cabin crew. Following UK CAA (Civil Aviation Authority) legislation, all commercial operations are required to carry a specific number of cabin crew on each flight to ensure adherence to safety regulations. The precise number will of course depend upon the size and type of the aircraft in question, and this could range from three to seven flight attendants on the smaller ones (i.e. Boeing 737) to 8 to 16 on the larger types (i.e. Boeing 747). The steady growth of the cabin crew profession seem to suggest that employee's actively seek to handle potential stressors evident by the bulk of crew who, on a broader level, manages to remain healthy. However, as pointed out in the available research, there is a wider issue of potential mental health problems amongst this population then, say, someone who is engaged in more regular patterns of 9-5 work or ground based shift work due to the somewhat atypical air crew lifestyle.

The literature on strategies of coping with threats to ill mental health suggest that any form of disruption to the psychological system is self limiting as the person will find a way to return to a stable state (see the 1984 study by Moos & Schaefer in Ogden, 2004). Once confronted with physical or psychological threats, the individual is thought to engage in three processes that constitute the coping process. Firstly, at the initial stage of disequilibrium, the person initially appraises the seriousness and the significance of the threat (cognitive appraisal). In keeping with the existing literature, such threats could be classified as either physical/biological such as jet lag and fatigue (Caldwell, 1997; Price & Holley, 1990; Sharma & Schrivastava, 2004), psycho social including low work control (Lauria et al., 2004), or disrupted social/personal relationships (Bor et al., 2002; Lauria et al., 2004; Levy et al., 1984; Rigg & Cosgrove, 1994). Secondly, the individual utilizes adaptive tasks as part of the coping process, and these can be seen as either general tasks or those specific to the threat itself. This is consistent with Caldwell's (1997) finding that sleep is the only remedy for fatigue, but this may not be easily achieved by air crew due to general demands such as the family or commitments outside of work. An explanation using psychological theory then, would suggest that general tasks such as sustaining relationships with family or friends may be utilized prior to, instead of, or as well as dealing with the core issue of sleep deprivation (Moos & Schaefer in Ogden, 2004). Thirdly, the person engages in a series of coping skills that are accesses to deal with the crisis. These coping skills can be divided into three categories: 1. appraisal-focused coping involving attempts to understand the threat and represent a search for meaning; 2. problem-focused coping such as confronting the problem and reconstructing it as manageable, and; 3. emotion-focused coping involving management of emotions and maintaining equilibrium. In Lauria et al., (2004) qualitative study of cabin crew, participants felt that talking to fellow colleagues normalized various difficulties associated with their work role. Explained in terms of psychological theory, it seems that seeking support and information from work colleagues allows crew to build a knowledge base by accessing any available information (problem-focused coping), which in turn allows them to redefine the situation in a realistic and acceptable way (appraisal-focused coping). According to this model, individuals use a variety of adaptive tasks and coping skills which in turn determine the outcome of the coping process. Such an outcome may be psychological wellbeing resulting from *healthy* adaptation or deterioration arising from *maladaptive* responses.

Whilst it is recognized that not all people respond to health threats in the same way due to individual differences in personal factors (age, sex, ethnicity, etc.) or social circumstances (available support networks), it is proposed that, for cabin crew, the unique work environment plays an important role in determining the use of adaptive tasks and coping skills (Michie, 2002). The operations of the airline industry mean that cabin crew are often removed from their home environment for long periods of time, and therefore, they are more likely to experience irregular contact with family or friends, which may exacerbate their capacity to, for example, maintain social and personal relationships.

Qualitative Research on the Effects of Work Stress on Cabin Crew Mental Health, and How They Deal With This

In a study consisting of in depth interviews with eight long haul cabin crew from major UK airlines, it was found difficult to separate the impact of physical and psycho-social work related stress on worker's mental wellbeing (Eriksen, 2005). Participants needed to have completed a minimum of seven years with no more than 15 years of in flight service, and a second requirement was that they had been operating on long-haul routes for a minimum of five years. The exclusion of short-haul staff was based on the wide difference in working requirements between the two populations with long-haul crew being more prone to jet-lag and health hazards.

The investigation utilized a qualitative paradigm with a particular focus on the strategies used by crew to cope with the range of stress arising from their profession.

The most stressful aspects of the cabin crew work role could be summarized into three main categories or themes as follows (Eriksen, 2005).

The Experience of Jet lag and Fatigue

The content of the first theme strongly supports the past literature, which concluded that the combined effects of repeated jet lag and shift work leading to accumulative sleep deprivation presents the most serious threat to ill health in aviation psychology (Price and Holley, 1990). This was reflected in participant's accounts with a majority of the interview time devoted to issues of tiredness and fatigue with limited reference to other physical challenges. One participant said:

> I think continuous time zone changes and also the tiredness, they take a toll on you, you find it hard to get up when you come home or sometimes the opposite you find it hard to get to sleep. You're always tired- If you're flying full-time you are probably tired ninety nine percent of the time... (Female, 32)

It also confirms the past findings that found reduced memory, attention span, alertness, and awareness to be common symptoms of both jet lag and fatigue (Caldwell, 1997; Sharma & Schrivastava, 2004). A 39-year-old female flight attendant said, "I feel that I'm not being as vivacious as I should be and you feel that you're doing things half-heartedly, not giving it your all."

Participants appeared to manage jet lag and sleeping difficulties by remaining on local time, administering relaxation techniques or physical exercise. However, they also talked about greater sleep disturbances occurring on longer trips characterized by larger time differences, and this seemed to generate concerns about emotional health such as mood fluctuations, anxiety or depression. A 37-year-old male participant said:

> You know, you get highs and lows quite a lot as well, sometimes you feel very sort of elated, just very happy and then other times emotions drop quite a bit, it's just finding it hard to deal with, your own personal emotions through tiredness and fatigue from work.

This apparent escalation seems to support the idea surrounding the influence of organizational structures beyond the control of the individual as participants felt they had little control over the random allocation of travel destinations (Michie, 2002). One participant said:

> Because (company) I work for, they have different schedules so you might have eight hours behind to get to LA and then the next week, you could go to Singapore and it's eight hours ahead so that sort of does throw you a bit... (Female, 30)

An explanation in terms of psychological theory (see the 1984 study by Moos & Schaefer in Ogden, 2004), would suggest a negative feedback loop between shift work through various time zones and lack of sleep, which over a period of time leads to accumulative sleep deprivation that appears to cause disruptions to participants emotional functioning. However, the extent to which sleep deprivation exacerbates cabin crew's emotionality, and thereby increases the risk of, for example, anxiety and depression, rests upon the individual's ability to self regulate, meaning the active (or non active) attempts to restore psychological equilibrium. The use of adaptive tasks, such as remaining on local time or relaxation techniques, creates a positive feedback loop that appears to ameliorate cabin crew's emotional state of mind, thus reducing the likelihood of psychological illness. In other words, the individual seeks to balance mental health by the means of healthy adaptations thought to "outweigh" the potential adversity of sleep deprivation. However, participants also spoke about the intensification of sleeping difficulties on longer trips, and it was within this context that three of the participants discussed infrequent use of sleeping pills or alcohol to force sleep.

Despite the potential negative effects of frequent and excessive use of such remedies, it appears that moderate and infrequent use may provide short-term amelioration in mental health. For example, a male cabin crew member spoke about "feeling absolutely exhausted" when "turning up" for a night flight (Sydney to Singapore) whilst having been unable to sleep all day, and as a result, he found it more difficult to function both emotionally and physically during the flight. Although he was adamant about the "emergency" use of sleeping pills, they appeared to help him fall asleep during the day, which also meant that he managed to avoid the above adversity associated with lack of sleep. It is also worth noting that more senior participants spoke about how, after years of flying, they had finally learnt to manage irregular sleeping patterns. This may indicate a potential transition in air crew's self regulatory system with a quantitative shift of higher tolerance of sleep deprivation. Alternatively, there may also be a qualitative shift such as the apparent change of rules illustrated by the following account:

> After I started to fly I realized that there's no point in trying to think that because there's an eight hour time difference there's no point getting up, I do just think go to sleep and generally I don't set an alarm... (Female, 43)

This indicates that there may be a supplementary stage involved in the coping process utilized by cabin crew. For example, it is possible that the initiation of a series of adaptive tasks, akin to a trial and error process, is necessary before workers eventually master the management of sleep deprivation. Although this does not directly contradict current psychological theory in that the individual is still keeping within the practice of self-regulation, it may indicate the need for further modifications to its present layout so as to cover the many holistic components that constitutes the holistic nature of air crew coping.

The Constant Strive to Balance Work with Life Outside of Work

This second theme overlapped with past findings of the complicatedness involved in balancing work and life outside of work (Lauria et al., 2004). The participants spoke about a constant alteration between adaptations to work life and home life with reference to the broader spectrum of daily living, and this created a sense of two existing worlds, a professional one and a private one. A 30-year-old female flight attendant said:

> As soon as you get into a pattern of doing something, you're off again, or you find a little bit of normality to your sort of routine, it's thrown out the window again... you detach yourself I think and sometimes you almost have two separate little worlds you live in.

There was strong agreement amongst cabin crew that lack of control over the work schedule appeared to constitute the root of the hardship of balancing professional and personal life, and once more, there seems to be a link with past studies of the connections between organizational structures and air crew stress (Michie, 2002). One 32-year-old female flight attendant said:

> I think the thing about working for an airline is that you don't have too much control over what you do; your life is decided for you by the schedule and department who decide where you go, when you go.

Participants appeared to utilize a diversity of individual techniques, mostly surrounding the use of organizational or time management skills such as, for example, arranging their home life around the working schedule in order to retain some structure and regularity in daily living: "My social life gets planned round my work…" (Male, 34).

Such adaptive tasks, according to psychological theory, could be seen as healthy attempts of self regulation with the ultimate aim of combating any potential imbalance on psychological functioning generated by the constant adjustment to both work life and home life. However, it is worth noting that both tiredness and private problems seemed to exacerbate participants' ability to cope with the constant disjointedness between professional and personal life as indicated by the following accounts.

> Jet lag for one is bad enough so I mean I really find that I don't function that well, physically and emotionally after a trip, you don't think straight and you don't think as logically as you would when you're rested or you might be going over things more than what they need to be… (Male, 34)

> You're annoyed with yourself cause you're actually wasting a day, and on the top of that if you've got stuff going on at home that can be buying a house or you've got problems with family or relationships, then that all takes a strain as well… (Male, 27)

This was characterized by the frequent discussions of increased emotional troubles such as ruminations, self-blame, and sadness in the wake of tiredness or private problems. This emergent finding is in accordance with the idea of a relationship between jet lag/fatigue and crew lifestyle, which taken together, may complicate cabin crew's scope for managing work related stress (Brown et al., 2001). In keeping with psychological theory, the operation of both physical and psycho-social *threats* may mean that those air crew are faced with a somewhat linear coping process, whereby successful self regulation rests upon their ability to initiate a chain of adaptive tasks pertaining to sleep management, private troubles and balancing the abnormal lifestyle. For instance, it appeared that talking to colleagues in the wake of personal problems at home provided short-term amelioration in mental health in that it appeared to prevent participants from excessive ruminations when left to their own devices.

> You might have something that's going on in your private life you just have to leave it on hold and go to work so you need to carry on. Sometimes I actually think it's better because sometimes I can dwell on things a bit too much and you might sort of like confide with the crew… (Female, 29)

More worryingly, however, was the finding of participants experience of having to "choose" between restoration of sleep or dealing with accumulative piles of urgent daily tasks within a time-limited framework, both of which could have potential negative consequences for the individual if not dealt with appropriately.

> There is a sort of frustration of being away, you can't get on with things that need to be done which are building up, you come back, you're tired and you have to deal with that quickly before you go away again… (Male, 37)

> I think with that you just try and keep going as much as you can because once you stop and think, I can't do it, I'm too tired, you miss out on whole days... (Female, 31)

This seemingly conflict of choice faced by cabin crew suggests that dealing with work-related health threats may perhaps involve an additional higher level process of decision making as to which threat requires immediate dealings and which one can wait. As such, the air crew coping process may be a somewhat more complex phenomenon than previously anticipated by both past literature in commercial aviation and existing psychological theory.

The Experience of Disruptions to Social and Personal Relationships

The third theme comprised the difficulties involved in both establishing and maintaining personal and social relationships (Bor et al., 2002; Lauria et al., 2004). Participants spoke about the constant forming and breaking ways of relating, and this appeared to also include relationships with colleagues.

> You've been on your own and you've been independent and then you have to become part of a couple or part of a family again and that takes some time and by the time you get into that role, you're going away again so it's continuously changing all the time... (Female, 32)

Participants attributed much of the complexity involved in socializing and co-habiting to both the irregular patterns of work and the forever changing crew composition with each new trip abroad: "I am always flying with new colleagues... You are with this group of people but, you know, essentially you are on your own as well..." (Female, 39).

More importantly, participants spoke about the significance of having partners, friends, or a family able to appreciate the strains of the job, particularly the erratic sleeping patterns associated with jet lag. For example, participants referred to feelings of guilt arising from the frequent absence from beloved ones, and according to one female cabin crew member, the experience of guilt seemed to worsen if a partner resented her frequent absence: "If you have a partner who resents you going away then your whole life you're made to feel guilty because you're going away..." (Female, 32).

The idea of a positive association between stable uncomplicated relationships and psychological wellbeing appeared to be a significant theme in earlier literature reviews in aviation psychology (Bor et al., 2002). In terms of psychological theory, there seems to be a negative feedback loop between lack of empathy from a partner and difficult relations, which in turn appeared to exacerbate the risk of psychological disturbances. The process of self-regulation seemed to include the use of preventative strategies with the aim of perhaps bypassing the negative consequences associated with compounded relations such as excessive stress leading to impaired performance at work and mental health risks (Michie, 2002). For example, participants spoke about attempts to select either partners with certain personality characteristics (independence and a high sense of security) or someone with a similar occupation. It was interesting to note that participants also talked about the avoidance of confrontations or attempts to over compensate for lack of physical presence by the means of excessive gifts: "When you have an issue with someone it's so easy to dodge that, it's very easy to disappear, it's very easy not to be accountable for certain actions…" (Male, 37). "I think when you are at home you overcompensate most definitely… it is an element of guilt you're sort of making up for not being there…" (Female, 30).

In keeping with psychological theory, such activities may lead to, apart from the stress of a potential build up of relationship difficulties, habitual maladaptive responses. In the long run such reactions may reduce the individual's ability to self-regulate, and thus increasing the likelihood of psychological imbalance. The lack of physical presence in home life makes it difficult to both establish and maintain personal relationships. When this is compounded by the lack of continuity and development in work relations, it can lead to a sense of loneliness and isolation. If less confrontation or buying gifts serves to keep friends, partners or the family happy or accepting of the crew lifestyle, the individual may simply protect him or her self from the potential greater risk of isolation or loneliness associated with lack or loss of friendships/ partnerships. This suggests that there may be a system of differing values placed upon competing "threats" to psychological ill health, perhaps indicating yet again the existence of higher order processes including the quick computation of the comparison of negative consequences pertaining to one threat in comparison to another.

The Inter-Connectedness between the Main Threats to Cabin Crew Mental Health

Cabin crew appeared to dedicate a great deal of their sayings to the difficulties involved in balancing their lifestyle through the wakes of tiredness and fatigue. On top of that they also voiced their concerns about the potential negative effects of problematic issues at home, such as an upset with a partner or a friend. The majority of the sample referred to the intensification of the reality of essentially "being on their own" through periods of large time differences coupled with either private troubles or difficult relations involving someone at home.

> It does happen that you don't find anyone on the crew, and you close that hotel room door and it's like one of the loneliest places that you've been, you just want to talk to somebody friendly at home, or if you've got a problem that's involving somebody else just to phone them up and kind of get it sorted, you can't even phone them, you just have to be conscious of all the time changes… (Male, 27)

This indicates that there may be a connectedness between physical, organizational, and psycho-social threats to air crew psychological wellbeing. It may be necessary to move beyond current thinking predominantly associated with the separate impacts of each of these events on workers mental health.

An alternative explanation that seems to accord with participants verbal accounts may be that the interactional effects of various threats could lead to a more complex coping process. It is possible that healthy adaptations pertain to, for example, achieving overall balance in the connections between physical and psycho-social health threats. This would avoid the onset of potential negative feedback loops that exceed the individual's threshold for self-regulation. For example, cabin crew talked about the negative effects of tiredness on their ability to both *think straight* and handle everyday tasks when returning from a night flight. However, for many of them, such activities were initiated prior to dealing with lack of sleep, which in turn may serve to exacerbate and further the issues of accumulative sleep deprivation. The end product then may be both accumulative sleep loss and a forever increasing list of mundane tasks, and taken together, the combined force of these may put even the most resilient crew member at serious risks of psychological disturbance.

The Role of the Central Executive in Goal Management

The new emergent theme of an apparent cyclic influence of, for instance, tiredness on participant's ability to cope with their lifestyle, and vice versa, suggests a negative association between tiredness and the ability to balance work with life outside of work. On the other hand, participants also made reference to a positive association between balancing lifestyle, jet lag and fatigue, and maintaining relationships. For example, two female cabin crew members spoke about how they were able to catch up on sleep (jet lag and fatigue) whilst away so as to preserve their energy for their time at home (balancing lifestyle) with friends and family (dealing with social and personal relationships). This indicates both positive and negative connections between the differing stressors as illustrated in Figure 13.1.

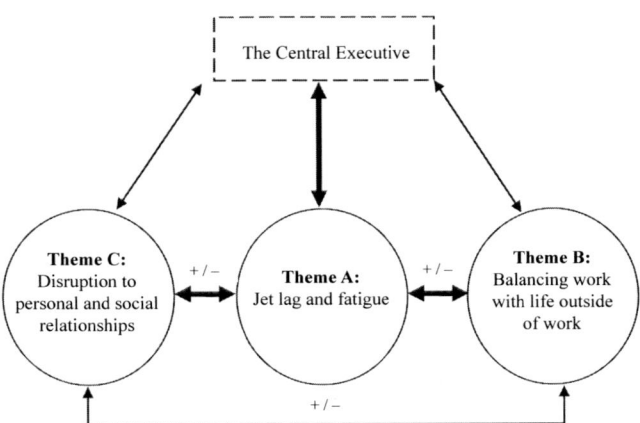

Figure 13.1 The Interactions between Lower Level Processes and their Connection to the Higher Level Central Executive

It seems that cabin crew's ability to maintain psychological equilibrium depends upon their ability to balance both the negative and positive connections with the aim of not exceeding their individual threshold and thereby cause imbalance to the self-regulatory system. More worryingly, however, is the tension arising from conflicting goals due to the frequent absence from home with little time in-between operations. For example, several cabin crew members spoke about having to make a choice between canceling social engagements and gaining sleep after or prior to a trip. Whereas the former could lead to feelings of guilt or loneliness, the consequences of the latter was, according to participants, reduced levels of daily cognitive functioning such as forgetfulness, clumsiness, or attentiveness. Faced with this apparent paradox, it seemed that cabin crew engaged in the process of decision making. For instance, one female participant spoke about feeling worse if letting her friends down than losing out on sleep. In contrast, the opposite was true for a differing cabin crew member, who had always been a bad sleeper. This indicates that there seem to be a higher order process, perhaps in the form of a central executive that takes an overall charge of goal management.

This may resemble Kahneman's (see the 1973 study by Kahneman in Gross, 2001) idea of a limited capacity processor that distributes resources to various tasks simultaneously so that, for instance, cabin crew can balance their lifestyle in the wake of jet lag as illustrated by the connections between lower level process and higher level process in the Figure 13.1. However, whether or not tasks can be performed simultaneously will depend upon how much demand they make on the limited capacity processor. In the case of the second crew member, existing sleeping difficulties may mean that her threshold for sleep deprivation is relatively low compared to normal sleepers. As such, the limited capacity processor compensates by allocating a majority of resources to the prevention of further sleep loss, thereby ensuring balance to the system. When demands are too high, such as the case of goal conflict, the central processor must decide how to allocate attention, and this of course, will depend upon the nature of the individual's self-regulatory system, that is the specific strength of connections between each stressor and the central executive. In keeping with both the existing literature and the verbal accounts given by participants, it is possible that the cognitive symptoms of jet lag and fatigue present the most serious threats to psychological imbalance for cabin crew. Apart from competing for the limited resources, reduced cognitive performance may also cause direct disruption to the higher order decision making process. This may suggest that making choices under the pressure of fatigue and tiredness is a rather more complex phenomenon than estimated by previous literature on the repeated effects of jet lag and shift work. One reason for this could be that the area of risk calculation under adverse conditions may essentially belong to the tradition of psychology as opposed to the school of medical research.

Conclusion

Drawing on the existing literature in commercial aviation there appears to be four major work related "threats" to cabin crew well-being. Firstly, shift work and frequent travels through differing time zones may cause physical health risks such as sleep loss, and symptoms of jet lag and fatigue even for the most resilient crew member (Boyd & Bain, 1997; Caldwell, 1997; Price & Holley, 1990; Sharma & Schrivastava, 2004). Secondly, the experience of irregular patterns of work could make it difficult for crew to balance their job with life outside of work (Lauria et al., 2004). Thirdly, the variable work schedule and long time spent away from home may come at a price in terms of stable social and personal relationships (Bor et al., 2002; Levy et al., 1984; Rigg & Cosgrove, 1994). Fourthly, the stress related to the managing of unruly or demanding passengers (Lauria et al., 2004). The literature also showed that stress comes into view once the resources of the individual are not sufficient to cope with the demands and pressure of the situation (Cooper & Marshall, 1976; Michie, 2002). This indicates that the total effect of work stress on psychological health will depend upon the individual's collection of means to deal with the various intimidations to healthiness. Whilst there has been much research into the effects of jet lag and fatigue upon cognitive abilities and performance (Caldwell, 1997; Price & Holley, 1990; Sharma & Schrivastava, 2004), there have been considerable fewer investigations into the coping strategies used by cabin crew to reduce the likelihood of adverse impacts of sleep disruptions.

Despite strong commonalities with past literature and research on the impact of work related stress on cabin crew health, the study found the process of coping with work related stress to be a complex and perhaps individual process (Eriksen, 2005). This would seem logical since each cabin crew's experience would have been completely unique, dependant as it was on their individual repertoire of coping skills, mental health, personality characteristics, social and personal relationship and so on. Added to this fact was the limited theoretical understanding available to explain the apparent interconnectedness between various stressors for this particular group of workers. This, in turn, seemed to bring the coping process to a whole new dimension with the possibility of differing levels of processing including a higher level central executive, which together, seeks to regulate the individual's emotional state of mind. In the light of these findings, it is recommended that mental health workers (psychologists, counselors etc.) may benefit from administering full mental health assessments with a specific focus on the connections between various stressors. This will help them to derive a better understanding of the complex, often conflicting challenges faced by cabin crew, and appropriate therapeutic approaches can be chosen accordingly, with more emphasis given to, for example, systemic or a gestalt orientations.

References

Banister, P. (1994). Report writing. In P. Banister, E. Burman, E., I. Parker, M. Taylor, & C. Tindall, (eds.) *Qualitative methods in psychology: A research guide* (pp. 160–179). Buckingham: Open University Press.

Beh, H.C. & McLaughlin, P. (1997). Effect of long flights on the cognitive performance of air crew. *Journal of Perceptual and Motor Skills, 84*, 319–332.

Bor, R., Field, G., & Scragg, P. (2002). The mental health of pilots: an overview of recent research. *Counselling Psychology Quarterly, 15*, 3, 239–256.

Boyd, C. & Bain, P. (1997). *Once I get you up there, where the air is rarified: Health, safety and the working conditions of airline cabin crew*. Oxford: Blackwell Publishers Ltd.

Brown, T., Rushton, L., Schucher, L., Stevens, J., & Warren, F. (2001). *A consultation on the possible effects on health, comfort, and safety of aircraft cabin environment.* Institute for Environment and Health, W5. Leicester: IEH.

Caldwell, J. (1997). Fatigue in the aviation environment: an overview of the cause and effect as well as recommended countermeasures. *Aviation, Space and Environmental Medicine, 68*, 932–938.

Cooper, C.L. & Marshall, J. (1976). Occupational Sources of Stress: a review of the literature relating to coronary heart disease and mental ill health. *Occupational health psychology, 49*, 11–28.

Cox, T. & Griffiths, A. (1995). The nature and measurement of work stress: theory and practice. In J.R. Wilson, J.R., & E. N. Corlett, (eds.). *Evaluation of Human Work: a practical ergonomics methodology* (2nd ed.) (pp. 783–803). London: Taylor & Francis.

Eriksen, C. (2005). *Long haul cabin crews account of how work-related stress affects their mental well-being, and how they deal with this*. Dissertation abstracts. UK: Department of Psychology, London Metropolitan University.

Haugli, L., Skogstad., A, & Hellesoy, O.H. (1994). *Health, sleep, and mood perceptions reported by airline crew flying short and long hauls*. Department of Occupational Health and Safety. Oslo: Scandinavian Airlines.

Kahneman, D. (2001). Attention and performance. In R. Gross (ed.), *Psychology: The science of mind and behaviour* (4th ed., pp. 183–198). Kent: Greengate Publishing Services.

Kagan, N.I., Kagan, H., & Watson, M.G. (1995). Stress reduction in the workplace: The effectivenss of psychoeducational program. *Journal of Counselling Psychology, 42*, 71–80.

Lauria, L., Ballard, T.J., Corradi, L., Mazzanti, C., Scaravelli, G., Sgorbissa, F., et al. (2004). Integrating qualitative methods into occupational health research: A study of women flight attendants. *Occupational and Environmental Medicine, 61*, 163–166.

Levy, D.E., Faulkner, G.L., & Dixon, R. (1984). Work and family interaction: The dual career family of the flight attendant. *Journal of Social Relations, 11*, 2, 67–88.

Lindgren, T., Andersson, K., Dammerstrom, B.G., & Norback, D. (2002). Ocular, nasal, dermal, and general symptoms among commercial airline crews. *Occupational Environmental Health*, *75*,7, 475–483.

Michie, S. (2002). Causes and management of stress at work. *Occupational Environment Medicine*, *59*, 67–72.

Moos, R.H. & Schaefer, J.H. (2004). The crisis of physical illness. In J. Ogden (ed.), *Health psychology: A textbook* (3rd ed., pp. 62–64). Buckingham: Open University Press.

Price, W. & Holley, D. (1990). Shiftwork and safety in aviation. *Occupational Medicine*, *5*, 343–377.

Rigg, R. & Cosgrove, M. (1994). Aircrew wives and the intermittent husband syndrome. *Aviation, Space and Environmental Medicine*, *65*, 654–660.

Samel, A., Wegman, H.M., & Vejvoda, M. (1997). Aircrew fatigue in long-haul operations. *Accident Analysis 29*, 4, 439–452.

Sharma, R.C. & Schrivastava, J.K. (2004). Jet lag and cabin crew: Questionnaire survey. *Journal of Aerospace Medicine 48*, 1, 10–14.

Williams, S., Michie, S., & Patani, S. (1998). *Improving the health of the NHS workforce*. London: The Nuffield Trust.

Chapter 14

Psychological Problems Among Cabin Crew

Chris Partridge and Tracy Goodman

Introduction

I want to work as part of a team and travel the world.

This common phrase, often heard at airline interviews, probably encapsulates the expectations and aspirations of airline cabin crew who apply for, what is perceived as, one of the most *glamorous* jobs in the world.

But what happens if, for some people, the glamor turns to grief and the travel to trauma? Behind the *plastic* smile, the groomed façade and the creaseless uniform can lie a conflicting balance of home and work life; professional and personal relationships and a demanding mix of interpersonal and intrapersonal skills.

In this chapter we aim to examine more closely the resulting psychological problems and complexes that can sometimes arise from these types of dichotomies. Speaking from the point of view of both an experienced senior cabin crew member, and professional counselor perspective, we invite the reader to travel more closely into this *alluring* world. We will look at the mental challenges faced when flying at 35,000 feet, whilst trying to keep one's feet firmly on the ground, and illustrate these challenges with hypothetical client case examples. We will also present an overview of the British Airways Crewcare counseling service, which has been an integral part of the support process for cabin crew throughout the last twenty years.

Background to Lifestyle

In order to present a realistic picture of the lifestyle of a cabin crew member, it is important to provide a brief overview of the initial cabin crew selection process.

Throughout the current British Airways recruitment campaign for cabin crew, the organization promotes the fact that working as a member of their cabin crew community is not so much a job, but in many ways a privileged life, with opportunities to visit places and cultures that are beyond most people's reach. However, the campaign carefully illustrates that the work is not all about jetting around the world and staying in exotic locations, it is about delivering outstanding customer service in all conditions. Ultimately, the key role of a cabin crew member is to be responsible for passenger safety.

British Airways is transparent about the job having glamorous elements whilst being an exhausting lifestyle that places tough demands on family and social commitments. It also advocates the fact that it is not a vocation for anyone who craves routine. A key element of the job requirement is that crew are expected to be totally flexible and prepared to work on a variety of routes across the British Airways network. For those crew members entering the profession from a more traditional 9-5 working environment, the inability to plan ahead and constantly working with different people can be particularly challenging.

The desired corporate skills sought for the role of a cabin crew member include a motivation to deliver excellent customer service, emotional resilience, developed interpersonal and intrapersonal skills, being a team player and a sound commercial awareness.

Required personable qualities include having a friendly and caring personality, competence in handling difficult situations, being a confident communicator, a good listener, supportive of colleagues, having an ability to remain calm and efficient under pressure and a willingness to treat everyone as an individual. In addition it is necessary to satisfy current British Airways and Civil Aviation health requirements and to take pride in personal appearance and grooming.

Following the recruitment process, which includes both group and individual interviews, medical and language testing, the new cabin crew member embarks on a four to six week training course. This covers statutory requirements such as safety and emergency procedures training, aviation medicine, hijack/security procedures, conflict management, restraining and breakaway techniques, all of which are required to conform to European statutory JAROPS standards (Joint Aviation Requirements Operations). This course also covers the "softer skills" such as communication and team working, service styles and standards and personal safety both on and off the aircraft.

The trainee is assessed and examined at regular intervals throughout the course and, as such, it is a fairly arduous and demanding period for the new employee. Once this is successfully completed, cabin crew then commence their flying roster, which is predominantly dictated by the company and operational requirements. For the initial six-month period the crew member is on "probation" and is regularly assessed on performance by a senior crew member. Having completed this six month probation the crew member earns their wings, and then returns briefly to the training school to consolidate the experience gained and to refresh their initial training.

Once out on the flying line the new recruit joins the ranks of the professional jet set and depending on fleet allocation, will fly on shorter, mainly European routes, involving day return trips or *nightstops* of up to three days away from base. Alternatively, worldwide crew will fly to destinations in the rest of the world, which may involve trips of up to thirteen days away from base.

In the initial *honeymoon* period, the opportunity to travel, to see the world and to meet a wealth of differing peoples and cultures contributes to the perceived glamor of the role. The opportunity to escape the routine and pending trays of many ground based jobs adds to this attraction. However, realities such as a lack of control of lifestyle, being away from family and friends and the difficulties in planning ahead, can soon begin to deflate the glamor *bubble*. The physical implications of the job, such as compounded sleep deprivation, jet lag and dehydration can also take their toll, both mentally and emotionally.

The constantly changing crews make it difficult to form stable and long lasting working alliances, resulting in a lack of continuity and development in work relations. A commitment to this kind of lifestyle also necessitates a consequential lack of investment in home life and often results in difficulties with establishing and maintaining personal relationships. When this is compounded by the fleeting nature of work relationships, it can lead to a sense of loneliness and isolation. In particular, the sense of loneliness within a crowd, or when working closely within a team of people, can often be the most difficult kind to bear. As one female crew member remarked on a recent BBC1 television programme, "Working in the Dark – Fly by Night;" "It can be a brilliant job, however if you have problems at home, it can be purgatory."

The lifestyle, which can also act as a psychological retreat, allows certain individuals continually to deny themselves the opportunity of addressing the heart of a particular problem. Inevitably, at some point, the problem will have to be faced. When this occurs, the British Airways Health Services offer a robust support mechanism and will assess an individual's physical and/or psychological fitness to fly should the need arise.

Cabin Crew Group Dynamics

A typical duty day for a crew member begins with a fifteen minute pre flight briefing at base to meet with a group of up to fourteen colleagues. Here, working positions are designated and crew are tested on emergency procedures and aviation medicine knowledge. Also specific information is given on the route and, in terms of security, the relative level of threat, from low to severe, designated to the destination by the Foreign Office.

When compared with Tuckman's (1965) model of group dynamics, the "forming, storming and norming," stages of the working alliance must be quickly established. The only opportunities for this are the fifteen-minute pre flight briefing, the bus journey to the aircraft and the seven minutes allocated to check all safety equipment and catering is onboard. Thus, the crew team needs to reach the "performing" stage of the model before the passengers board.

Ideally, there would be a longer period for these group formation processes to be completed, but commercial pressures mean that the performing stage of the Tuckman model needs to be reached relatively quickly. The requirement to reach this stage often leads to a number of ways of acting out by individuals in order to achieve this necessary, but often fragile, group harmony. Typically these include succumbing to the perceived pressure to be accepted and valued by the crew group through the projection of the attractive side of one's personality. Of course, how one judges attractive traits is largely subjective but, for example, the use of humor or personal anecdotes are just two methods employed to aid individual inclusion within the crew grouping.

Once in the performing role the crew member must quickly adapt to the group dynamic and often bypass individual needs in order to meet the needs of the customer/passenger. Ultimately, of course, the main purpose for the group forming is not to satisfy individual requirements, but for larger organizational objectives. Primarily, these are the need to move the customer/passenger from point A to point B efficiently and safely, whilst caring for their basic physiological needs.

Once the trip or duty day is over, the group then reaches the "mourning" stage of the Tuckman model. Here, there is a sudden break in both working and social relationships, with a realization that crews will not fly with each other again within the same grouping. Even for the more robust personality type, this constantly formed and then broken way of relating leads to a lack of stability and continuity. For the more vulnerable, this dynamic can have more serious personal consequences.

Company Culture

As previously discussed, the role of a cabin crew member requires a certain type of individual who is able to combine a whole range of qualities. These include being emotionally available, open, flexible, adaptable, empathic, dependable, reliable, independent, and self sufficient. Frequently, people who display characteristics from this range of qualities require a degree of a sense of emotional nurturing from the organization too. Indeed, themes of British Airways being perceived as the "nurturing parent" are sometimes apparent when cabin crew employees present with personal difficulties at work.

This parental role assumed by the organization can be seen to take responsibility for the control of the employee by looking after the physiological needs of the individual once onboard the aircraft and whilst overseas. For example, meals are provided for crew members on the aircraft; hotels allocated overseas; allowances available to cover food; porters to take cases to rooms and coaches to transport individuals to the airport, etc. It follows that a good parent will unquestionably take care of these needs and so, in this way, the organization becomes more than an employer. Often a large dependency is established and/or cultivated by the crew member as the organization provides more than just financial remuneration to the employee; it provides a way of life.

However, with any *family type* relationship there will be times when domestic harmony has to be compromised. Given commercial and resource pressures, employees are required to work harder and cope with additional stressors, such as increased security under the threat of terrorism, within the constantly changing dynamic of the industry. This often leads to some resentment building toward the organization at large, as the employee may sense that they are somehow less cared for. Sometimes the employee can regress to an infantile state where they perceive that the parent organization is not providing for all their needs. This dynamic can often be witnessed in clients presenting to the Crewcare counseling service.

The culture of the organization is thus largely subjective, since it depends on the relative emotional maturity of the employee in determining how much responsibility they take for satisfying their own needs. In a job, however, which can be deemed to be a lifestyle choice, there will be a greater tendency to expect the organization to be more nurturing of personal needs, since much control and personal power is sacrificed by the employee.

We accept Bor's view (2004) that, on an unconscious level, the cabin crew in turn might also be regarded as maternal substitutes for the passenger. They are, after all, in constant view and take responsibility for providing food and drink. Passengers sit on board an aircraft in a non-participative role and can yield to a regressive "pull," exaggerated by the fact that they are sat helplessly at 35,000 feet, cared for by parental figures of unknown experience.

Thus cabin crew have to demonstrate these parental qualities in their everyday working lives. Consequently, they themselves will sometimes need to feel *parented* by the organization. Given the service orientated airline industry, this hierarchical culture of caring has obvious benefits. However, for the caring dependent employee, the consequences of subsequently not feeling cared for are often tangible.

How Cabin Crew Manage These Psychological Anomalies

In recent times the emphasis in the recruitment process has focussed on people who can be empathic, and therefore emotionally available, where required. However, in contrast this employee group also needs to show resilience and independence of character, in order to deal with the more "isolating" aspects of the job and the various pressures presented, particularly under a heightened state of aviation security.

So what personality traits are employed to manage these contrasting requirements? Here we expand on some of the perhaps less psychologically healthy ways of defending and acting out, often utilized by cabin crew, in order to survive in this kind of contradictory environment:

- Rescuer types – sums up a large percentage of cabin crew. The job attracts "caring" individuals who gain self-value primarily by helping others. They may also be termed "selfless" types who have a tendency to avoid their own issues by concentrating on others' needs.

Case No. 1

Background Stewardess who had had no quality time to grieve having been closely involved with supporting her husband's family after his death, and then acting on a life coach's advice to return to work as soon as possible:

> My husband was killed in an accident two years ago and I've been flying for a year and a half now. However I'm finding that I'm suffering flashbacks and nightmares, particularly when I'm away and alone in my hotel room. I have taken advice from people and tried to get on with my life again as quickly as possible but I just can't seem to move on. I am getting obsessive about my personal safety, always checking my door is locked and feeling that something terrible is going to happen.

Counseling Outcome

This client was indicative of the rescuer type quality so often present in our client group. She had selflessly taken care of other people's needs following her husband's death, probably in an unconscious attempt to avoid her own sense of existential loss and grief. This had resulted in her own cycle of grief being interrupted and as she tried to "move forward" with her life she found that this avoidance manifested in obsessive compulsive tendencies. Through follow up care and counseling we were able to bring these issues into awareness and refer this client onto a bereavement counseling service.

- Ain't it awful types – a traditionally distinctly "British trait." Certain individuals will describe how badly they have been treated or how "victimized" they feel. This is often used to aid group cohesion, as people around are often "drawn" into sharing similar "ain't if awful" type stories.
- Personal disclosure types – perhaps in order to make fleeting and brief but intense relationships hold more meaning, many crew share a great deal of personal information relatively quickly.
- Holding pattern types – as working relationships are brief but intense, some crew will use the same tried and tested formula for gaining group acceptance and attention. This means that various psychological complexities are played out again and again, but in constantly changing relationships they can be happily repeated many times, without the expectation to "move on."
- Uniform used as facade – given the fact that working alliances must be forged alongside social interaction whilst away from home/base, some crew will literally change personas when in uniform. There is obviously a requirement to behave in certain expected ways whilst representing the company, but if this is too far divorced from the essence of the real person, then problems can occur.
- Stuck in limbo syndrome types – as crew's feet are never literally "on the ground" for great periods of time, there is a temptation that life issues never quite have to be faced. Just when domestic or personal problems become

too intense the next rostered trip is waiting to provide an escape from the impasse.
- Avoidance types – similar to the previous trait, but includes people who are attracted to the job as a form of "escapis" from difficulties at home, this being their primary motivation for flying. When problems with this type of person occur, it is often worth investigating what problems are being "left behind" on the ground. What are they using the job to avoid?

Case No.2

Background Stewardess who used to work as a secretary in her home town and has been having an affair with a work colleague:

> When I'm away I find I can really be me. The job has expanded my horizons and I've met people who are so different from those back home. My family are very traditional and I've only been married for 2 years but already I've realised that I've given up my independence too soon. I've never spent much time on my own, always having been in a relationship and adapting myself to other people's needs. Now I want to spend some time on my own to see where I want to go with my life.

Counseling Outcome

Leaving her home town and traveling the world had resulted in this client discovering new parts of herself which had previously been repressed by background and culture. Stepping back from her adapted self and realizing that she had previously compromised her own independent needs resulted in her rebeling and acting out by having an affair at work.

It took the personal crisis of having a work affair for this client to realize how the job had previously provided a "smokescreen" for her domestic issues.

- Workaholic types – these types often see the job as their primary focus. They invest little in their lives outside of this single work dimension and so, when fulfilment wanes or retirement nears, they may realize that they have defined themselves too narrowly within the role.
- Infantile types – this type happily regresses to an infantile state with the expectation that many personal needs will be provided for by the organization. However, as previously discussed, at some point the "parent" organization will be perceived to let them down and this type of personality will assume little responsibility in seeking resolution through emotionally mature means. This often manifests in a general avoidance of responsibility for the self and leads to an unhealthy degree of responsibility being transferred from the individual to the organization.
- Glamor interject types – there are some crew members who carry out the job as a result of significant others' encouragement and/or expectations. For example, friends, family members or partners may persuade the individual

that the job is wonderful and glamorous. Consequently, they may interject this belief, often at the expense of their own true inner, and perhaps contrary, experience.
- Addictive types – finally many of the above ways of acting out may be totally avoided or at least "numbed down" by the types who turn to readily available alcohol, or even drugs, in order to escape the inherent dilemmas.

Psychological Effects of the Cabin Crew Role

> Air travel disrupts human relationships and behaviours, as well as bodily functions and systems (Bor, 2004, p.8).

Given the conflicting and contradictory demands inherent in the cabin crew role it is not surprising that certain "splits" can occur in the personality in order to try to appease these. An awareness of when it is appropriate to "switch" between various public facades is healthy and is demonstrated by individuals who can reach into their psychological toolbox of coping "mechanisms." However, problems can emerge when an individual over identifies with one or more of these "masks" and begins to lose a sense of their core self. For example, the requirement for group inclusion, a basic prerequisite for human happiness, can lead to an over reliance on the plastic veneer. Consequently, vulnerabilities and psychological issues not representative of that "smile" become suppressed and ignored. Hughes, writing in *Counselling at Work* similarly comments:

> Potential psychological damage can be caused by the forcing or falsifying of emotions in order to carry out the required tasks of a job, i.e. the explicit and implicit demands for flight attendants to 'display' a friendly, courteous disposition at all times, reflecting the mission statement of the company. (Hughes, 2000, p.7)

Thus, as the individual attempts to deal with the splits that can occur, detrimental manifestations may include anxiety, panic attacks, stress, manic episodes and other depressive states. Some crew members will present to health services with more directly related work problems, such as sleep deprivation and dietary issues. However, further discussion often reveals a deeper level of psychological angst associated with this falsifying of emotion.

The availability of alcohol within the job can compound and blur deeper routed personal problems. The lifestyle can act as a psychological retreat and alcohol affectively used to aid this process. However, the recent introduction of the statutory drugs and alcohol policy is acting as a support for the early disclosure of underlying addictive symptoms. The consequences for not disclosing such problems within a safety critical role can lead to disciplinary action. We are thus finding that crew with addictive type issues are more readily presenting themselves for help and the organization has support policies in place for this kind of proactive approach.

Case No. 3

Background A steward with self esteem issues who previously operated shorter European routes and has been flying to longer worldwide destinations for the past few months:

> I'm drinking a bottle of wine a day now. I find since I've been on long haul it's the only way I can get to sleep. Also it's such a good way to socialize with people when I'm away as normally I'm very shy around people I don't know well. I've realized now that it's getting out of control.

Counseling Outcome

This client presented his concerns regarding drinking to the company. The introduction of the Drugs and Alcohol policy now means that a network of support is put in place if and when an employee should proactively present with such a concern.

It transpired that this client had started to use alcohol as a means to rest and sleep since operating longer worldwide routes and that this was becoming a problem for him. Further counseling revealed that this client had come from a strictly religious family background and that he had found himself in a "shame/guilt" cycle since using alcohol to help him sleep. Supportive and reparative work here helped this client to break his shame/guilt associative cycle.

British Airways also has an attendance management policy in operation. This aims to effectively manage absence by offering support and an understanding of the reasons why employees cannot report fit for work. Similar to the drugs and alcohol policy, this provides a forum for crew to speak more openly about any personal problems as they arise and before they become too unmanageable.

For a certain number of crew members, particularly when they have been employed in the role for ten years or more, there is sometimes a resultant unconscious reliance on the job which indirectly supports an avoidance of dealing with underlying emotional struggles. As previously discussed, cabin crew can remain in a "holding pattern" until certain life crisis strike which serve to wake up the individual from their psychological slumber, or to provoke them into exploring deep seated emotional problems.

The Impact of 9/11

The attacks in America on September 11 served to nudge many individuals' deep-seated existential fears into awareness and heightened the need to support airline cabin crew across the world. Consequently, these events had an impact on the volume of contacts received by the Crewcare counseling helpline. There were a significant number of cabin crew who had to remain stationed overseas for several days, unable to return home. Following the hijacking incidents some crew were desperate to talk about how they were feeling and what it meant to be away from loved ones in a world forever changed by terrorism.

Possibly, for the first time, individuals' existential anxieties had a plausible "outlet." The inherent and/or latent fears around flying were dramatically given license for expression through the mass outpouring of emotions associated with this event. As many questioned their own mortality an increase in self-responsibility was evident and some crew members, particularly those with families and small children, began questioning their motivation for the job. The terrorist attacks gave license for cabin crew to talk more candidly about their other deep fears and anxieties as their psychological *retreat* became less viable.

Case No.4

Background Mother with children facing existential guilt about leaving them particularly after September 11th.

> I've been flying for 10 years now and I keep getting these anxiety rushes when I'm driving into work. It's crazy and I can't explain why it's suddenly started to happen. I find that I don't want to talk to people at work because I feel if I begin to talk about it, it may get out of control. For a while I've managed to keep a lid on it because it seems so crazy…

Counseling Outcome

This client presented with minor panic attacks when she was getting ready to go to work and often actually in the car traveling to work. During the counseling process she became aware of the heightened importance of her own physical safety following the terrorist attacks of September 11th, particularly in relation to her responsibility for taking care of her children. Her suppressed existential guilt over leaving her children to go away manifested in physically anxious symptoms.

Even prior to this current world crisis, the role of a cabin crew member could be stressful. Crew have an awareness of their responsibility to maintain an air of calmness under pressure and to present themselves in a positive, welcoming manner. As previously noted, their outward confidence may not always represent their internal feelings and subsequently true emotions are often suppressed. Within a working community comprising of approximately 14,000 cabin crew, the issues raised in the counseling service are broad and varied and often represent a snapshot of societal issues at large.

The Crewcare Counseling Service

The British Airways Crewcare counseling service is a unique, confidential, airline counseling service which, has been in operation for almost 20 years. The department is operated by a team of 23 trained counselors, all of whom are crew members themselves, who have been flying as cabin crew for at least two years. They combine flying duties with working a six-day counseling roster every four to five weeks. The peer group aspect of the service is something that users value greatly. The service receives positive feedback relating to how reassuring it is to speak to someone who understands the lifestyle and uniqueness of the job. If problems raised by the client cannot be resolved in a single counseling session, Crewcare have the facility to refer a crew member to the British Airways' Employee Assistance Provider, the Independent Counselling and Advisory Service, for ongoing counseling. Additional referrals will be offered via The British Association of Counselling and Psychotherapy (of which Crewcare is an organizational member, adhering to their Codes of Ethics and Practice), or the United Kingdom Confederation of Psychotherapists.

Throughout the last 20 years, Crewcare has continued to develop a range of services and support systems. These include a 24-hour freephone helpline; a drop in counseling facility located on site; follow up help and guidance for crew involved in critical incidents; a support group for working parents who are combining family and working life; an opportunity to re-orientate crew returning from long term sick; awareness campaigns on issues facing the community; workshops on counseling skills for managers; stress management programs; a web presence on the company intranet and the continual updating of an extensive resource directory.

Critical incidents can range from a passenger's death on board to air rage, and any work related event that may be traumatic for the individual. In these cases, the role of the Crewcare counselor is primarily a defusing one; listening to the crew member's experiences and responding to their immediate needs.

Quarterly statistics are provided to the relevant managers within British Airways. This data illustrates current trends and issues which are raised without compromising the position of the service as a confidential facility. Inevitably, as commented by Kutek (1996), "We are change agents whether we like it or not. If people come to us and talk about bullying and harassment, that means pointing to something that needs changing in that organization" (p.285).

The Crew Perception of Crewcare

Following 20 years of 24-hour support, the Crewcare function is now firmly embedded in the support systems offered to cabin crew both at home and overseas. During this time utilization rates of the service have increased substantially.

Statistics for 2001 show us that the service received approximately 1082 contacts from an employee group of approximately 14,000. This figure comprises information and referral contacts through to active crisis counseling calls and/or personal visits. This represented 7.7 percent of our then total client base – a high take up rate when compared to most Employee Assistant Programmes.

Given this level of usage amongst the department's client base there is evidence that it positively impacts the whole of the cabin crew culture. Indeed the service can be seen as a "safety net" into which not everyone will fall, but who's presence is nevertheless valued. We concur with Briner when he says:

Changes in a client's mood and emotional state may also play a significant role in changing the feelings and behaviours of co-workers. There is plenty of evidence to suggest that we can "catch," through a process known as emotional contagion, the feelings of other people. (Briner, 2000, p. 3)

During this same year we also canvassed our client base and asked for feedback on the service provided. The following comments are a cross section of the many anonymous responses received:

Positive "Having used Crewcare I feel this is a service that is well needed and it helped me so much to get back on track when I felt there was nowhere else to turn. I am truly grateful."

As a Safety Net "I think it is an excellent service – even knowing that it's there often provides support. The times I have used it the counselors have been discreet, professional and efficient."

Individual Perception "I think people who haven't used the service tend to think it's only for suicidal, depressed alcoholics; more could be done to let people know about the kind of problems you can help with- big or small."

Implications of Peer Support Service "Crew life can be fatal for some people sometimes and Crewcare is helpful because they are crew themselves, so they can understand crew lifestyles. None of my immediate friends ever understand why I am tired, lonely or jet-lagged. As far as they are concerned I am on a permanent holiday."

Effect on Organizational Culture "I'm glad that BA as a company can offer this service. I have never had this service at previous companies."

The Success of Crewcare

Overall the success of the Crewcare service is probably due to the fact that it evolved from a genuine need from within the flying community. The concept was originally proposed to management by two crew members following a suicide overseas.

Subsequently, the whole ethos of the Crewcare service has grown from a place of empathic understanding and realization of the often covert issues that exist within the flying lifestyle. The fact that all the counselors recruited are also crew members continues to build on this central tenet of a peer group who has insight, understanding, and empathy with its client base.

The flying community is, by its nature, largely an absent workforce, so the on-sight, no appointment needed facility located at the operation center at Heathrow Airport is further vital to Crewcare's success. In this way the support functions are perceived by the client group to be readily available, reliable, and spontaneous to their immediate, often "in crisis" needs.

The Crewcare service is placed in a unique but important autonomous position within the overall support facilities available to flying crew. Thus it works closely with occupation health, the trade unions, manpower services and scheduling departments but maintains its identity and framework of confidentiality within this support system. In this way it is perceived by its client group to offer a unique kind of service, which is not duplicated by any other function and therefore valued in its own right for its confidentiality and autonomy.

The Future of the Crewcare Counseling Service

Learning from the past has helped the Crewcare counseling service build for the future. As the industry is constantly changing, the Crewcare counseling service recognizes that it cannot be complacent. It is constantly fine-tuning the services on offer to the cabin crew community.

With the continued need to offer a duty of care in the increasingly litigious working environment, to the constant evolution of the job in the modern world and as a sanctuary of retreat in a dynamic organization, we like to think that our services will be valued well into Terminal 5 and beyond.

Note

The information in this is confidential and propriety to British Airways Plc and has been obtained from a number of sources, is edited, checked where possible and believed correct at the time of going to print. The accuracy of source material cannot be warranted and British Airways is under no obligation to update the information. This information is for general information only and is not intended to be relied upon. The case studies illustrated in this chapter are purely fictitious to preserve anonymity of individuals. However, they are a good illustration of the types of issues presented to the Crewcare counseling facility.

References

Bor, R. (2004). *Anxiety at 35,000 feet. An introduction to clinical aerospace psychology*. London: Karnac.
Briner, R. (2000). Do EAPs work? A complex answer to a simple question. *Counselling at Work, 29*, 1–3.
Hughes, R. (2000). The impact of emotions at work. *Counselling at Work, 30*, 7–9.
Kutek, A. (1996). Managing stress at work. *Counselling, November*, 284–286.
Tuckman, B. (1965). Developmental sequence in small groups. *Psychological Bulletin, 63*, 384–399.

PART 3
Related Themes in Aviation

Chapter 15

Psychological Aspects of Astronaut Selection

David M. Musson

In all countries that have a human spaceflight program, the application process to become an astronaut is extremely competitive. In each case, the number of highly qualified applicants greatly exceeds the limited number of available openings for this extraordinary job. At NASA, for example, astronauts are selected approximately every two years, and for each selection opportunity, typically, between 2,000 and 4,000 individuals will apply. Formal application includes a personal statement and detailed biographical data including academic achievement, extracurricular interests, flight history including aircraft type and experience, ancillary qualifications and medical history. Of the thousands who apply every two years, about 120 are eventually invited to the Johnson Space Center (JSC) in Houston, Texas, for detailed evaluation. From this group, only about 20 will be offered entry into the Corps as an Astronaut Candidate. Once selected, attrition from the candidate pool has essentially been nil (Santy, 1994), so that all applicants can expect to eventually qualify as astronauts following one to two years of specialized training.

The astronaut selection process begins by reviewing applications to identify individuals who meet specific minimum qualifications of academic achievement, flight experience, and medical history. A subset of applicants are then brought to a central testing facility (JSC in the case of NASA) for further assessment that includes extensive physiologic and medical evaluations, psychological testing, and a series of clinical and non-clinical interviews. Some authors have criticized the limited role played by psychological evaluation during selection, citing a culture that has a deep rooted aversion to psychology and psychiatry (Helmreich, 1983; Santy, 1994). This chapter will present an overview of the history of astronaut and cosmonaut selection, and discuss how the process of psychological assessment is related to the overall selection strategy. The recent shift away from short duration spaceflight missions to longer duration expeditions, as well as the current multinational nature of spaceflight, has significant implications for astronaut/cosmonaut selection. Recent experiences in extended-duration orbital missions are reviewed, along with a discussion of what research conducted in terrestrial "space analogue" environments has revealed. Current practices in selecting and training of long duration crews are also presented, along with a discussion of issues relevant to future research and astronaut selection for longer duration lunar and interplanetary missions.

The First American Astronauts

When the National Aeronautics and Space Administration (NASA) was formed in 1958, one of its first tasks was to select the seven men who would serve as the astronauts for Project Mercury. The Mercury capsules that were the United States first step in the exploration of space were designed to hold only one human being and President Eisenhower decreed that these first astronauts would be experienced military test pilots. Although NASA was a civilian agency, military pilots were thought to be the most appropriate candidates for reasons of both individual aptitude and national security (Santy, 1994). From an initial pool of over 500 military test pilots, 110 were identified as meeting the requisite experience and qualification level for the Mercury Program, and from those, 69 agreed to volunteer for further assessment (Atkinson & Shafritz, 1985). The medical and psychological testing employed in selecting the astronauts for project Mercury was extensive, and psychological assessment included most of the tests that were available at that time, including the Minnesota Multiphasic Personality Inventory (MMPI), Rorschach Inkblot Test, the Thematic Apperception Test (TAT), and the Wechsler Adult Intelligence Scale (WAIS). The purpose of this testing was to identify individuals who possessed the characteristics theorized to be most appropriate for the mission. These included: high intelligence, drive, and stability; low levels of impulsivity; and the ability to depend on others when needed yet also be capable of functioning independently if required. From the pool of 69 pilots, seven men were selected to become the first American astronauts.

The decision to consider only military test pilots for the first NASA astronauts had several implications for future astronaut selection and training. This was a highly screened population, and successful military pilots had already demonstrated above-average degrees of achievement motivation, an aptitude for complex operating environments, and the ability to perform well under stress. They had also all succeeded in military careers, and by inference could be presumed to possess certain other important characteristics, such as a willingness to follow orders, adherence to the concept of national service, and the ability to thrive in squadron-like operational environments. Whether NASA was aware of it or not, these attitudes were implicitly part of the selection process, even though they were not overtly being tested. The culture of military pilots would become the founding culture for the future astronaut corps, bringing along with it specific attitudes and beliefs that would later become hallmarks of astronaut culture. As described by Santy (1994), viewing physicians and psychologists as an unwelcome threat to an astronaut's continued flight status were key elements of this professional culture.

Perceiving medical and psychological assessments as potential threats to one's flying career, both astronauts and pilots typically prefer to spend as little time as possible in the flight surgeon's office, and even less time within sight of a psychologist or psychiatrist. According to some accounts, this dislike of behavioral sciences led to the premature cancellation of an extensive study of astronaut psychological stress and performance undertaken by NASA scientists during the early years of project Mercury (Santy, 1994). Astronaut selections conducted in 1962 and 1963 in support of the future Gemini and Apollo programs were expanded to include civilian pilots, and in

1965 the first scientist pilots were chosen from the fields of natural science, medicine, and engineering (NASA, 2003). Although psychological evaluations continued as part of the selection procedure, no formal programs to validate the existing criteria or collect data to advance the psychological selection process were in place. Broadening the selection base to include non test pilots, non pilots, and civilian personnel meant that the implicit selection of military and pilot characteristics that existed during Mercury was no longer in place. By the end of the Apollo program, the norms of astronaut culture had been well established, and included a fundamental aversion to psychological assessment and interventions. This would prove problematic for later efforts to study and revise existing selection standards, and to accommodate the needs of longer duration and more psychologically demanding missions.

The First Soviet Cosmonauts

Few details are readily available regarding the early years of Soviet Cosmonaut selection, though Santy (1994) provides an excellent summary based largely on translated works from that period and personal interviews. In many ways, the selection of the first cosmonauts paralleled the US selection of astronauts – both countries decided that military jet pilots were the most suited for this new role. Yuri Gagarin, the first man in space and a Soviet national hero, recounts his experience of selection in his memoirs (see Gagarin (1978), as cited in Santy, 1994). He describes a similar process to that of the Mercury applicants – days of medical and psychological tests that winnowed down the large applicant pool to a small group of cosmonauts. Psychological tests employed during this process were known to include Soviet variants of many Western tests, including a modified version of the MMPI (referred to as the SMIL), the Thematic Apperception Test, the Eysenck Personality Inventory, the Rorschach, and the Personality Scale of Taylor, along with a number of other testing batteries. In general, the Soviet program was selecting for factors such as stress tolerance, concentration, and stability in these applicants (Santy, 1994).

There are two important factors regarding psychology that differentiate the Soviet era cosmonaut selection from that of its American counterpart. The first was that while the Soviets used many Western psychological testing instruments, there were significant differences in psychological theory and application in the Soviet Union compared to Western psychology. Soviet psychology of that era has been severely criticized as being overly political and less than scientific, and its avoidance of Western theory meant that much of the science of cosmonaut selection was based on the evaluation and matching of biophysical and biorhythmic patterns – a practice viewed with considerable skepticism in the West. The second factor was that while Soviet psychology differed greatly in its theoretical foundations, the role of psychology in the Soviet space program was much more prominent that at NASA. Unlike NASA, which sponsored little or no psychological research until recently, the Soviet space program was characterized by a much more visible role for psychology and psychiatry. It appears that an awareness of the potential problems posed by stress, isolation, and confinement were well known to mission planners and cosmonauts alike. This may be partly due to the fact that the Soviets had focussed their space program on increasingly long duration orbital space station missions while the US pursued short duration missions well into the Shuttle era. With a focus on long-duration space station missions, it would have been hard to ignore psychological factors. However, despite the higher regard given to psychological issues in the Soviet space program, anecdotal accounts of psychological problems on board Soviet stations over the past two decades are not rare. Regardless, the Russians are to be commended for recognizing the potential difficulties in behavior and performance in these missions when their counterparts in the West did not.

The Space Shuttle – Two-Week Missions in Space

In the United States, the Mercury, Gemini, and Apollo programs of the 1960s and early 1970s focussed on getting to the moon. With the development of the space shuttle in the late 1970s and the launch of the first shuttle Columbia in 1983, the US human spaceflight program shifted exclusively to short-duration orbital missions. These missions would be limited to 14 days in duration due to engineering limitations posed by the massively complex spacecraft, and the expected shift to a more permanent presence in space would not occur for many years.

The first shuttle crew was named in 1978, and since then, more than 150 astronauts have been selected into the NASA astronaut corps. In the shuttle era, astronauts are classified as either Pilot Astronauts or Mission Specialists. Pilot Astronauts are typically (but not exclusively) pilots of military background whose role is to command and pilot the Shuttle. Mission Specialist Astronauts are scientists, engineers, and physicians whose primary responsibilities include a mix of onboard science and mission operations. Astronaut selection for the Shuttle program was modeled on that of the earlier Mercury, Gemini and Apollo programs, though psychological contributions to the selection process were advanced by proponents such as psychiatrist Patricia Santy and later by psychiatrist Chris Flynn and psychologist Al Holland. Selection was advanced to two clear phases – that of "select out" and "select in." Select out refers to eliminating those applicants who are found to possess some degree or significant risk of psychopathology, whereas select in refers to the process of identifying the most appropriate individual from a pool of qualified and healthy applicants.

During this period, the European Space Agency (ESA), Canadian Space Agency (CSA) and the National Space Development Agency (NASDA) of Japan have adopted similar approaches to the "select in" and "select out" method described above (Sekiguchi, Umikura, Sone, & Kume, 1994; Ursin et al., 1991). In particular, ESA expended considerable efforts in the late 1980s and early 1990s to define desirable psychological profiles for its astronaut selection process, with emphasis on traits related to emotional stability, task motivation and social empathy (Goeters & Fassbender, 1991).

While the criteria for identifying someone with real or potential psychiatric difficulties may be relatively straight forward, formally identifying those healthy individuals who would be most appropriate for the job is less so. Later in this chapter we will look at recent and current efforts to identify what those factors are.

Shuttle-Mir or Mir-Shuttle: The US Returns to Long Duration Missions

In the late 1990s, a series of seven US astronauts joined Russian crews on the space station Mir for periods ranging from 115 to 188 days.[1] Mir had been launched several years earlier, and represented the crowning achievement of the Soviet (and later Russian) era of manned orbital space flight. An entertaining and relatively thorough account of this experience appears in Brian Burrough's book, *Dragonfly* (Burrough, 1998), and a number of astronauts have written autobiographical accounts of their experiences aboard Mir during this program.

Unlike the Soviet and Russian space program, the US program until this time had consisted of primarily short duration missions of 10 to 14 days in length. The exception to this was the Skylab program in 1973 (Skylab 1 – unmanned, Skylab 2 – 28 days, Skylab 3 – 59 days, and Skylab 4 – 84 days). While the US had experienced some crew problems on Skylab, most notably a 24-hour work stoppage by exhausted astronauts (Cooper, Jr., 1996), it is reasonable to say that NASA was unprepared for the psychological difficulties experienced by its crews during the course of the Shuttle-Mir program. Through Shuttle-Mir NASA learned that psychological factors were key to mission success in long duration missions – this of course was already known to the Russians who had been in the business of long duration spaceflight for much longer.

Psychologists had long warned of potential mood, behavioral, and social issues that would be experienced by crews during long duration spaceflight (Helmreich, 1983). At NASA, astronaut selection had focussed on choosing astronauts who were task focussed, highly skilled, and able to assimilate the mountain of technical information required to perform the requisite tasks during the one or two weeks in orbit. There was a general sense among the astronauts that they could work with anyone for 14 days, which was probably true. Six months of isolation and confinement was a different story, and the separation from family, friends and home was certainly difficult for all of the astronauts, though undoubtedly more for some than others. There was little to guide the selection of those who would cope well and those who would not. Selection, at least in the US, had never been designed to identify those who would perform well under such conditions.

---

1 This program was named Shuttle-Mir in the United States, though it was referred to as Mir-Shuttle in Russia. The inability to find a common name was perhaps prophetic of the difficulties in coordination that were to be experienced by these two space superpowers in managing this complex program.

Nick Kanas and his research team at the University of California at San Francisco along with Russian collaborators Salnitskiy, Gushin, and Kozerenko studied the Shuttle-Mir (or Mir-Shuttle) crews while on orbit, and have reported on a host of psychological difficulties during that mission (Kanas et al., 2001). Even some of the American astronauts that flew aboard Mir have publicly discussed some of the psychological difficulties of those missions, including depression, loneliness, and socio-cultural isolation (Florida Today – Space Online, 1997). The experiences of Shuttle-Mir served to raise awareness at NASA of the potential psychological problems involved in supporting crews for prolonged periods in space. These included not just the problems that could occur in orbit, but also the need for coordinating support during multinational missions and the demands placed on personnel who would train for extended periods in foreign cultures before the flight. Compared to earlier missions, being an astronaut now seemed to include cultural flexibility, the ability to tolerate prolonged period of low stimulation, and the ability to coexist with others in confined spaces for months at a time.

International Space Station – A Permanent Presence in Space

Most recently, the construction of the International Space Station (ISS) has been the focus both the Russian and American space programs as well as space agencies from Europe, Canada, Brazil, and Japan. Since 1999, the station has been continually occupied by crews spending up to six months at a time on board the orbiting facility. While originally planned to house seven multinational crew members, budget cuts early in the ISS program led to the cancellation of the development of an adequately sized escape vehicle, which in turn led to the reduction of the maximum number of ISS crew to just three astronauts. The grounding of the shuttle fleet following the loss of the Columbia in February of 2003 further reduced that number to only two, and at the time of writing this chapter, it is unclear if and when the ISS crew compliment will be increased.

Current selection for the NASA astronaut program has not changed significantly from how selections were conducted during the shuttle program, though psychological assessment now appears to figure more prominently in that process (Galarza & Holland, 1999b). NASA psychologists are reluctant to publish which specific psychological tests they are presently using due to concerns of test compromise. They have, however, generally described what they are looking for in terms of psychological profiles. In keeping with operational experiences and findings from terrestrial research, current selection attempts to identify high levels of emotional stability, performance under stress, group living skills, teamwork, conscientiousness, and effective team leadership (Galarza & Holland, 1999a; Kanas & Manzey, 2003).

In preparation for missions aboard ISS, astronauts are put through a series of exercises and experiences designed to both train and assess skills for living and working in close confinement and under stress. Such exercises include a short (1-2 week) arctic survival program, currently conducted in the Canadian North and loosely based on Air Force arctic survival training. NASA has also been experimenting with short duration exercises in which ISS astronauts live and work with marine scientists onboard the Aquarius undersea habitat located in 60 feet of water off the Florida Keys. While the primary purpose of this program is to train and provide some familiarization with living in confined environments, it is likely that all of these experiences are used, at least informally, as an additional level of selection for space station assignment.

The Right Stuff, Wrong Stuff, and No Stuff

The idea of the *Right Stuff*, a term made popular by Tom Wolfe's account of test pilot Chuck Yeager and the seven Mercury astronauts has become a commonly used phrase for describing those individuals who possess the special qualities required by astronauts and other high performing individuals. The impression given by Wolfe's account was that these men knew no fear, triumphed under adversity, possessed superior intellect and demonstrated a natural inclination towards heroic tendencies. They were also described as rugged individualists, stoics, and completely self reliant – traits believed to be critical to withstanding the dangers and stresses of such a dangerous undertaking. These were the attributes that were to become known as the *Right Stuff*, and they were probably appropriate characteristics on which to base the first astronaut selections. However, despite these assumptions, there was little scientific evidence to guide exactly what should be looked for in selection. There can be little argument that the astronauts and cosmonauts that were chosen in the past have performed their duties with utmost distinction. But for many, there has been a nagging sense that as human space flight progresses to longer and more socially complex missions that the psychological make up of tomorrow's space explorers may need to be significantly different than those who came before.

Drawing from the work of psychologists such as Janet Spence and Robert Helmreich, psychologists and psychiatrists at NASA began looking at validating existing theories of personality and performance in astronauts, beginning in the late 1980s. In particular, Helmreich, along with his former students, most prominently Tom Chidester and Clay Foushee, had conducted a series of studies on airline pilots looking at the link between personality, attitude, and flight performance in multi crew cockpits. Using the fundamental constructs of Instrumentality (or Achievement Motivation) and Expressivity (or Social Orientation), they identified three distinct clusters of traits that appeared to predict superior performance in this population. The highest performers possessed elevated scores in both trait families, along with low levels of hostility. These individuals were not only task-focussed, but highly competent at coordinating their crews and building a sense of teamwork. They labeled these pilots the real *Right Stuff*. The second cluster consisted of pilots who were also highly instrumental, but were less socially competent and tended to score higher on measure of hostility. Helmreich and Chidester labeled these individuals as possessing the *Wrong Stuff*. Interestingly, subsequent research has suggested that while these pilots may be hard to fly with, crews eventually compensate for these factors and can perform quite well after a period of adjustment. A third cluster was characterized by low levels of task motivation, and these pilots did not fare as well as either of the two previous groups, and were labeled the *No Stuff* cluster (Chidester & Foushee, 1991; Chidester, Helmreich, Gregorich, & Geis, 1991).

These findings from aviation became the basis for conducting studies on NASA astronauts in the late 1980s and early 1990s. Robert Rose, Helmreich, Santy and others administered personality tests to active astronauts at NASA, and correlated trait scores with peer and supervisory assessments from that same population. While the findings were limited by both small sample size and restricted ranges on such scales as Achievement Motivation (on which all astronauts scored very highly), initial evidence was found to replicate some of the airline pilot findings in this population. In an extension of that study, the Spence and Helmreich testing battery was administered to all NASA astronaut applicants for the following five years as part of a long term study of astronaut performance that is still underway. Analysis of these data has demonstrated that applicants who fell into the *Right Stuff* categorization were no more likely to have been selected, suggesting that the formal selection process in place from 1989–95 did not screen specifically for those trait clusters (Musson, Sandal, & Helmreich, 2004).

Analogue Environments

Opportunities to conduct research with astronauts are exceedingly rare, and when obtained, the findings are limited by the problem of small subject numbers. As a result, NASA and other agencies have funded extensive research in settings considered to be analogous in one or more ways to spaceflight (Harrison, Clearwater, & McKay, 1991; Stuster, 2000).

Research into human performance with relevance to spaceflight has included studies of aquanaut performance in undersea habitats (Radloff & Helmreich, 1969), sealed-chamber space station simulations (Sandal, 2001; Sandal, Vaernes, Bergan, Warncke, & Ursin, 1996), and expedition performance at Antarctic research stations (Suedfeld & Weiss, 2000). Broadly, these settings are often referred to as "space analogue" environments, though they are also at times referred to as "extreme environments" or ICE settings (Isolated and Confined Environments) in various literatures. Aviation is one of the most obvious analogue environments for the conduct of psychological research relevant to astronaut performance and selection, and the work of Helmreich, Chidester and others that figures so prominently in current selection strategies discussed earlier is based largely upon studies of airline pilot performance. The batteries developed by Spence and Helmreich have been used in many space analogous settings, most prominently Antarctic research stations and confinement simulations. The *Right Stuff* cluster of traits has proved to predict performance in a number of these settings, including Antarctic teams (Sandal, 2000), submarine crews (Sandal, Endresen, Vaernes, & Ursin, 1999), and confinement studies (Sandal, Vaernes, & Ursin, 1995).

In addition to the Spence and Helmreich scales, other inventories have been used in these "space analogue" settings in attempts to predict different aspects of human behavior under conditions of isolation and confinement. These include the Hogan Personality Inventory (HPI) (Biersner & Hogan, 1984), the Fundamental Interpersonal Relations Orientation-Behavior (FIRO-B) Scale (Palinkas, Gunderson, Johnson, & Holland, 2000), and the NEO Five Factor Inventory (NEO-FFI) (Steel, Suedfeld, Peri, & Palinkas, 1997). Key findings from this body of research have been that low levels of *emotional instability*, low *extraversion* and low *conscientiousness* have been positively correlated with performance (Palinkas, Gunderson, Holland, Miller, & Johnson, 2000). The finding that low levels of *conscientiousness* correlate positively with supervisor and peer-assessed performance is at odds with research from other employment settings where *conscientiousness* has consistently been positively correlated with multiple indices of performance. A proposed explanation for this finding is that individuals characterized by higher levels of *conscientiousness* may have high needs for achievement, and are frustrated by equipment failures, schedule delays, and inadequate supplies which are often a reality of living and working under the harsh conditions of an isolated research station. Individuals with lower needs for achievement and perfection may show superior psychological adjustment under such circumstances (Gunderson, 1974; Palinkas et al., 2000). A major concern in the extrapolation of research findings from space analogue settings to the spaceflight environment is that there appear to be fundamental population differences between astronauts and analogue populations. Astronauts are characterized by unusually high levels of *achievement motivation, emotional stability (low neuroticism), and conscientiousness* that are not uniformly shared by the research subjects in analogue settings (Rose, Fogg, Helmreich, & McFadden, 1994; Musson, Helmreich, & Sandal, 2004). These differences can be worrisome when findings in analogue settings are unanticipated and counterintuitive, such as the correlation between low *conscientiousness* and superior coping in the Antarctic, particularly when those are the very traits on which the two populations differ (Musson, 2003). Concerns aside, these environments are still unquestionably valuable settings for conducting studies that will guide the future selection of astronauts for the long duration missions of the future. Research into human performance in analogue environments remain a critical means of identifying those traits critical to mission success on the long duration missions planned for the near future.

Back to the Moon and On to Mars

At present, the future of the space station is somewhat uncertain, as is the future of NASA's fleet of shuttles. Regardless of the current state of the various global national space programs, there is a general consensus that the future of human spaceflight will involve long and increasingly complex and dangerous expeditions back to the moon, and then on towards Mars and the other planets in our solar system. Psychological issues are currently listed among the top biomedical concerns to be overcome if these missions are to be successful. Existing plans to return to the Moon (by 2020 at present estimate) involve prolonged stays of weeks to months. Even the shortest timelines for an expedition to Mars place a crew of four–six in transit to the red planet for six months, surviving on the surface for one–three years, then at least another six months in transit on the way back. The current understanding of bioastronautic problems is most clearly laid out in NASA's Critical Path Roadmap – a continually revised assessment of biomedical limitations for human spaceflight. Along with lethal doses of cosmic radiation and the serious phenomenon of continual bone loss, psychological issues and crew performance are the most severe human factors limitations posed by extended duration missions beyond low Earth orbit (NASA, 2005). Psychological selection, of course, is only part of the solution to the problem of maintaining the mental health of crews and preserving performance capabilities on these projected missions. Selection of the entire crew, not just of individual astronaut and cosmonaut will become a major consideration. Historically, our Russian colleagues have always considered the problem of crew composition to be a major focus for long duration missions. However, the inevitable multinational and multilingual composition of future crews will pose additional issues beyond those previously addressed in long duration space flight. Crews will need to be more self-sufficient and autonomous than has been the case previously in spaceflight and selection will need to include such considerations. There has been speculation that crews on long missions may need to have some sort of crew counselor or therapist on board to help deal with the inevitable problems that will arise between crew members. While this responsibility would likely fall to the crew medical officer, all crew members would have to possess some degree of skills specifically aimed at prolonged confined living. Lessons from previous long duration expeditions, findings from analogue environments, and lessons learned from current crews aboard ISS will all serve to inform the process for selecting the next generation of astronauts.

References

Atkinson, J. D. & Shafritz, J. M. (1985). *A history of NASA's astronaut recruitment program: The real stuff.* New York, NY: Praeger Publishers.

Biersner, R. J. & Hogan, R. (1984). Personality correlates of adjustment in isolated work groups. *Journal of Research in Personality, 18*, 491–496.

Burrough, B. (1998). *Dragonfly: NASA and the crisis aboard Mir.* New York, NY: Harper Collins.

Chidester, T. R. & Foushee, H. C. (1991). Leader personality and crew effectiveness: a few mission simulation experiment. In R. Jensen (ed.), *Proceedings of the 5th International Aviation Psychology Symposium Vol. II.* Columbus, OH: Ohio State University.

Chidester, T. R., Helmreich, R. L., Gregorich, S. E., & Geis, C. E. (1991). Pilot personality and crew coordination: Implications for training and selection. *International Journal of Aviation Psychology, 1*, 25–44.

Cooper, H. S., Jr. (1996). The loneliness of the long-duration astronaut. *Air Space, 11*, 37–45.

Florida Today - Space Online (1997). Blaha candid about battling depression during early days aboard Mir. Florida Today - Space Online March 3 1997 [On-line]. Retrieved on February 24, 2005 from http://uss001.infi.net/flatoday/floridatoday/space/explore/stories/1997/030397b.htm.

Gagarin, Y. A. (1978). *The road to space*. Moscow: Nauka Press.

Galarza, L.& Holland, A. W. (1999a). Critical astronaut proficiencies required for long-duration space flight. *Proceedings of the International Conference on Environmental Systems, Denver, CO, July 12-15*.

Galarza, L. & Holland, A. W. (1999b). Selecting astronauts for long-duration space missions. *Proceedings of the International Conference on Environmental Systems, Denver, CO, July 12-15*.

Goeters, K. M. & Fassbender, C. (1991). *Definition of psychological testing of astronaut candidates for Columbus missions ESA 8730/90/NL/IW*. Hamburg, Germany: European Space Agency.

Gunderson, E. K. (1974). Psychological studies in Antarctica. In E.K. Gunderson (ed.), *Human Adaptability to Antarctic Conditions* (pp. 115–131). Washington, DC: American Geophysical Union.

Harrison, A. A., Clearwater, Y. A., & McKay, C. P. (1991). *From Antarctica to outer space: Life in isolation and confinement* (1st edn.). New York: Springer-Verlag.

Helmreich, R. L. (1983). Applying social psychology in outer space: Unfulfilled promises revisited. *American Psychologist, 38*, 445–450.

Kanas, N. & Manzey, D. (2003). *Space Psychology and Psychiatry*. (1st ed.) El Segundo, CA: Microcosm Press.

Kanas, N., Salnitskiy, V., Weiss, D. S., Grund, E. M., Gushin, V., Kozerenko, O. et al. (2001). Crewmember and ground personnel interactions over time during Shuttle/Mir space missions. *Aviation Space and Environmental Medicine, 72*, 453–461.

Musson, D. M. (2003). *Personality determinants of professional culture: Evidence from Astronauts, Pilots and Physicians (Doctoral Dissertation)*. The University of Texas at Austin.

Musson, D. M., Helmreich, R. L., & Sandal, G. M. (2004). Baseline personality comparisons between astronauts and Antarctic personnel: Implications for generalization of psychological research findings. In International Civil Aviation Organization (ed.), *Proceedings of the 55th International Astronautical Congress 2004*, Vancouver, Canada.

Musson, D. M., Sandal, G. M., & Helmreich, R. L. (2004). Personality characteristics and trait clusters in final stage astronaut selection. *Aviation Space and Environmental Medicine, 75*, 342–349.

National Aeronautics and Space Administration (2003). *The Astronaut Fact Book NP-2003-07-008JSC*. Washington, DC: Author.

National Aeronautics and Space Administration (2005). Critical Path Roadmap for Bioastronautics. http://bioastroroadmap.nasa.gov/ [On-line]. Available at http://bioastroroadmap.nasa.gov/.

Palinkas, L. A., Gunderson, E. K., Holland, A. W., Miller, C., & Johnson, J. C. (2000). Predictors of behavior and performance in extreme environments: the Antarctic space analogue program. *Aviation Space and Environmental Medicine*, *71*, 619–625.

Palinkas, L. A., Gunderson, E. K., Johnson, J. C., & Holland, A. W. (2000). Behavior and performance on long-duration spaceflights: evidence from analogue environments. *Aviation Space and Environmental Medicine*, *71*, A29-A36.

Radloff, R. & Helmreich, R. L. (1969). Stress under the sea. *Psychology Today*, *3*, 28–29.

Rose, R. M., Fogg, L. F., Helmreich, R. L., & McFadden, T. J. (1994). Psychological predictors of astronaut effectiveness. *Aviation Space and Environmental Medicine*, *65*, 910–915.

Sandal, G. M. (2000). Coping in Antarctica: is it possible to generalize results across settings? *Aviation Space and Environmental Medicine*, *71*, A37-A43.

Sandal, G. M. (2001). Crew tension during a space station simulation. *Environment and Behavior*, *33*, 134–150.

Sandal, G. M., Endresen, I. M., Vaernes, R., & Ursin, H. (1999). Personality and coping strategies during submarine missions. *Military Psychology*, *11*, 381–404.

Sandal, G. M., Vaernes, R., Bergan, T., Warncke, M., & Ursin, H. (1996). Psychological reactions during polar expeditions and isolation in hyperbaric chambers. *Aviation Space and Environmental Medicine*, *67*, 227–234.

Sandal, G. M., Vaernes, R., & Ursin, H. (1995). Interpersonal relations during simulated space missions. *Aviation Space and Environmental Medicine*, *66*, 617–624.

Santy, P. A. (1994). *Choosing the right stuff: The psychological selection of astronauts and cosmonauts*. Westport, CT: Praeger Publishers.

Sekiguchi, C., Umikura, S., Sone, K., & Kume, M. (1994). Psychological evaluation of Japanese astronaut applicants. *Aviation Space and Environmental Medicine*, *65*, 920–924.

Steel, G. D., Suedfeld, P., Peri, A., & Palinkas, L. A. (1997). People in high latitudes: The "Big Five" personality characteristics of the circumpolar sojourner. *Environment and Behavior*, *29*, 324–347.

Stuster, J. W. (2000). Bold endeavors: behavioral lessons from polar and space exploration. *Gravitational and Space Biology Bulletin*, *13*, 49–57.

Suedfeld, P. & Weiss, K. (2000). Antarctica natural laboratory and space analogue for psychological research. *Environment and Behavior*, *32*, 7–17.

Ursin, H., Olf, M., Sandal, G., Warncke, M., Maki, P., Pettersen, R. et al. (1991). Final definition of ESA test battery. In K.M. Goeters & C. Fassbender (eds.), *Definition of Psychological Testing of Astronaut Candidates for Columbus Missions for European Space Agency* (pp. 288–316). Hamburg, Germany: German Aerospace Research Establishment (DLR).

Chapter 16

Occupational Factors in Pilot Mental Health: Sleep Loss, Jet Lag, and Shift Work

Jim Waterhouse, Ben Edwards, Greg Atkinson, Thomas Reilly, Mick Spencer, and Adrian Elsey

During the course of the last hundred years or so, the task of flying a plane has changed enormously; no longer is it the responsibility of a single pilot but rather of a battery of computers and a whole air crew. Modern aviators are required for their bravado and sense of adventure much less than in pioneering days; for much of the time, at least during the cruise phase of a flight, they are computer-minders. One of their main problems is to remain awake and alert so that they can diagnose and respond to the remote possibility of an emergency or malfunction. To interpret the wealth of information that the complex electronic systems in the cockpit generate requires specialized training; however, the individuals have, from physiological and psychological viewpoints, remained unchanged. Does this mean that humans, with their limited abilities to remain vigilant and their difficulties in functioning effectively when deprived of sleep, have become the weak link in this situation?

In the first part of this chapter, the factors responsible for sleep and mental performance in subjects living normally will be considered. In the second, the abnormalities in the air crews' lifestyles produced by their flight schedules will be considered. The effects of these abnormalities upon sleep and performance in air crew will be considered in the third part. Finally, and based upon current understanding of the causes of the difficulties, advice to minimize these difficulties will be given, in order to maintain the highest levels of safety for the passengers and air crew.

Factors Controlling Sleep and Mental Performance

Modern flights can be long, they can take place at times that would not coincide with normal working hours, and they can result in crossing time zones. In order to understand the effects that these abnormal circumstances exert upon sleep and performance, it is necessary to consider first the changes in these variables that are likely to occur to an individual living a normal, regular lifestyle – awake and active

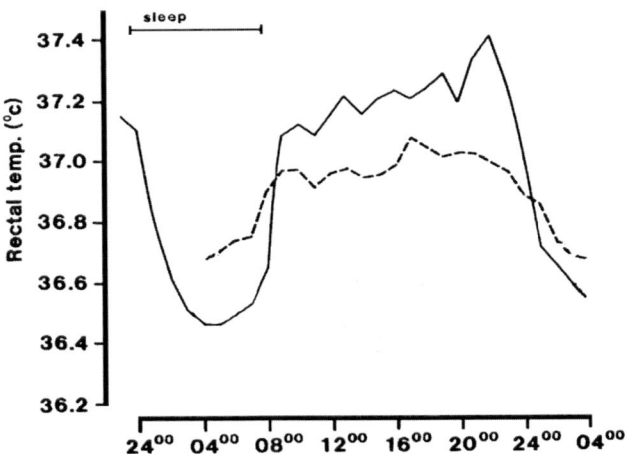

Figure 16.1 Core Temperature v. Time[1]

during the day-time and asleep at night. To do this requires an understanding of the body clock.

The Body Clock

When subjects living normally (day-time activity and night-time sleep) are studied, body temperature shows higher values in the day-time and lower values at night (Figure 16.1). Many other variables, including mental performance, show rhythms that are timed similarly. By contrast, for many hormones – melatonin, for instance, a hormone released into the bloodstream by the pineal gland deep within the brain – the observed rhythm is the inverse, with nocturnal concentrations being higher than those in the day-time (Reilly, Atkinson, & Waterhouse, 1997).

These rhythms are not wholly because the body is responding to a day-orientated society, with opportunity at night for sleep and recuperation. This fact can be deduced from studies of individuals during a "constant routine" protocol. In this protocol, the subject is required: 1. to stay awake and sedentary for at least 24 hours in an environment of constant temperature, humidity, and lighting; 2. to engage in similar activities throughout, generally reading or listening to music and; 3. to take identical meals at regularly-spaced intervals. In spite of the fact that this protocol means that any rhythmicity due to the environment and lifestyle has been removed, the rhythm of body temperature (and other variables that have been studied) persists,

---

1 The mean rectal temperature rhythm of 9 subjects living normally and sleeping midnight-08:00 h (full line), and then performing a "constant routine" starting at 04:00 h (dashed time). (From Waterhouse et al. 2002).

even though its amplitude is decreased (see Figure 16.1). Three deductions can be made from this result:

1. The rhythm that remains must arise within the body; it is described as an endogenous rhythm, and its generation is attributed to a "body clock."
2. Some effect of the environment and lifestyle is present, since the amplitude of the rhythm has decreased; this component of the rhythm is termed "exogenous." In the case of body temperature, it is raised by light and mental and physical activities during day-time waking, and decreased by darkness, sleep and inactivity during the night.
3. In subjects living a conventional lifestyle, these two components are in phase. During the day-time, the body temperature is raised by the body clock acting in synchrony with the environment and activity; and, during the night, the clock, environment and inactivity all act to reduce body temperature.

Humans have paired groups of cells, the suprachiasmatic nuclei (SCN), in the base of the hypothalamus, a region of the brain closely associated with the regulation of body temperature, the release of hormones, appetite, and sleep (Moore, 1995). One of the many pieces of evidence that the SCN is the site of the clock is that cells from this area show rhythmicity in nerve activity when they are cultured *in vitro* in constant conditions; no other region of the brain has cells with this autonomous activity. The genetic and molecular mechanisms responsible for generating the activity of the clock have now been described in the scientific literature (Clayton, Kyriacou, & Reppert, 2001).

When humans are studied in an environment in which there are no time cues – in an underground cave, for example – the daily rhythms of sleep and waking, body temperature, hormone release, and so on, continue (Waterhouse et al., 2002). This fact confirms their endogenous origin, but it is observed that the period of such rhythms is closer to 25 than 24 hours and the subject becomes progressively more delayed with respect to the outside environment. The clock-driven rhythms measured in such circumstances are called circadian (from the Latin for "about a day").

Such circadian rhythmicity implies that the body clock needs to be continually adjusted for it to remain synchronized to a solar (24-hour) day, which is what normally happens. Synchrony is achieved by zeitgebers (German for "time-giver"). These are rhythms resulting, directly or indirectly, from the environment. In humans, the most important zeitgebers are the regular alternation of the light-dark cycle (Czeisler et al., 1989; Minors, Waterhouse, & Wirz-Justice, 1991) and release at night of melatonin (Haimov & Arendt, 1999). These changes are normally underpinned by the whole social structure of individuals' lifestyles, including social, mental and physical activities, and mealtimes (Reilly et al., 1997).

The effect of light depends on the time of exposure. Pulses of light that are centred in the 6-hour "window" immediately after the trough of the body temperature rhythm (the trough normally being 03:00–05:00 h, see Figure 16.1) produce a phase advance; those centred in the 6-hour window before the temperature minimum, a

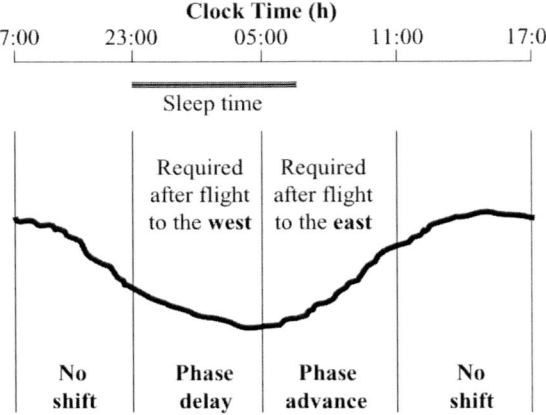

Figure 16.2 Shifts of Body Clock caused by Light[2]

phase delay; and pulses centred away from the trough by more than a few hours have little effect (Figure 16.2). Bright light such as is found outdoors or indoors near a window produces larger phase shifts than does domestic lighting. In practice, most humans normally have little exposure to natural daylight.

Therefore, exposure to light on waking in the morning, and even to light passing through our eyelids when asleep at this time, will cause a small advance of the body clock. This will result in the body clock showing a period of 24 hours, synchronized to the solar day. Ingestion of melatonin, in the form of pills, can also exert different effects upon the body clock according to the time of day at which ingestion takes place. The shifts produced are in the opposite direction to those produced by light exposure at the same time. Thus, ingestion of melatonin in the afternoon and evening tends to advance the body clock, and in the morning tends to delay it. Receptors for melatonin are present in the SCN. Since bright light also inhibits melatonin secretion, the phase-shifting effects of light and melatonin reinforce each other. Bright light early in the morning just after the temperature minimum advances the phase of the body clock not only directly but also indirectly, since it suppresses melatonin secretion and so prevents the phase-delaying effect that melatonin would have exerted at this time (Lewy & Sack, 1996).

The Role of the Body Clock

The body clock produces daily rhythms in body temperature, plasma hormone concentrations, the outflow of the sympathetic nervous system, and activity in

---

2 Diagram to illustrate times relative to the rhythm of body temperature in a subject living normally when light causes a phase advance, a phase delay, or no phase shift (dead zone).

the sleep centers of the brain, all of which exert effects throughout the body. The implications of this for sleep and mental performance will be considered later.

The general effects of the body clock are two-fold. Firstly, those actions involving physical and mental activity and their associated biochemical and cardiovascular changes are promoted in the day-time, and those involving recovery and restitution during a period of inactivity are promoted at night. The second role is to enable preparations to be made for the switches from the active to the sleeping state, and *vice versa*; individuals have to prepare biologically for going to sleep and for waking up. Such changes require an ordered reduction or increase in activity of a whole series of biochemical and physiological functions – and this has to be set in motion before the actual events of falling asleep or waking up take place.

To achieve the above aims, the body clock needs to be stable and robust, rather than ready to respond to transient changes in the environment or lifestyle of the individual. A clock that rapidly adjusted would compromise the phasing of the circadian rhythms of an individual who woke transiently in the night and switched on the light, or who took a nap in the day-time. The observation that the body clock is slow to adjust to changes in lifestyle therefore makes sound ecological sense.

However, such stability means that, for those who have to work at night or undergo time-zone transitions, the normal synchrony between the endogenous and exogenous components of the circadian rhythms will be lost. It is this lack of synchrony between the body clock and the outside world that causes these individuals to suffer the negative subjective and objective effects that are known, respectively, as "shift workers' malaise" and "jet lag." These effects will disappear only when adjustment of the body clock to the changed sleep-wake pattern or local time has taken place. They will be considered below in more detail.

Two of the most important effects of the body clock are in regulating sleep and altering mood and mental performance.

The Body Clock and Sleep

The ease of getting to sleep shows a circadian variation that is linked to the rhythm of body temperature (Åkerstedt & Gillberg, 1981). Sleep onset is easiest to initiate at, or just before, the temperature trough and most difficult at, or just before, the peak. Another time when sleep is transiently easier to initiate is at about 13:00–15:00 h. This is called the "post-lunch dip." This dip in wakefulness (increase in sleepiness) is present even in fasting subjects, though it is accentuated by a meal and alcohol intake.

Staying asleep is most easy when body temperature is low, and most difficult when it is high. Spontaneous waking shows a daily propensity that is the opposite of that for falling asleep; it is most likely to occur when body temperature is rising and least likely when temperature is falling.

If these findings are combined, the best time for taking a full sleep, of 6 hours or more in length, is when the sleep will span the time of the trough in body temperature. Such a sleep would normally start at the beginning of the falling phase of body temperature, about 23:00 h, and would last until the rising phase is well under way, about 07:00 h. This is, of course, the time when sleep is normally taken (Dijk & Czeisler, 1995). Sleep might be initiated more easily in the middle of the night, when body temperature is at its lowest, but it would then be shorter and curtailed by the rising body temperature. Therefore, when the body temperature is at its lowest, and during the so-called post-lunch dip, are the times when short sleeps or naps (½ – 2 hours in length) are most easily taken. The value of these naps will be considered later.

When the quality of sleep is considered, measurements by the electro-encephalogram, EEG, indicate that sleep is associated with continual changes in the frequency and amplitude of electrical activity (Reilly et al., 1997). By convention, this continuum of electrical activity is divided into several components, some of which are sleep stages 1–4 and REM sleep. A normal night's sleep consists of a mixture of "light sleep" (stages 1 and 2), "deep sleep" (stages 3 and 4, also known as "slow wave sleep," SWS) and "paradoxical sleep" (rapid eye movement sleep, REM, when the EEG is very active and muscle tone decreases). The role of light sleep remains unclear, but the amount of SWS reflects the amount of prior wakefulness and so is generally assumed to represent the dissipation of some factor that has accumulated during the time that has been spent awake. It is also thought that REM sleep is recuperative in some way; it is associated with dreaming, and the propensity for it to occur is inversely related to body temperature.

During the course of a normal night's sleep, there are about four nonREM-REM cycles (where nonREM refers to stages 1–4 of the EEG), each lasting about 90 min. The first nonREM-REM cycle has most SWS and least REM sleep; in subsequent cycles, there are falls in the amount of SWS and rises in the amounts of REM sleep and light sleep. Spontaneous waking is generally from light sleep.

The Role of Naps

If time does not permit a full sleep, or the body temperature is too high to enable this to take place, then a shorter sleep or nap is an alternative (Campbell, 1992; Dinges, 1992; Naitoh, 1992; Rogers et al., 1989; Waterhouse et al., 2001). It is not unusual for adults, particularly elderly ones, to nap during the day-time but it seems that all subjects benefit from naps during prolonged work hours. Thus a 2-hour nap after a vigil of 45-53 hours results in a considerable degree of recuperation, and a 4-hour nap can substantially reverse the day-time decrement following a night without sleep. Naps shorter than this can be of some use also; a 1-hour nap taken at 02:00 h during a night without sleep has been shown to reduce the overnight decrement in auditory vigilance.

There have been attempts to investigate if a routine of more frequent, shorter sleeps is as effective as a single, long sleep in maintaining mental performance throughout the waking period (Naitoh, 1992). In other words, is the conventional 8-hour sleep/16-hour awake the ideal, or would two equal "days" (each of 4-hour sleep/8-hour wake), or even shorter days (involving multiple "days" lasting 8, 6 or 4 hours, for example), be equally effective? In favor of such schemes, it has been found that the shorter the waking period, the less is the decrement in mental performance that takes place during the shorter time awake (see below). However, there is another effect of sleep that tends to offset the advantage of many short sleeps; this is "sleep inertia" (Naitoh et al., 1993). Its effect is that, immediately after waking from a nap, performance *falls* for some minutes after being woken. The effect seems to be more marked if the subject wakes, or is woken, from SWS. Sleep inertia tends to be less after spontaneous awakening since this is rarely from SWS. The greater part of this inertia has worn off about 10 min after waking up but improved mental performance (due to the loss of fatigue) is achieved only after a longer period, up to about 1 hour, of having been awake. There are clear implications here for air crew sleeping during a duty period.

In summary, it appears that "days" that are too short show, overall, poorer performance because effects due to sleep inertia offset the decrease of fatigue due to having had a sleep. Nevertheless, a routine involving multiple naps might be useful in those contexts where sleep loss has occurred.

Naps may be described as "recuperative," "appetitive" or "prophylactic" (Nicholson et al., 1984). Recuperative naps can be taken to relieve fatigue due to lost sleep or time awake, and this is the sense in which they are most commonly used and has been used above. Appetitive naps are taken at the same clock time due to habit; they can be regarded as acting as "anchor sleep." This is the regular timing of a portion of sleep, generally at a time coincident with night on home time. It seems to stabilize circadian rhythms, though whether it is the regular sleep itself or other regularities of lifestyle that it will produce (particularly the light/dark cycle) is not known. Whatever the exact explanation of the effect, its value to air crew is that it will help them maintain rhythms adjusted to home time, if they so choose (see below). Prophylactic naps are taken before a period of work in order to start duty feeling as refreshed as possible.

It is clear that the suitable use of naps is an important part of the armory against sleep loss and its effects.

Mental Performance

Safety and reducing accidents to a minimum are paramount when the performance of any workforce is concerned (Dinges, 1995; Folkard, 1990; Graeber, 1982; Winget et al., 1984). Accidents are rare, and they must be investigated retrospectively, with all the interpretative difficulties that this entails. As a result, attention tends to be directed instead to critical incidents, that is, occasions when errors are made, one outcome of which could be an accident, though this is not necessarily the case. Even though such critical events are more common than accidents, they are still comparatively rare, and their complex causal nexus does not make them easily amenable to scientific study. In practice, therefore, it is more expedient and common to attempt to measure mental performance, subjective assessments of mood, and so

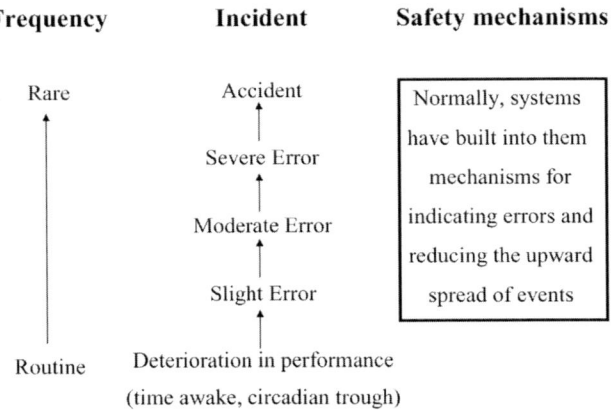

Figure 16.3 Causal Links between Routine Deteriorations[3]

on, in order to find those circumstances where performance is poorer or mood is worse. The argument – that, by minimizing circumstances that lead to a deterioration of mood or performance, the frequency of accidents will be reduced – is an indirect one (Figure 16.3). It also ignores the fact that safety mechanisms can be put into place by which the sequence – poor performance → increased frequency of errors → increased frequency of incidents → increased likelihood of an accident – can be broken. Even so, it is generally the case that accident prevention involves maximizing attention and mental performance in the first place.

Much of the relevant work has been performed in the laboratory and has made it clear that there are many facets to "mental performance" and "mood." A huge variety of tests of mental performance and mood have been devised, differing in sophistication, cost, and the aspect of performance that they measure. There is always debate about the relevance of a particular test to a real-life situation. For example, how much about the efficiency of an air crew can be learned by considering reaction time, speed and errors in mental arithmetic, vigilance, and so on? One solution would be to use flight simulators in assessments of the effects of jet lag, of time awake, of sleep loss, and so on. However, the time and cost of their use tend to be prohibitive, and there is always the problem of deciding if the results from a "virtual world," no matter how realistic it might appear to be, can be compared with those from the real one.

In spite of these difficulties, many of the results produce the same general conclusions. It is these that will now be considered.

---

3 Diagram to illustrate the assumed causal links between routine deteriorations in mental performance and accidents.

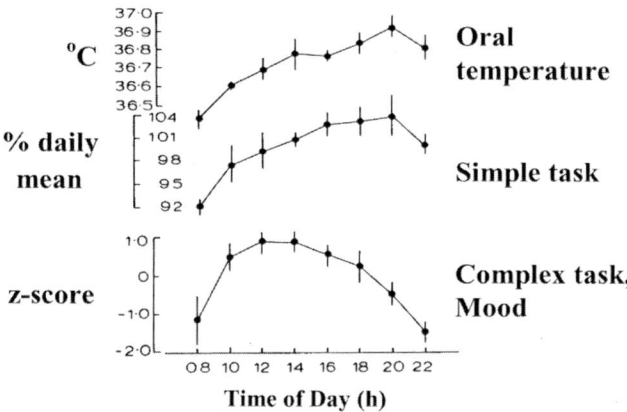

Figure 16.4 Day-time Rhythms[4]

Circadian Rhythms in Mental Performance and Mood

Figure 16.4 shows the rhythms of oral temperature, a simple mental performance task and mood (or a more complex performance task), all measured in the day-time in healthy subjects living a conventional sleep-wake schedule. Clearly there are rhythmic changes with lowest values in the early morning and late evening (Folkard, 1990). When these measures are extended through the night, they are, like body temperature, lowest at this time. This general parallelism between temperature, mood, and mental performance is often interpreted to indicate a causal link. Conceptually, it is easy to consider that a rise in body temperature will increase brain activity, and that a fall is associated with a tendency towards feeling tired and falling asleep.

A closer inspection of Figure 16.4 shows that the time-course of performance at a simple mental task differs from that when mood or performance at a complex task are considered. These differences can be seen as variations in the rate at which a decrement occurs as time awake increases. As a general rule, simpler performance tasks deteriorate less quickly than complex ones. This deterioration is often referred to as "fatigue" and has been found in field as well as laboratory studies.

A common conclusion is that, whatever other factors might exist, mood and mental performance are determined to a large extent by the interaction between circadian rhythms and time awake and, moreover, that circumstances where both factors are unfavorable – "double negatives" – should be avoided. These two factors dominate much of the modelling of mood and performance that has been carried out (see below).

---

4 Day-time rhythms, in subjects living normally, of oral temperature, a simple mental performance task, and mood or a complex performance task (based on Folkard, 1990). Used by permission.

Other Factors Affecting Performance and Mood

Other factors include the amount of time spent performing a particular task, the environment, and sleep loss (reviewed in Waterhouse et al., 2001).

Time on Task

Performance deteriorates with the amount of time spent at a particular task. This is especially so when the task is repetitive or boring, or if it is too complex and imposes too high a workload. When the task is boring, efforts should be made to increase the interest it arouses in the individual, since lack of motivation renders the individual prone to making errors. When the workload is too high, this normally implies an inadequacy somewhere else in the system. In such circumstances, individuals might be forced to sacrifice accuracy for speed, and confidence and motivation suffer, particularly if the workload is externally imposed rather than self-selected.

In these cases, a brief rest allows time for recuperation. Alternatively, changing the nature of the task being carried out might result in an improvement in mood and performance, and this raises the possibility of "job-sharing."

The Environment

In studies assessing human performance, the effects of the environment are generally controlled or corrected for. Nevertheless, the role of the environment when it is not controlled should not be underestimated. Lighting and noise both play roles in this regard. Light that is too bright can be annoying and make it difficult to read dials with ease; too little light is soporific and can also render other tasks difficult to perform. Noise produces similar effects, with too much distracting an individual and making it difficult to focus upon the task in hand, and too little sound predisposing towards inattention and sleepiness. Mood is susceptible to abrupt changes due to environmental factors and these can occur at any time of the day.

Sleep Loss

Sleep loss leads decrements in mood (Reilly & Pierce, 1994) and mental performance, subject to the following generalizations:

a. Sleep loss of as little as 2 hours can produce decrements in performance.
b. Sleep loss often affects the speed of performance rather than its accuracy.
c. Sleep loss affects complex and vigilance-requiring tasks most.
d. Sleep loss causes nocturnal performance to be particularly affected, with "lapses" and "micro-sleeps" – extreme deterioration to the point of inactivity – occurring.
e. Effects due to sleep loss and time awake or time spent on a task interact. Thus, towards the end of a 16-hour work period, the accident rate has been measured

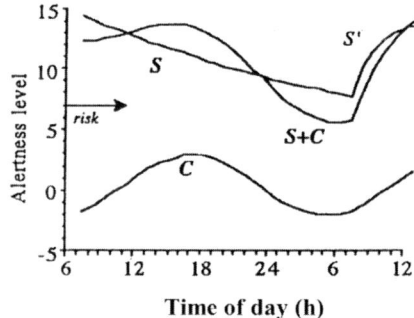

Figure 16.5 Alertness v. Time of Day[5]

to begin to rise steeply, and deterioration with time on complex tasks can occur after as little as ½ hour if sleep loss has occurred also. These are further examples of "double negatives," and even "triple negatives" (a combination of sleep loss, performance at the circadian trough, and extended time on task) can be envisaged.

Modeling Human Performance and Mood

There have been several attempts to construct mathematical models that would predict performance and alertness (see Spencer, 1987, for example), and the results of a recent workshop to discuss and compare several models have recently been summarized (Neri, 2004). The models are generally based on a circadian component (parallel to body temperature) combined with a homeostatic component that reflects the increasing amount of fatigue associated with the amount of time spent awake. The main role of such models is to enable prediction of those circumstances where performance would be compromised, particularly the "low points" and the "double negatives" that will occur in any work schedule.

One such model is the "three-process model of alertness regulation" (Folkard & Åkerstedt, 1991; Åkerstedt & Folkard, 1995). Using subjective alertness data from a number of experiments on altered sleep/wake patterns, alertness could be predicted from the combination of a circadian and a homeostatic component – plus a component for sleep inertia. The model was validated against the EEG and measures of sleepiness in the laboratory and the field.

The simplified version of the model consists of two components: C and S (Figure 16.5). Process C represents sleepiness due to circadian influences and has a form

---

5 A model to predict alertness from a knowledge of its circadian (C) and time awake (S) components. Alertness is predicted from the sum of S + C, and predicted values less than 7 indicate a risk of inadequate performance. S' indicates the recovery of alertness during sleep. For more details, see text (from Waterhouse et al., 2001). Used by permission.

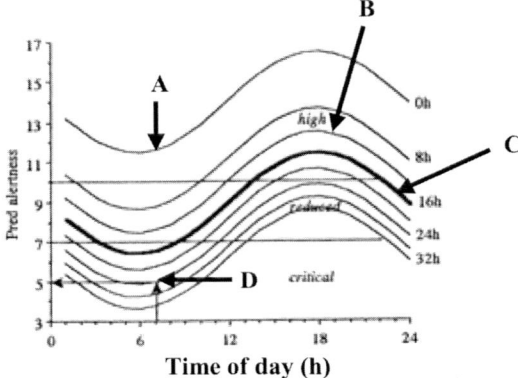

Figure 16.6 Predicted Alertness v. Time[6]

parallel to the rhythm of body temperature. Process S is a function of time-since-awakening; it is high on awakening, and then falls exponentially, initially rapidly and then progressively more slowly. At sleep onset, process S reverses (S'), and recovery occurs in an exponential fashion, initially increasing very rapidly but subsequently leveling off towards an upper limit. Recovery is usually accomplished in 8 hours. Process S' has been linked to the presence of slow wave activity (mainly occurring during SWS) during sleep. (There is also a third component in the full model, the exponential wake-up process W, or sleep inertia, which shows a steep initial rise during the first minutes after awakening and then levels out towards a limiting value).

The inputs to the model are the times of rising and going to bed, and the output from it is the predicted alertness. This output is the arithmetic sum of S+C. Alertness ranges from 1-16, with 3 or less corresponding to extreme sleepiness, 14 or more to high alertness, and values of 3-7 to an increased risk of sleep onset and performance failure.

Figure 16.6 shows predicted alertness on waking at 07:00 h after a full 8-hour sleep (A), at the end of a day-time work period (B), in the evening before going to bed (C), and when wakefulness is extended by 8 hours (to a total of 24 hours), as is frequently the case with the first night shift (D). The output shows that alertness falls during the night shift to values in the early morning that are in a "critical" zone when there is a risk of failing to respond to the demands of the task in hand and even of falling asleep. During the following sleep, the steep recovery of factor S', together with the rising phase of C, cause a rapid increase in alertness – though this is normally latent, since it occurs during sleep.

When this model is used to predict alertness under different circumstances, some of the main findings have important implications for air crew and are as follows:

---

6   Adapted from Waterhouse et al., 2001. Used by permission.

a. In a subject adjusted to the normal waking day, the fall due to S after waking is offset by the rise in C until the evening. This maintains alertness during the day-time, and then enables preparation for sleep to be made in the evening, when C also begins to fall.
b. If C, the phase of the circadian rhythm of body temperature, is suitably adjusted, then alertness does not fall below the threshold associated with an increased risk of sleep onset during the course of a night shift.
c. If one or more naps are taken during a period of extended wake time, then alertness is maintained above the critical zone for longer. In practice, if such a nap were to end more than about an hour before the start of a work period, then the effects of sleep inertia would be effectively over; such a nap would be "prophylactic" in nature.

Effects of Short- and Long-Haul Flights upon Lifestyle

In air crew involved in short-haul flights, the hours of work can be irregular from day to day, both due to the schedule being worked and to delays and diversions that arise in the several sectors that are involved. Often the duty period requires an early time of awakening. As a result, the individual starts the duty period with some loss of sleep since it is difficult to go to bed and sleep sufficiently early the night before. This loss can accumulate over successive days, particularly if duty hours are different each day, so preventing the individual from getting into a routine. Sleep loss occurs also if duty hours encroach upon the evening or the first part of the night. In the longer term, such schedules become debilitating, particularly if they are extremely irregular and involve high workloads.

For long-haul flights, although levels of workload might not always be very high, the problems associated with physiological adaptation are increased for several reasons (Cabon et al., 1993; Dinges, 1995; Graeber, 1982, 1989; Lowden & Åkerstedt, 1998; Nicholson et al., 1984, 1986; Samel et al., 1995, 1997, Spencer et al., 1991; Stone et al., 1993; Wegmann & Klein, 1985). The flights are necessarily longer than normal; they often take place during the night, whether this is reckoned by destination or by home time; and, unless the flight is in the north-south or south-north direction, there will be a time-zone transition to contend with when the air crew arrives at its destination. With aircraft development, the duration of flights has increased steadily over the years, and flights to the antipodes, involving a single stopover *en route*, typically of 24 hours or 48 hours, are now commonplace. It is likely that, in the future, there will not be such a stopover on many of these journeys.

Such changes to an individual's daily routine will also be experienced by the passengers, but there are some crucial differences. Firstly, even the more ardent globetrotter is unlikely to undergo such disruptions as frequently as does the air crew. Secondly, whereas the passengers can relax throughout the flight and choose to sleep whenever they wish to, this option is unavailable to the air crew who, collectively, must be vigilant throughout the flight. Passengers can plan their rest on the flight to optimize their alertness on arrival, whereas air crew need to plan their rest on the ground to optimize their alertness during the most demanding parts of the flight (take-off and landing).

There are very many possible "tours of duty" currently being practised by air crew, and it would not be possible to describe them all. Even so, the journey can be split into several sections. In the following account, the kinds of problem encountered will be outlined; a more detailed description of them and their causes has already been given, or will be given later.

The Outward Flight

This can last for 10 hours or more, though more than one air crew might be involved in the longer flights. Since the aim is to land the passengers in the day-time by local destination time, this flight might take place during the day-time or overnight as measured by local time at the place of departure. These flights will cause fatigue, due to their length, and might also give rise to the further difficulties associated with night work ("shift workers' malaise"). If the flight starts early in the morning, the air crew might have some degree of sleep loss due to decreased sleep the night before. If, by contrast, the flight does not start until the evening then the total amount of time awake by the end of the flight will be extended beyond the normal period of about 16 hours. Unless the flight is in the north-south or south-north direction, a time-zone transition will be encountered, with local times differing at the points of departure and arrival. This difference in local time will give rise to "jet lag."

The Layover Days (Time Spent in the New Time Zone before Returning Home)

The length of time spent on layover can vary from the minimum consonant with the regulations regarding flight hours to several days. To maximize the utilization of the aircraft, the layover period is normally close to a multiple of 24 hours, and is frequently either 48 hours or 72 hours at the more remote destinations. This time is primarily intended for rest and recuperation after the outward flight. During this time, individuals can try to catch up on lost sleep, but they have the problem of deciding whether or not to attempt adjustment of their sleep-wake cycle and body clock to the new time zone.

Economic considerations mean that this time need not be spent only relaxing and sightseeing; if the layover lasts several days, it is not uncommon for local, shuttle flights to be undertaken. These can take place either during the day or at night but, generally, do not involve crossing time zones. Not crossing more than two time zones is important to the airlines since, at least with the UK's flight-time limitations scheme, an individual is adjudged to be on "local time" after having spent three local nights on the ground in the same time zone, and the maximum duty hours depend on whether or not the pilot is living on "local time" (Civil Aviation Authority, 2004).

Also, local positioning flights may be undertaken which may interrupt the air crew's planned sleeping schedule.

The Return Flight or Subsequent Flight(s)

If the tour of duty ends with a return flight to the original point of departure then the problems involved are very similar to those that were described, above, for the outward flight. If the individual has been successful in attempts to stay on home time when away from home (see above), then re-adjustment of the sleep-wake cycle and the body clock to home time should not be required once he/she has reached home. However, if he/she attempted to adjust to destination time, then the process of adjustment back to home time must be accomplished after arrival home. If the flight is to another destination rather than back home, then the issue of whether to adjust to the new time zone or remain on home time recurs.

In summary, air crew members undergoing long-haul flights can experience one or more of the following, often inter-related, abnormalities when compared with their normal routine:

- extended hours of work;
- extended time awake;
- the symptoms of "jet lag" due to time-zone transitions;
- the symptoms of "shift workers' malaise" due to night work;
- loss of sleep due to duty hours, jet lag and/or night work.

These abnormalities are illustrated by examples of flights practised by the airline KLM (Figures 16.7-9).

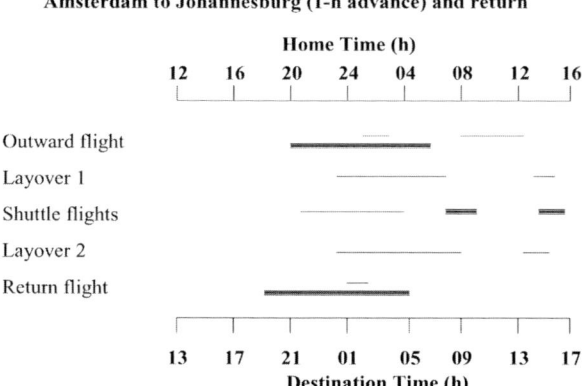

Figure 16.7 Times of Sleep and Naps, 1 Hour Advance[7]

---

7   Duty hours indicated by thick lines, sleeps and naps by thin lines. Carvalho, Bos et al., 2003. Used by permission.

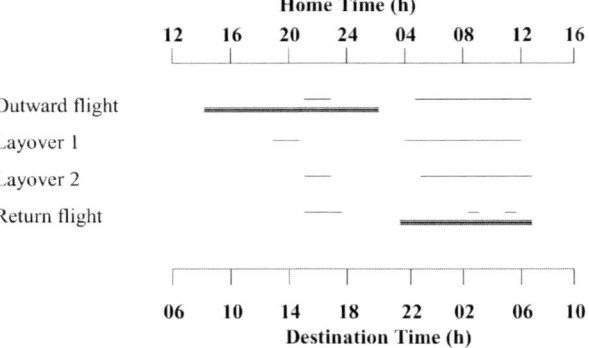

Figure 16.8 Times of Sleep and Naps, 6 Hour Delay[8]

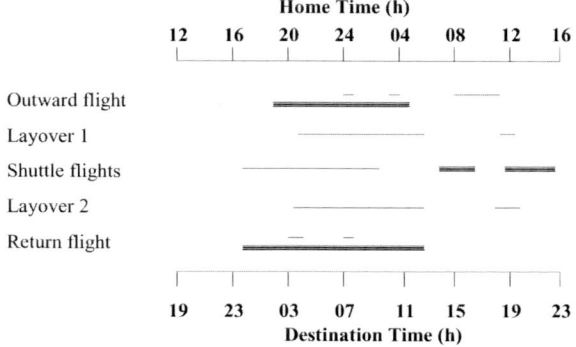

Figure 16.9 Times of Sleep and Naps, 7 Hour Advance[9]

Alterations in Mood, Sleep and Performance in Air Crew

The above abnormalities have a direct effect on the sleep, mood, and performance of air crew. In any particular situation, there will be a considerable variability in the effect on an individual. This variability is both inter-individual (i.e. different people may respond differently in the same situation) and intra-individual (the same person may respond differently in the same situation on different occasions). Nevertheless, considering groups of individuals, there are general trends which can be reliably

---

8   Data presented as in Figure 16.7, Carvalho, Bos et al., 2003. Used by permission.
9   Data presented as in Figure 16.7, Carvalho, Bos et al., 2003. Used by permission.

demonstrated and which represent the underlying effect upon air crew of a particular pattern of duty. These trends are summarized below.

"Jet lag" (Haimov & Arendt, 1999; Waterhouse et al., 1997; 2000) is the term commonly applied to an assemblage of symptoms that include:

- feeling tired in the new local day-time, and yet unable to sleep at night;
- feeling less able to concentrate or to motivate oneself;
- decreased mental and physical performance;
- increased incidence of headaches and irritability;
- loss of appetite and general bowel irregularities.

The problems encountered cannot be attributed to differences in climate or culture between the destination and the home country. For example, they are likely to be marked for European air crew traveling to Australia, New Zealand or the west coast of America, but slight for those travelling between Europe and Africa. They are due to the slow adjustment of the body clock, as a result of which the normal synchrony between the exogenous and endogenous components of a circadian rhythm is lost.

Jet lag is an inevitable accompaniment to time-zone transitions, though its severity and its detailed nature depend upon the individual. It does not become less marked with experience of crossing time zones, though coping mechanisms are developed (see below). The negative effects are generally more marked and last longer after eastward flights, and up to one day per time zone crossed is a rough estimate of the time taken for jet lag to disappear (a reflection of the time taken for the body clock to adjust). There can be a large variation between individuals in this rate of adjustment. It is possible that age is one of the factors responsible for such variation (Gander et al., 1993; Moline et al., 1992); if so, then there are possible implications for air crew that remain to be investigated. Mood and mental performance are both likely to deteriorate because the trough of the temperature rhythm falls in the new local day-time and also because sleep quality is poor (again associated with the unadjusted temperature rhythm).

Night Work and Shift Workers' Malaise

The general malaise, poorer mood and decrement in performance experienced by those who work at night are similar to the effects of jet lag. This is because the "shiftworkers' malaise" is also caused by following a sleep-wake schedule that is out of synchrony with the body clock (Reilly et al., 1997; Waterhouse et al., 1992). Mood and performance will be poorer at night, because the individual is working during the circadian trough of body temperature. The individual feels unmotivated and performance is more difficult to sustain. The problems will be more marked if the individual is suffering also from sleep loss and/or is working extended hours (examples of "double" or "triple-negatives"). The negative effects abate much more slowly on successive night shifts due to the contradictory information from different zeitgebers – the timing of the sleep-wake cycle, for example, might promote adjustment whereas the timing of the natural light-dark cycle during flights that do not cross many time zones might not.

Night work entails day-time sleep, and this is poorer than normal since it is being attempted during the day-time when body temperature is high. There is the possibility that sleepiness will mean that an individual takes an unplanned sleep during a duty period, and there is both subjective and objective evidence that this does happen. Moreover, a form of sleep paralysis has been observed, during which the individual is apparently awake but becomes immobile and unable to respond to the environment, possibly missing important information.

The Effects of Workload and Evidence for "Stress"

Studies of workload of air crew during flights have shown that the times of peak load are during take-off and landing, with the load during the cruise phase being lower and less demanding. If the flight begins in the early morning, the take-off phase might coincide with a time of low body temperature, associated with which might be a decline in crew alertness; if it begins late in the day, then the landing phase will correspond with extended hours awake. In both cases, there might be added difficulties due to previous sleep loss and to working at "night" by "body time," that is, by the time zone to which the body is adjusted.

The low workload associated with the cruise phase of a flight is not without its problems. Boredom and monotony can lead to inattention and even short sleeps, particularly if the individual has, for whatever reason, some degree of sleep loss.

There is indirect evidence to indicate that, in some circumstances, an air crew's lifestyle can be regarded as "stressful." For example, in some circumstances, there is an increased frequency of menstrual disorders in female cabin attendants (Preston et al., 1973; Suvanto et al., 1993). More directly, in military flights, there is an increased secretion of catecholamines – the "stress hormones" – in air crew following missions. Clearly, the relevance of this to civilian flights is unknown, but it might have some relevance when flying conditions are more difficult than normal – with poor weather, marked air turbulence or technical problems during the flight, for example.

Problems with Decrements in Sleep, Mood, and Mental Performance

These problems have been mentioned before, but they are important enough to warrant further discussion.

It is the fall in alertness, associated with which is a tendency to fall asleep, that is the main concern of those dealing with biological difficulties experienced by air crew. The tendency to doze off is one of those most frequently mentioned by air crew when asked about difficulties associated with long-haul flights. Some 20–30 years ago, such reports were made only in confidence but, since that time, it has become accepted that naps and sleepiness are a reflection of human physiology rather than indolence and irresponsibility. Laboratory-based simulations of the irregular schedules worked by air crew found a decreased efficiency of sleep, desynchronization between circadian rhythms, and declines in performance. These declines in performance depended on the type of mental task being performed and were attributed to the combination of working near the temperature trough, working extended hours, and cumulative sleep loss. However, the relationship between such results and those in the field has been questioned, particularly when the nature of the tasks required of pilots (a large amount of vigilance) and the relationship between poor performance in a laboratory test and the possibility of error or accident are considered.

In his review of the reports from the confidential reporting system for airborne "incidents," Green (1985) stressed the problems of sleep loss and sleep disruption, as have all reviews of the area (see, for example: Dinges, 1995; Gander et al., 1993; Graeber, 1982; 1989; Nicholson et al., 1984, 1986; Samel et al., 1995; 1997, Spencer et al., 1991; Stone et al., 1993; Wegmann & Klein, 1985; Winget et al., 1984). Sleep during layovers has been assessed from air crews' logs but also by objective measures, including the EEG, measurements of eye-blinking and wrist activity. In all studies, it is found that fatigue increases, that sleep is poorer and fractionated, and that sleep efficiency is decreased. The EEG measurements have also shown that "micro-sleeps" are taken during the flight if the individual is feeling tired. This is of more concern than the case of the individual who announces that he/she will take a nap, and it is an active area of research to try and warn individuals when unintentional lapses into sleep are about to happen. The problem is exacerbated by the observation that individuals do not seem able to predict reliably when the change from a feeling of tiredness – but the ability to continue to respond to the environment – will drift into a nap or micro-sleep.

The problem of sleep loss is worse after an eastward flight than after one to the west. Two factors are thought to contribute to this. Firstly, an eastward flight presents the opportunity of a sleep that is too early, and air crew are undecided whether to take this opportunity, to nap, or to forego sleep altogether. After a westward flight, by contrast, a delayed sleep is invariably taken (this is shown very clearly in one large study, Graeber et al., 1986). Secondly, adjustment to an eastward shift requires an advance of the body clock that is more difficult, possibly because the natural period of the body clock is slightly in excess of 24 hours. Moreover, during a tour of duty, which might involve several flights and days away from the home time zone, sleep loss tends to accumulate with successive days, and such "cumulative sleep loss" is more marked with age (Gander et al., 1993).

One component of the negative impact of sleepiness upon mood is an increase in irritability of individuals. This might affect work performance as well as personal relationships; an irritable crew might cooperate inefficiently whereas a motivated and friendly crew (unless it becomes too familiar and careless) will act the converse.

Performance testing is rare, for reasons of practicality, though studies made during air crew layovers or in a simulator tend to support predictions about decrements in performance following sleep loss and/or when measured at the trough of the circadian rhythm of body temperature. In one study made during flights (Samel et al., 1995), EEG measurements gave evidence of "critical decrements" in air crew during long-haul flights, particularly ones to the east.

In summary, air crews are victims of the combination of their human nature and duty schedules. The difficulties can be summarized as reductions in motivation and alertness, and problems in maintaining performance standards. These difficulties arise due to changes from normal in their sleep-wake cycles, and these are caused by the work schedules themselves and by the jet lag that results from time-zone transitions.

Advice to Air Crew

Given that the difficulties arise from the combination of flight schedules and human biology, there would be an argument for considering each in turn. Flight-time limitations are placed on air crew with regard to the number of hours worked, whether this is per duty period or for some other period of time (a week or year, for example). Current understanding of the human factors that are involved can form the rational basis for giving advice to air crew with regard to behavioral changes that can reduce the difficulties. Crew Resource Management training includes modules covering stress and fatigue and crew interaction. Also, the use of Standard Operating Procedures assists in ameliorating the effects of mood, stress and fatigue.

Just as experienced flight passengers and shift workers are aware of many of the difficulties involved and have often devised ways to deal with them, so, too, air crew learn by experience how to reduce the inconvenience that they suffer. The account that follows summarizes this body of information, as well as summarizing the ways of decreasing the difficulties that have been described, implicitly or explicitly, above. Many airlines are now beginning to make available to air crew handbooks that contain some of this information, and the whole area is one in which research is being actively pursued.

Maintaining Motivation and Performance on Duty

As time spent on a single task increases, so does the likelihood of boredom and of making an error, particularly if the task is repetitive or monotonous. The problems will be more marked if the individual is working in the middle of the night (by body time) and/or is sleep deprived. When the task is particularly demanding, then continued performance is likely to become stressful, again with an increase in the number of errors made.

In all cases, a rest from the task of an hour or so will improve performance when returning to the task. During the rest time, relaxation, or even a nap (see below), appear to be beneficial. However, often a task cannot be left; in this case, it is worth considering if it is possible to exchange tasks with somebody else – "a change is as good as a rest."

The Role of Naps, Caffeine, Melatonin, Alcohol, and Other Drugs

Sleep does not have to last for 6 hours or more to be beneficial. Shorter sleeps or naps of ½-2 hours are valuable in replacing lost sleep (Bonnet & Arand, 1994; Nicholson et al., 1985; Rogers et al., 1989; Rosekind et al., 1992, 1995). After the individual has fully woken up, there is an improvement in performance and mood. The time at which a nap is taken can vary, and this will influence the role that it plays, though, in practice, these roles often overlap considerably.

The scientific literature indicates that air crews regularly take naps. These naps can be taken just after a duty period (a "recuperative" nap), just before one (a "prophylactic" nap), when it is night by "body time" (the normal time for sleeping), or during the night on home time (an "appetitive" nap acting as "anchor sleep" to prevent adjustment of the body clock). Whether or not a nap should be taken as part of a scheme to prevent adjustment of the body clock will be considered below.

Caffeine (at a dose comparable to that obtained from about 2 cups of coffee) has been shown to produce a temporary improvement in alertness during a night without sleep in laboratory studies (Walsh et al., 1995; Wright et al., 1997). Coffee is routinely drunk by air crew, but there do not seem to be data on the normal frequency of intake.

The use of melatonin, both as a natural soporific (or hypnotic) and as a chronobiotic (a substance that promotes adjustment of the body clock), is problematical. There is much scientific literature attesting to the efficacy of this substance (Haimov & Arendt, 1999), but this is based upon studies of passengers rather than air crew. The frequency of its use by air crew is unknown. At least some airlines have refused to support regular use of it by air crew. For example, an unpublished communication prepared for air crew in 1994 included the following statements: "In the light of what is now known about melatonin, its use by flight and cabin crew is not recommended. Because melatonin will cause sleepiness and impair performance immediately after ingestion, and because the after-effects could be detrimental to operational efficiency, the use of this compound less than 12 hours before the start of a flying duty period and on board an aircraft should be forbidden."

Possible reasons for the lack of support for the use of melatonin include (Waterhouse et al., 1997):

a. Pure melatonin is difficult to obtain in many countries except by prescription.
b. Whilst melatonin might promote sleep, it is uncertain if its effects have worn off by the next day. That is, is mental performance improved the next day, or is there still some residual decrease in alertness and mental performance?
c. The long-term effects of melatonin ingestion are unknown.

d. Since it is common for air crew to return to their home time zone after only a short layover at their destination, and they tend to retain sleeping habits appropriate to their home time zone (see below), the chronobiotic properties of melatonin might be inappropriate.
  e. Whenever melatonin is to be used as a chronobiotic, its time of ingestion relative to the phase on the body clock is critical. Unless the individual has been in the same time zone and not working at night for at least seven consecutive days, the phase of the body clock might not be synchronized with the environment, and there is no easy way to assess what this phase might be (Turner & Stone, 1997).

Alcohol has several negative effects upon performance and mood and an individual's ability to perform a complex task reliably. Its use by air crew during and just before flights is forbidden.

Various drugs, including amphetamines, pemoline, and modafinil, have been shown to increase alertness after waking and sustain performance during extended periods without sleep (reviewed by Åkerstedt & Ficca, 1997). Their use, even for military operations, is controversial, and would not be appropriate in civil aviation. Apart from their ability to promote wakefulness, they have several disadvantages. Firstly, their use can be counterproductive since they also reduce the ability to initiate and sustain sleep (and it is sleep loss that is the civil air crew's main concern). Secondly, their extended use can lead to drug dependence. Thirdly, the full implications of their effects on mood and mental performance have not been investigated in sufficient detail.

The unpublished communication (of 1994, referred to above) concluded: "The use of sleeping pills and alertness enhancing drugs appears not to be significant among commercial pilots."

Napping on the Flight Deck

It is now recognized that naps are also taken during the cruise phase of flights. At this time, the workload is smaller and so air crew will have more difficulty in shrugging off the effects of tiredness, particularly if the flight is during the night by body time and/or the individual has lost sleep previously. Unambiguously informing the rest of the crew is imperative, of course. Indeed, it is essential that, if one pilot on a two-crew operation plans to take a nap, procedures are in place to ensure that the other pilot does not fall asleep, and that the risks associated with sleep inertia are also taken into account.

The benefits for air crew of scheduled in-flight sleeps were investigated by Rosekind et al. (1992). Such naps improved subjects' performance at tests of reaction time and attention, and there were fewer attention lapses (the test were performed in-flight during a period of low workload) in comparison with air crews who took no naps.

In this last study, the problem of "sleep inertia" was described, and it seems important that provision for its effects is built into any nap that is taken while on board or immediately before a flight. At least 15 min is required after a nap, and longer after a full sleep. In practice, therefore, it would be advisable for an in-flight nap to end with at least 15 min of being awake and relaxing before duties are resumed.

In summary, based solely on physiological considerations, naps can be extremely beneficial and, in general, the longer the nap the greater the benefit; once the effects of sleep inertia have worn off. However, operational considerations may dictate that naps are limited, or even that they are not permitted, for non-augmented crews. Indeed, if there are situations where a minimum crew is required to plan in-flight napping in advance, this raises the question whether such flights should routinely carry an additional crew member, with the provision of in-flight rest facilities.

Crew Augmentation

On the longer, more demanding flights, the crew is often augmented by an additional crew member. Indeed, such provision may be required under flight-time regulations, depending on the duration of the flight duty period. Typically, the main crew members will operate during the take-off and landing but, during the cruise phase, they will alternate with the additional, "relief" crew member. Under this arrangement, it is important that the allocation of rest to the individual crew members ensures that they are sufficiently alert during their time at the controls.

Often an equal division of the cruise phase into three rest periods will be appropriate, with the pilots taking their periods of rest alternately. If the relief pilot takes the first rest period, this will ensure that the main crew will have had a more recent opportunity for rest prior to the critical landing phase. However, sometimes an unequal allocation of rest may be preferable, for example when one of the rest periods occurs at a time when none of the pilots is sufficiently sleepy to benefit from the rest. As an alternative, the cruise phase may be divided into six sections, with the crew members each taking two rest periods. The advantage of this arrangement is that each pilot has a second opportunity to obtain some sleep, if the first was unsuccessful.

On the longest flights, a double crew is carried, and the crews alternate their rest in pairs. The cruise phase may be divided into two or four sections, depending on whether the crews take one or two rest periods. Otherwise, similar considerations apply as for the three-crew situation. On some very long flights, such as those with the new Airbus 340-500, it may even be appropriate for one or both of the crews to take three different rest periods.

One question that is often asked is whether the organization of the rest periods, which is the responsibility of the Captain in Command, should be planned in advance of the flight. This would enable the pilot(s) to whom the first rest period is assigned to restrict their sleep prior to departure, with the aim of achieving better sleep on the flight. This may work well in theory, but there is a risk that the pilots may still fail to achieve any sleep on the aircraft, and become extremely sleepy during the remainder of the flight when they are back at the controls.

In-flight sleeps may be taken on the flight deck, in passenger seats or in bunks separated from the rest of the air crew and the passengers. The choice will depend in part upon the number of crew and the amount of augmentation that is present.

To Adjust or Not to Adjust to Destination Time?

This is a complex issue. In most cases, flight schedules do not allow sufficient time for adjustment to the new time zone to take place. Accordingly, many air crews choose to remain on home time as much as possible during the tour of duty (Carvalho Bos et al., 2003; Lowden & Åkerstedt, 1998), particularly if the tour is of short duration and the destination is to the east of the home time zone (so requiring a phase advance of the body clock). This is illustrated by Figure 16.9, for example. To retain home-time whilst away requires at least some sleep in a darkened room to be taken during the night by home-time. Such a sleep acts as an "anchor sleep"; the rest of the sleep requirement can be naps at times that are convenient to the individual. Exposure to bright light outdoors should also be limited where possible to times when there is daylight in the home-time zone.

This regimen is unlikely to be compatible with sightseeing or normal social activities, and it is not possible if flight schedules (generally, day-time shuttle flights) are incorporated into the layover days. Moreover, there are often difficulties obtaining restful sleep during the day in the hotel accommodation provided. In these circumstances, at least partial adjustment of the body clock is a possibility. This will occur following the appropriate timing of exposure to, and avoidance of, bright light. Theoretically, adjustment could be promoted also by ingestion of melatonin at suitable times (Haimov & Arendt, 1999), and sleep and performance could both be enhanced by the use of hypnotics and drugs that enhance alertness (Åkerstedt & Ficca, 1997; Stone & Turner, 1997; Walsh et al., 1995). In practice, however, possibly due to current ignorance about possible side-effects and long-term effects of these substances, airline authorities do not recommend them (see above).

When attempting to adjust the body clock, bright light in a 6-hour "window" before the temperature minimum delays the body clock, and in a 6-hour window after this minimum, advances it (Figure 16.2). In addition, light should be avoided at those times that produce a shift of the body clock in a direction opposite to that desired (Czeisler et al., 1989; Minors et al., 1991). Implementing this information requires knowledge of body time and, particularly, of the time of the minimum of the body temperature rhythm. It can be assumed to be about 05:00 h (see Figure 16.1) if the individual has been in the time zone that is being left for at least the last seven days. Table 16.1 translates body time into those local times in the destination time zone when light should be sought or avoided on the first day after a time-zone transition. After the first day, when partial adjustment of the body clock has occurred, the individual is then advised to alter the timings of light exposure and avoidance

Table 16.1   Body Clock

	Bad local times for exposure to light	Good local times for exposure to bright light
Time zones to the west		
4h	01:00-07:00[a]	17:00-23:00[b]
8h	21:00-03:00[a]	13:00-19:00[b]
12h	17:00-23:00[a]	09:00-15:00[b]
16h	13:00-19:00[a]	05:00-11:00[b]
Time zones to the east		
4h	01:00-07:00[b]	09:00-15:00[a]
8h	05:00-11:00[b]	13:00-19:00[a]
10-12h	Treat these as 14-12h to the west, respectively[c]	

Notes
a this will advance the body clock (see Figure 16.2)
b this will delay the body clock (see Figure 16.2)
c this is because the body clock adjusts to large delays more easily than to large advances

towards the local light-dark cycle by one hour per day. In this way, the individual's exposure to light gradually becomes synchronized with that of the locals.

Intuitively, adjusting as fully as possible to the lifestyle and habits of the locals in the new zone would seem to be most appropriate. For westward flights, this intuition is correct, and can be illustrated by the example of a westward flight through eight time zones. To delay the clock requires exposure to bright light at 21:00-03:00 h *body time* and avoidance of it at 05:00-11:00 h *body time* on the first day. By new local time (see Table 16.1), this schedule becomes equal to 13:00-19:00 h for bright light exposure and 21:00-03:00 h for bright light avoidance, staying indoors in dim light. (On the second day in the new time zone, light exposure at 14:00-20:00 h, and light avoidance at 22:00-04:00 h are required, and so on). Exposure to the new local day-night cycle – by getting "out and about" and mixing with the local people – would satisfy these requirements. That is, exposure to the new environment promotes adjustment of the body clock in the appropriate (delaying) direction, and the times of minimum temperature, poorest performance and maximum sleepiness all become later and soon coincide with the evening or night on local time. This is one reason why adjustment to westward time-zone transitions is not found to be too arduous. (However, such light exposure is undesirable if the individual wishes the body clock to retain home time, as considered above).

By contrast, after a flight to the east, immediately adopting the times of exposure to and avoidance of light in the new environment can be the wrong approach. Consider a flight to the east through eight time zones, when a phase advance of the body clock is being attempted. This (see Table 16.1) requires bright light exposure at 05:00-11:00 h *body time* (13:00-19:00 h local time) and dim light (light avoidance) at 21:00-03:00 h *body time* (05:00-11:00 h local time) on the first day. It also requires these times to be advanced by 1 hour per day on subsequent days. It can be seen that morning light for the first day or so would be unhelpful and tend to make the clock adjust *in the wrong (that is, in a delaying) direction* – though light in the afternoon and evening is fine. The problem is particularly marked on those occasions when the crew has flown overnight, flies into the sunlight of the new day, and lands in the new time zone in the morning by local time. Indeed, laboratory simulations point to the likelihood that the body clock will be delayed by this schedule, unless suitable precautions are taken (see below).

Accepting that, after an eastward flight, exposure to the natural environment will tend to produce a phase delay (if the landing is in the morning by local time), it follows that full adjustment will require a substantial phase delay (of 16 hours after an eastward flight across 8 time zones, for example) and that this will take some days to achieve. The alternative, indicated in Table 16.1, is to attempt to adjust light exposure/avoidance to promote a phase advance of the body clock. As indicated by this table, and discussed above, a phase advance requires avoidance of bright light in the new morning. The effects of arrival in daylight in the morning by local time can be opposed if the bright light is avoided by wearing dark sunglasses until the shelter of a dimly-lit hotel room is reached, and sleep in a darkened room is attempted. It must also be realized that, if such a phase advance of the body clock is to be attempted, the time of minimum body temperature (and, therefore, of minimum alertness and maximum fatigue) will advance through the afternoon and morning on subsequent days. The effects will be noticed more in this case than in the case where a delay is attempted since, with a delay, the time of poorest performance will delay through the afternoon and evening, and fit in better with relaxing at these times.

It is advised that times when outdoor light is to be sought can be coupled with bouts of activity, brisk walking, swimming or jogging, for example. This helps some individuals combine the structuring of their day with appropriate light exposure.

Clearly, the air crew is presented with a set of options that are equally challenging. This, coupled with the greater difficulties associated with sleep, explains why adjustment to eastward flights is more difficult than to westward flights, and why eastward flights cause more difficulties for those involved.

The issues of adapting to local time are more complex for air crew than for passengers who may spend several days or weeks in their new destination. Air crew are likely to have only a few days back at base before undertaking another long-haul flight, possibly to an entirely different destination. If crew members have adapted, even partially, to local time, they may not have time to readapt to home time before they depart once again. In these circumstances they could be in an almost perpetual state of circadian desynchronization. The advantages of not adapting to local time are therefore considerable and for this reason a remote eastern destination, where adaptation is difficult, may be preferable to one in the west. It must be remembered that, if individuals succeed in adapting to local time (more likely after a westward flight, see Figure 16.8), they then face the problems of readapting after the return eastward flight.

If flights include a substantial shift in latitude, particularly flights towards the Arctic and Antarctic, then there will be an added problem due to the changing number of daylight hours. Furthermore, if the flight crosses the equator, there is a change in season and whether or not daylight saving time is in operation. Exposure to the climate and light-dark schedule in the new environment can cause disorientation, therefore, whether or not there has been a time-zone transition.

References

Åkerstedt, T. & Ficca, G. (1997). Alertness-enhancing drugs as a countermeasure to fatigue in irregular work hours. *Chronobiology International*, *14*, 145–158.

Åkerstedt, T. & Folkard, S. (1995). Validation of the S and C components of the three-process model of alertness regulation. *Sleep*, *18*, 1–6.

Åkerstedt, T. & Gillberg, T. (1981). The circadian variation of e x p e r i m e n t a l l y displaced sleep. *Sleep*, *4*, 159–169.

Bonnet, M. & Arand, D. (1994). The use of prophylactic naps and caffeine to maintain performance during a continuous operation. *Ergonomics*, *37*, 1009–1020.

Cabon, Ph., Coblentz, A., Mollard, R. & Fouillot, J. (1993). Human vigilance in railway and long-haul flight operation. *Ergonomics*, *36*, 1019–1033.

Campbell, S. (1992). The timing and structure of spontaneous naps. In: C. Stampi (ed.), *Why we nap* (pp.71–81). Boston: Birkhauser.

Carvalho Bos, S., Waterhouse, J., Edwards, B., Simons, R. & Reilly, T. (2003). The use of actimetry to assess changes to the rest-activity cycle. *Chronobiology International*, *20*, 1039–1059.

Civil Aviation Authority (2004) *The Avoidance of Fatigue in Air crews, Guide to Requirements*, CAP 371, 4th edition, Civil Aviation Authority, Gatwick, West Sussex.

Clayton, J., Kyriacou, C. & Reppert, S. (2001). Keeping time with the human genome. *Nature*, *409*, 829–831.

Czeisler, C., Kronauer, R., Duffy, J., Jewett, M., Brown, E. & Ronda, J. (1989). Bright light induction of strong (Type 0) resetting of the human circadian pacemaker. *Science*, *244*, 1328–1333.

Dijk, D.-J. & Czeisler, C. (1995). Contribution of the circadian pacemaker and the sleep homeostat to sleep propensity, sleep structure, electroencephalographic slow waves, and spindle activity in humans. *Journal of Neuroscience*, *15*, 3526–3538.

Dinges, D. (1992). Adult napping and its effect on ability to function. In: C. Stampi (ed.), *Why we nap* (pp. 118–134). Boston: Birkhauser.

Dinges, D. (1995). An overview of sleepiness and accidents. *Journal of Sleep Research*, *4 Suppl. 2*, 4–14.

Folkard, S. (1990). Circadian performance rhythms: some practical and theoretical implications. *Philosophical Transactions of the Royal Society of London, Series B*, *327*, 543–553.

Folkard, S. & Åkerstedt, T. (1991). A three process model of the regulation of alertness and sleepiness. In: R. Ogilvie & R. Broughton (eds.), *Sleep, Arousal and Performance: Problems and Promises* (pp. 11–16). Birkhauser, Boston.

Gander, P., Nguyen, D., Rosekind, M., & Connell, L. (1993). Age, circadian rhythms and sleep loss in flight crews. *Aviation, Space and Environmental Medicine, 64,* 189–195.

Graeber, R. (1982). Alterations in performance following rapid transmeridian flight. In: F. Brown & R. Graeber (eds.), *Rhythmic Aspects of Behaviour* (pp. 173–212). Hillsdale and London: Lawrence Erlbaum Associates.

Graeber, R. (1989). Jet lag and sleep disruption. In: M. Krugger, T. Roth & C. Dement (eds.), *Principles and Practice of Sleep Medicine,* (pp. 324–331). Philadelphia/London: WB Saunders Co.

Graeber, R., Lauber, J., Connell, L., & Gander, P. (1986). International air crew sleep and wakefulness after multiple time zone flights: a co-operative study. *Aviation Space and Environmental Medicine, 57, Suppl.,* B3-B9.

Green, R. (1985). Stress and accidents. *Aviation, Space and Environmental Medicine, 56,* 638–641.

Haimov, I. & Arendt, J. (1999). The prevention and treatment of jet lag. *Sleep Medicine Reviews, 3,* 229–240.

Lewy, A. & Sack, R. (1996). The role of melatonin and light in the human circadian system. *Progress in Brain Research, 111,* 205–216.

Lowden, A. & Åkerstedt, T. (1998). Retaining home-base sleep hours to prevent jet lag in connection with a westward flight across nine time zones. *Chronbiology International, 15,* 365–376.

Minors, D., Waterhouse, J. & Wirz-Justice, A. (1991). A human phase-response curve to light. *Neuroscience Letters, 133,* 36–40.

Moline, M., Pollack, C., Monk, T., Lester, L., Wagner, D., Zendell, S., Graeber, R., Salter, C. & Hirsch, E. (1992). Age- related differences in recovery from simulated jet-lag. *Sleep, 15,* 28–40.

Moore, R. (1995). Organisation of the mammalian circadian system. In: D. Chadwick & K. Ackrill (eds.), *Circadian Clocks and their Adjustment* (pp. 88–106). Ciba Foundation Symposium, 183. Chichester: Wiley.

Naitoh, P. (1992). Minimal sleep to maintain performance: the search for sleep quantum in sustained operations. In C. Stampi (ed.), *Why We Nap,* (pp. 199–216). Boston: Birkhauser.

Naitoh, P., Kelly, T. & Babkoff, H. (1993). Sleep inertia: best time not to wake up? *Chronobiology International, 10,* 109–118.

Neri, D. (2004). *Proceedings of the Fatigue and Performance Modeling Workshop,* June 13-14, 2002, Seattle, Washington. *Aviation, Space and Environmental Medicine, 75 (3), Section II, Suppl.,* A1-A199.

Nicholson, A., Pascoe, P., Roehrs, T., Roth T., Spencer, M., Stone, B., & Zorick, F. (1985). Sustained performance with short evening and morning sleeps. *Aviation, Space and Environmental Medicine, 56,* 105–114.

Nicholson, A., Pascoe, P., Spencer, M., Stone, B., & Green, R. (1986). Nocturnal sleep and day-time alertness of air crew after transmeridian flights. *Aviation, Space and Environmental Medicine, 57,* B42-B52.

Nicholson, A., Stone, B., Borland, R., & Spencer, M. (1984). Adaptation to irregularity of rest and activity. *Aviation, Space and Environmental Medicine, 55,* 102–112.

Preston, F., Bateman, S., Short, R., & Wilkinson, R. (1973). Effects of flying and of time changes on menstrual cycle length and on performance in airline stewardesses. *AerospaceMedicine, 44,* 438–443.

Reilly, T. & Piercy, M. (1994). The effect of partial sleep deprivation in weight-lifting performance. *Ergonomics, 37,* 106–115.

Reilly, T., Atkinson, G., & Waterhouse, J. (1997). *Biological Rhythms and Exercise.* Oxford: Oxford University Press.

Rogers, A., Spencer, M., Stone, B., & Nicholson, A. (1989). The influence of a 1 hour nap on performance overnight. *Ergonomics, 32,* 1193–1205.

Rosekind, M., Graeber, R., Dinges, D., Connell, L., Pountree, M., Spinweber, C., & Gillen, K. (1992). Crew factors in flight operations: IX effects of planned cockpit rest on crew performance and alertness in long-haul operations. *NASA Technical Memorandum, 103884.*

Rosekind, M., Smith, R., Miller, D., Co, E., Gregory, K., Webbon, L., Gander, P., & Lebacqz, J. (1995). Alertness management; strategic naps in operational settings. *Journal of Sleep Research, 4 Suppl. 2,* 62–66.

Samel, A., Wegmann, H., & Vejvoda, M. (1995). Jet-lag and sleepiness in air crew. *Journal of Sleep Research, 4 Suppl. 2,* 30–36.

Samel, A., Wegmann, H., Vejvoda, M., Drescher, J., Gundel, A., Manzey, D., & Wenzel, J. (1997), Two-crew operations: Stress and fatigue during long-haul night flights. *Aviation, Space Environmental Medicine, 68,* 679–687.

Spencer, M. (1987). The influence of irregularity of rest and activity on performance: a model based on time since sleep and time of day. *Ergonomics, 30,* 1275–1276.

Spencer, M., Stone, B., Rogers, A., & Nicholson, A. (1991). Circadian rhythmicity and sleep of air crew during polar schedules. *Aviation, Space and Environmental Medicine, 62,* 3–13.

Stone, B., Spencer, M., Rogers, A., Nicholson, A., Barnes, R. & Green, R. (1993). Influence of polar route schedules on the duty and rest patterns of air crew. *Ergonomics, 36,* 1465–1477.

Stone, B. & Turner, C. (1997). Promoting sleep in shiftworkers and intercontinental travellers. *Chronobiology International, 14,* 133–143.

Suvanto, S., Härma, M., Ilmarinen, J., & Partinen, M. (1993). Effects of 10 h time zone changes on female flight attendants' circadian rhythms of body temperature, alertness and visual search. *Ergonomics, 36,* 613–625.

Turner, C. & Stone, B. (1997). *Melatonin: recommendations concerning its use by air crew.* DERA report DERA/CHS/PP5/CR/97093/1.0, QinetiQ, Farnborough.

Walsh, J., Muehlbach, M., & Schweitzer, P. (1995). Hypnotics and caffeine as countermeasures for shiftwork-related sleepiness and sleep disturbance. *Journal of Sleep Research, 4 Suppl. 2,* 80–83.

Waterhouse, J., Edwards, B., Nevill, A., Atkinson, G., Reilly, T., Davies, P., & Godfrey, R. (2000). Do subjective symptoms predict our perception of jet lag? *Ergonomics, 43,* 1514–1527.

Waterhouse, J., Folkard, S., & Minors, D. (1992). *Shiftwork, Healthy and Safety. An Overview of the Scientific Literature 1978-1990.* London: HMSO.

Waterhouse, J., Minors, D., Åkerstedt, T., Reilly, T., & Atkinson, G. (2001). Rhythms of human performance. In: J. Takahashi, F. Turek, & R. Moore (eds.) *Handbook of Behavioral Neurobiology: Circadian Clocks* (pp. 571–601). Kluver Academic/Plenum Publishers, New York.

Waterhouse, J., Minors, D., Waterhouse, M., Reilly, T., & Atkinson, G. (2002). *Keeping in Step with Your Body Clock.* Oxford: Oxford University Press.

Waterhouse, J., Reilly, T., & Atkinson, G. (1997). Jet-lag. *Lancet, 350,* 1611–1616.

Wegmann, H. & Klein, K. (1985). Jet-lag and air crew scheduling.In: S. Folkard & T. Monk (eds.) *Hours of Work* (pp. 263–276). Chichester: John Wiley.

Winget, C., deRoshia, C., Markley, C., & Holley, D. (1984). A review of human physiological and performance changes associated with desynchronosis of biological rhythms. *Aviation, Space and Environmental Medicine, 55,* 1085–1096.

Wright, K., Badia, P., Myers, B., & Plenzler, S. (1997). Combination of bright light and caffeine as a countermeasure for impaired alertness and performance during extended sleep deprivation. *Journal of Sleep Research, 6,* 26–35.

Chapter 17

Legal Aspects of Aviation Health: The Changing Landscape[1]

D Anthony Frances

Introduction

Aviation and airlines have a unique history. Having just celebrated 100 years of powered flight, aviation has been one of the most intriguing and interesting modes of transport in the modern world. It has been stated by many governments and professional organizations that it is *"one of the safest forms of transport,"* although recent events, most notably the loss of Helios Air B737-300,[2] the incident of the Air France A340-300[3] and the September 11 attacks on the World Trade Center in New York have changed people's attitudes towards traveling by air and most importantly the inherent risks of flying.[4] These dramatic events have in turn, caused great concern among flight crews and passengers about the dangers of flight.

Adjusting our perception of safety on the ground and in the air appears to form the basis of recent media stories that suggest there are potentially increasing dangers to passengers and air crew. These potential dangers include Deep Vein Thrombosis (more commonly known as "DVT"), poor air circulation due to aircraft operations and the effects of cosmic radiation. This physiological hyper-awareness, when added to the danger of mechanical failure and in particular terrorist-induced injury, has produced within passengers and crew psychological conditions that have made some afraid to fly and others reluctant to put themselves at risk. Others have appeared to have insulated themselves against the physiological and psychological stresses by resorting to alcohol misuse or drug abuse both on the ground and while in flight. Too often, passengers, and on occasion flight crew have been affected by alcohol and drugs before boarding an aircraft. Since pilots and those who hired them have a higher duty of

---

1 The views expressed in this chapter are those of the author and do not necessarily reflect that of the partnership of Clyde & Co.

2 Incident of a Helios Air B737-800 near Athens on 14 August 2005, with the loss of all aboard.

3 Incident of an Air France A340-300 at Toronto Airport, August 2005 where the aircraft slid off the runway and crashed into a nearby ditch – all passengers and crew survived.

4 See article entitled "Is flying still the safest way to travel?" www.bbcnews.com 24 August 2005.

care for those they serve, evidence of psychological issues and their resulting effects on both professional and private lives among air crew has on occasions, sparked liability suits in the courts. While these types of cases do occur, they are not widely publicized or reported by the popular media.

This chapter outlines the legal issues regarding flight crew and associated passenger health and its relationship with commercial aviation and airline operations. It is not intended to be an exhaustive exposition of all the relevant laws and regulations with respect to carriage by air, but rather an indication of the legal liabilities and scenarios that modern aviation has had to deal with since the unfortunate loss of Pan Am 103 *Maid of the Seas* over Lockerbie, Scotland in 1988 and more recently, the events of September 11, 2001. Like most industries, aviation has had to adjust to the advent of new technologies and an ever-changing society, as more people use aviation as a common form of transportation. Despite the ever-increasing amount of air traffic, further pressures have been placed upon flight crews, air traffic controllers, and all associated staff with respect to airline and aircraft operations, which has on isolated occasions, led to fatal outcomes.

Airline's Perspective and Expectations

Aviation has changed remarkably in the last 20 years. As the old-style predominantly state-run airlines operations (such as Olympic Airways[5] and Alitalia) begin to make way for the newer low-cost carriers, such as Easyjet, Southwest, Virgin Blue, Flybe, BMIBaby and Ryanair, the stresses and strains on flight crew, passengers and those associated with the industry have become more prominent. In addition, crews have had to cope with the need to maintain and hopefully increase profitability of the airline, while at the same time, ensuring that their services operate efficiently and on time.

Despite the portrayal of airline travel as an *experience*, the modern role of the airline is simply to carry passengers from point A to point B. The flying experience appears to be virtually the same for pilots, flight attendants, and passengers, whether one travels on scheduled operators, such as British Airways or Qantas Airways, or charter operators such as Britannia Airways or Air 2000. However, a brief analysis of these two categories of flight may be important, when one considers the matrix of potential legal liability of operators for health and behavior issues regarding flight crew and passengers.

As will become apparent in this chapter, crew and passenger health and behavior varies depending upon the circumstances of flight – ranging from the *internal* conditions of the aircraft during flight, to the additional extras that airlines provide for both crew and passengers. Factors such as internal cabin space and seat pitch, flight deck set-up, air flow through the cabin and the general well-being of crew and passengers are factors that lead to determining the way crew and passengers may

---

5   The national airline of Greece, which is in the process of a privatization process.

behave aboard aircraft. In addition, the flight itself will often have an effect on the health and behavior of crew and passengers. All too often, severe turbulence, engine fires, technical failures or other occurrences make flight crew as well as passengers, fearful and in turn, prone to behavior changes, mood swings and the *perceived loss of control*, due in large part to unfamiliar conditions that are inherent parts of air travel. When considering the health of flight crew and passengers, one must also consider the potential and actual psychological effects of the relationships between the individuals on the flight and cabin crew, and those most relationally significant to them. Relationships between the parties involved, as they apply to the circumstances of a flight incident or accident, may affect the liability matrix in the event of a claim. It is therefore important to consider the operational aspects of any aircraft incident, in order to identify the key causal nexus.

When considering the health and behavior of air crew and passengers, one must consider the expectations, services and experiences of the commercial airliners or operator as these relate to the well-being of air crew and passengers. Certain carriers have good reputations for treatment of their staff, flight crews and passengers. Notwithstanding the importance of health and behavior issues, one must also consider carefully how promoting good health and appropriate behavior can be achieved while not ignoring the principal objective of flight safety.

It is to these points that this chapter now turns in greater detail.

Flight Crew's Perspective

The airline crew's role has significantly changed in recent years from one of concentrating on providing an air transportation service to ensuring a higher degree of safety and security for air and ground operations of commercial aircraft. This additional dimension has had an effect on the health and behavior of air crew.

It is arguable that in the last 20 years and as airlines have become more financially tighter, more strain has been placed upon flight crew to perform and maintain the quality and service, despite more rigorous schedules and demands placed upon them by airlines and aircraft operators.

Hostilities from Passengers

There have been a number of recent instances where flight and cabin crew members have been involved in incidents that have forced these employees to perform their duties in an extremely hostile environment. Incidents such as attempted interference with an aircraft while in flight,[6] frequent and persistent bomb threats, such as those experienced by Olympic Airways (for flights predominantly bound for London

---

6   See incident on board a British Airways B747-400 near Nairobi, Kenya 2003 where a passenger attempted to interfere with the aircraft during flight.

from Athens, Greece), disorderly and rude conduct by intoxicated passengers[7] and the ever occurring risk of hijacking and air piracy have all contributed to increased anxiety and an ever-increasing strain on pilots and flight crew. These factors have in turn, arguably interfered with the flight crew's ability to safely perform their flight duties and in some case, have contributed to aviation accidents or incidents.

Hostilities from within the Airline

There have also been a number of incidents involving flight crew where the alleged misconduct was directly related to internal hostility between airline employees. One prominent example of this type involved the loss of Egyptair 990 (a Boeing 767-300 aircraft) en route from New York to Cairo in 1999. The alleged deliberate placing of the aircraft into an uncontrolled dive leading to the destruction of the aircraft was thought to be suicide or related to the financial problems of the First Officer, Gemeel El-Batouty. In contrast, the *Los Angeles Times* reported[8] on this same incident that, "... *a former Egyptair captain suggested a possible motive other than suicide ... the crash was an act of revenge against an Egyptair executive who was on the flight and had reprimanded El-Batouty for sexual misconduct.*" If these allegations are proven to be accurate, then arguably more stringent psychological profiling and health awareness of pilots by airlines and regulators may need to be considered and where appropriate, more regulated. This may in turn, become more of an issue in instances where an airline may be facing unlimited liability situations and the possibility of pecuniary damages. It is well accepted by airlines that contesting liability claims in the United States is generally a very expensive and complicated business.

Hostilities from Terrorists

There have also been a number of terror scares involving airlines that have led to further operational and occupational stresses on flight crew and passengers. For example, in August 2004, two Russian civilian aircraft were brought down by explosives believed to be carried on board by terrorists posing as innocent passengers. There have also been attempts to destroy in-flight aircraft using small explosives hidden on board.[9] Too often, there have been reports of flights being escorted by F-16 military aircraft in circumstances where there has been either a report of an attempted hijacking or some communication difficulty.[10] These instances would no doubt have effects upon the health of flight crew and passengers, but may also have

---

7   See article entitled "*Briton Faces 20 years for Jet Mayhem*", Daily Mail, 14 July 2005 located at www.aviation-health.com.

8   See article entitled "*Egyptair co-pilot crashed plane*" at www.bbcnews.com 21 March 2002.

9   See for example, the "shoe-bomber" Robert Reid who threatened to blow up an aircraft bound for the United States from Europe.

10 See article entitled "*Threats no pilot can ignore*" www.bbcnews.com 6 October 2004.

damaging effects on an airline's bottom line, due to that airline's reputation in the marketplace. There has been media speculation that some airlines more than others, are particularly susceptible to terrorist attention. In addition, numerous liability suits brought by the survivors or estate in fatal accident cases do also affect an airline's ability to operate and remain financially viable. There have been recent examples, especially in respect to the loss of the Pan Am B747-100 over Lockerbie, which, over time, affected the airline's bottom line.

Hostilities from the Work Environment and Onboard the Aircraft

Aside from those instances set out above, there are a number of other *occupational stresses* that flight crew endure. For example, the increased exposure and risk to contracting cancer through exposure to cosmic radiation. The question then becomes, what is management's responsibility (if any) to flight crew exposed to cosmic radiation?

Cosmic Radiation Medical researchers have pointed in research that sustained flight operations at high altitude may lead to an increased risk of cancer and eye damage, due to inadequate protection. These matters raise additional occupational health and safety issues for the airline or aircraft operator. For example, exposure to cosmic radiation has been rated to be equivalent to regular x-rays.[11] While there is no *instant* cause for concern from a liability viewpoint, as flight crews, airlines and their regulatory authorities become more aware of the hazards of their occupation and the potential threat of grounding due to the side-effects, it is anticipated that there will be flight crew demand for greater protection for their personal safety as part of their health and safety working environment. The European Union and relevant regulatory authorities (most notably the Civil Aviation Authority in the United Kingdom) have as a result, implemented regulations to ensure that flight crew are exposed to such radiation to a minimum.[12]

Air Quality of the Cabin There have also been issues regarding in-flight air circulation and toxic fumes being inhaled during flight.[13] A very recent tragic example was a Helios Air flight from Larnaca to Prague, via Athens, where the main cause of the crash appears to have been a loss of oxygen in the cabin.[14] There have been also a

---

11 See article entitled *"What Air crews should know about Occupational Exposure to Ionizin Radiation";* W Frieberg & K Copeland. Civil Aerospace Medical Institute at www. cami.jccbi.gov.

12 See European Union Council Directive 96/29 and Air Navigation Order 2000 SI 2000/1562 (UK).

13 See articles entitled *"In-Cabin Trace Chemicals and Crew Health Issues"*; Dr J C Balouet (Aerospace Medical Association 20 May 1998) at www.aopis.org and article entitled *"Leaking oil fumes threat to air crews"* The Sunday Times 10 April 2005 located www. aviation-health.com.

14 Accident near Athens, Greece: 14 August 2005. The was the main alleged cause, although no official report at the date of writing this chapter had been concluded.

number of reports of fume inhalation and toxic chemical affects on pilots, especially with respect to certain aircraft types, most recently reported by pilots flying the BAe 146 fleet.[15] Once again, these issues can have knock-on effects for flight crew health and can lead to liability claims against the airline and manufacturers, especially with respect to what actions and precautions the airline and manufacturers took in respect to these matters.

A recent example of further difficulties facing flight crews and the potential liabilities for airlines are infectious diseases brought on board by passengers who may or may not know they are so infected. Such was the case during the 2002–2003 SARS pandemic that quickly migrated from the Far East to Canada and other cities. Flights were canceled between known and potentially infected areas, profoundly upsetting airlines and their operations on a global level. The motivation to quarantine areas by cutting off air transportation was due to the health threat, but one cannot rule out the influence of the persistent liability issues attached to the health threat, those affecting flight crews and passengers alike.[16]

Crew Selection and Resource Management

Aside from potential foreign threats by way of disease or technical issue, one of the other difficult issues facing an airline is undertaking proper selection procedures for pilots. Issues such as operational imbalances, difference in gender and age and the cultural differences between crew need to be considered. Too often in the past, accidents have been assessed to have been caused by flight crew members failing to work as a team during flight emergencies. This is especially apparent where the Captain is very senior with many thousands of hours of flying time and the First Officer and/or Engineer are relatively junior in flying status.

This concept, also known as *cockpit* or *crew resource management* has been a flight issue since 1916. As human factors scientists became more interested in flight crew interaction in the late 1970s and 1980s, airlines and aircraft operators started requiring flight crew to participate in Crew Resource Management (also known as CRM) training courses. The urgency for the new CRM training was directly related to a number of accidents where the Captain had overruled other flight deck opinions voiced by the First Officer, Flight Engineer or other qualified flight crew. In those cases, where the flight crew did not properly manage their cockpit resources, the result was often tragic as a result of the senior pilot or pilot in command failing to appreciate that they may be wrong and in turn putting the aircraft and its passengers in danger.[17] Two

---

15 See article entitled "*Leaking oil fumes threat to air crews*" *The Sunday Times* 10 April 2005 located www.aviation-health.com.

16 See for example World Health Organization Circular, 15 March 2003 at www.who.int.

17 See for example, "*Making air travel safer through Crew Resource Management*" 2004 at www.psychologymatters.org and R. L. Helmreich et al. "*Culture, Error and Crew Resource Management*" University of Texas at Austin, Human Factors Research Project, 2001.

dramatic examples of failure to properly manage cockpit resources were the loss of the Air Florida B737[18] in Washington DC and the collision of the KLM and Pan Am B747s at Tenerife in 1977. A more recent example was the loss of the Korean Airlines Boeing 747 cargo freighter near London Stansted Airport in December 1999. In the Korean Airlines case, the reports[19] suggest that there was confusion on the flight deck regarding the reading of instrumentation, whereby the Captain overruled other members of the flight crew, which led in turn to the loss of control of the aircraft and subsequent crash into a hillside. All crew perished. Many commentators were of the view that authoritarian leadership styles and cultural reluctance to question authority are factors, which must be considered in any loss.

CRM training became an important matter for liability reasons, as airlines attempted to overcome the problems associated with pilot error and dealing with proper cross-checking techniques and practices. To this day, proper pairing of crew members is still an important personnel decision for airlines. Airline operators consider carefully these aspects, as they cannot hide behind excuses that they are unaware of the problems of mismanaging crew pairings. It is appreciated that flight crews do undergo special psychological screening and exhaustive flight training to ensure that only the best pilots are allowed to operate commercial aircraft and that only those pilots and cabin crew who embrace cooperative management of flight deck resources are allowed to form flight crews. In fact, many airlines make psychological profiling an important ingredient to a successful application for employment as an airline pilot.[20] Some airlines have made strategic decisions not to employ ex-airforce pilots, as their mindset is often incompatible with that of a commercial airline or operator. This active assessment of crews and vigilant pursuit of safe flying practices has helped air carriers with their risk management practices and in turn, reduce the potential exposure to third parties. The common practice amongst airlines of confidential reporting of instances whereby flight crew report on others for deficiencies or flouting procedures or rules is a way to regulate the behavior of flight crew to not only weed out those who are not performing to performance criteria, but also to address and identify operational and practical deficiencies of a crew member.

Not only must flight crews deal with their own environmental and occupational issues, but also the welfare and in rare cases, the mental health of the passengers in order to ensure the safe operation of the aircraft. It is to this, that we now turn.

---

18 The aircraft crashed into the Potomac River in 1982.
19 See article "*Safety of Korean Air under a Microscope*" Air Safety Week 17 January 2000.
20 See for example, "selection criteria" located on the Eurowings (an European airline) at www.eurowings.com.

Passenger's Perspective

Passengers on aircraft are mainly divided between those who are traveling for business reasons and those who for whatever reason desire air transportation over other modes to arrive at their destination. Of course, there are numerous sub-categories of passengers, but for the purposes of this chapter, only two types passengers will be considered. For the present discussion, non-frequent passengers tend to travel in economy or coach class, whereas frequent or business travelers tend to fly in business or first class.

Economy or Coach Class

For most people who do not travel often, a flight is an *adventure* and it is a chance to take advantage of the amenities, which in practical terms means food and drink, possibly including alcohol. The amount of freely moveable space one has, coupled with mandated confinement to one's assigned seat and obedience to security restrictions and when mixed with frustrated expectations of the flight, can often cause an array of problems for the flying passenger, other passengers, and affect the cabin crew tending to the needs and desires of the flying public. Long-haul flights are the most problematic, as passengers' minor annoyances often become blown out of proportion.[21] Those who suffer behavioral upset when in ideal settings, find the restriction and confinement of an aircraft to be unbearable. When alcohol is mixed with volatile mood swings, passengers can become combative and even violent. If the cabin crew are not in a position to quell the problem, other passengers will volunteer to *fix* the problem, which in turn can have further effects upon the safety of the aircraft.

It is noted that trans-oceanic flights are not the only breeding grounds for behavioral upset and violence among passengers. Even before the events of September 11, passengers have been suspicious of troublesome persons who threaten others. For example, during a domestic flight in the United States several passengers subdued a nineteen-year-old male who allegedly posed a threat to the lives of the passengers and the crew as he attempted to break into the cockpit on a Southwest Airlines flight. The young man allegedly died of suffocation as a result of the intervention and excessive use of force by crew and passengers.[22]

Business or First Class

Passengers who are frequent flyers and who travel mostly on business class have a different experience of flying. Such passengers tend to have the added space for

---

21 *United States v. Cordova* (1948), where two men brawled on a Flying Tigers, Inc. aircraft, with no deaths involved.
22 *Las Vegas Review Journal, Sept. 17, 2000*, Mr Burton dies after struggle with passengers.

seating, for moving around the aircraft and generally are busy with work commitments. To them, issues of flight revolve around ensured *smooth* service and arrival on time. It is not often that such passengers are the cause for concern for the flight crew. The real problems often erupt in the economy cabin, where cramped conditions and contaminated air confined within a comparatively smaller space are common complaints and experiences for those who fly, especially for long distances.

Alcohol also plays an important role in the flight experience. It is appreciated that economy passengers who do not fly often tend to become (on average) more intoxicated and disruptive due to the circumstances of flight and the fact that the alcohol is readily available. In addition, drunkenness, no smoking rules, presenting mental illnesses, abuse of flight crew, and interference with fellow passengers, as well as disputes with the flight crew are common problems leading to in some cases, assault charges and reckless endangerment of aircraft, especially where the passenger needs to be restrained. Such behavior is subject to and governed by the relevant aviation security and navigation legislation.[23]

Special attention to new security rules, taken together with the typical passenger problems that present during many flights, has arguably added further strain on flight crews. This behavior has arguably increased stress and anxiety that psychologically obstruct flight crew attention. In turn, these result in furthering endangerment of passengers on board, as the flight crew become distracted from their required employment responsibilities. This new in-flight battleground has alarmed airline insurance and risk managers in terms of limiting risk and prevention, when considering the airline's exposure and the underlying liability of the flight crew when performing their flight duties.

Specific Cases of Crew and Passenger Behavior

Flight crew and passenger health and behavior as outlined above also may lead to incidents on board aircraft, which often become very public.[24] Many airlines wish to avoid such incidents and the resultant publicity and in order to do so, have instigated a system of procedures for incidents involving flight crew and disruptive passengers. Such procedures have been set out in in-flight magazines and often are publicized on television and radio to ensure that passengers understand the legal consequences of their actions.

It has been common for airlines to even *red card* passengers who have offended on more than one occasion for not only disorderly conduct on board the aircraft, but also endangerment and safety violations during flight. These passengers are often then banned from that airline or in some cases, all airlines by way of court order.

---

23 See Aviation Security Act 1982 (as amended) and the Air Navigation Order 2000 (UK) for regulations and sanctions.

24 See for example, an incident involving Peter Buck from REM on board a British Airways flight from the USA to London.

Liability of Airlines/Passengers

The legal regime for airlines for the carriage of passengers where they suffer injury or death is made up of both domestic and international rules and regulations. In the United Kingdom, international carriage by air is set in the primary document, known as the Warsaw Convention (1929). This document outlines the liabilities of a carrier and also the extent to which a carrier (or airline) may be liable for damage sustained by a passenger during flight or during embarkation and disembarkation. This legislation has been incorporated into local law in England, by way of the Carriage by Air Act (1963) as amended.

The Warsaw Convention has been modified by way of additional Protocols and Conventions, which have clarified and amended certain provisions that were initially set out in the Warsaw Convention. In the United Kingdom, the Hague Protocol of 1955 has amended the Warsaw Convention and is the applicable legal regime for international carriage by air. In addition, the Montreal Convention 1999 has been recently ratified by the European Union and is therefore in operation. It applies to international carriage by air only between signatory states.

Although the Montreal Convention 1999 will only apply to two States who have ratified the Convention, the following are the main points in that respect:

a. Strict liability to passengers (for death or bodily injury) upon the carrier up to 100,000 Special Drawing Rights (SDRs) without proof of fault against the carrier;
b. Unlimited liability if a carrier cannot show that it (or its agent) was not at fault (in negligence) for death or injury to passengers;
c. The carrier is required to pay an interim amount to the passengers or their dependants in the case of death or injury in order to meet economic needs;
d. Entitlements for passengers to bring proceedings against a carrier where the passenger (suffering death or injury) had its principal or permanent place of residence ("the fifth jurisdiction");
e. New rules and regulations with respect to cargo, baggage and additional matters such as delay including their defences.

In the late 1990s, additional legal instruments have been incorporated into English law relating to carriage by air. The European Union has also issued Directives with respect to airline liability for European carriers and those who are flying to and from a European airport. In addition, some airlines have agreed amongst themselves to vary the liability limits above what has been agreed by way of international Convention. In these cases, special contracts have been drawn up and are issued to passengers by certain carriers through the issuance of their tickets. These variations to liability and their limits also need to be considered when ascertaining the legal liability of an airline to passengers and third parties.

With respect to disorderly behavior by passengers, there are a number of statutory rules that regulate the conduct of aircraft. The Aviation Security Act 1982 (UK) as amended and the Air Navigation Order 2000 (UK) regulate *inter alia*, the operation of an aircraft, powers of the commander of the aircraft, the offences for conduct and the sanctions for such behavior. In the current climate of terrorism, the sanctions are severe.

Unruly Passengers and Air Rage

In the last decade, the influx of a number of low cost carriers and the ease and convenience of travel has exposed many more people to airline travel than were envisaged. In this day, airline travel is not for those who can afford it, but for anyone. Access simply, is not an issue.

As a direct result of the increased ease and access of flight, coupled with the environmental influences of flight as set out in this chapter, there is a higher probability that the flying public will be involved in an incident, whereby unruly passengers act out in ways that put at risk the passengers and the flight crew. It is viewed that these incidents, which in turn have often been contributed by the peculiarities of flight, have led to the following string of cases and liability issues for consideration. These are a practical demonstration of the types of behavior that can affect a flight crew's health from an occupational and operational point of view.

a. In *R v Heather Susan Tagg* (2001),[25] the defendant appealed against a conviction and costs order as a result of being drunk while on a flight from New York to London. The appeal was refused by the court on the basis of the court's interpretation of the relevant legislation, which is not directly relevant to this discussion. However, this case should be considered in light of the court's comments regarding passenger behavior and aircraft safety. It was the court's view that a passenger's behavior where induced by alcohol in an aircraft environment must be considered in the light of aviation safety – this is a paramount consideration.

b. In *R v Julian Ayodeji* (2000),[26] the defendant appealed against a conviction for being drunk and endangering the safety of an aircraft in flight. The court stated that in the circumstances, the defendant embarked upon a course of action to make himself drunk and therefore he had no reasonable excuse for his actions. The court also considered the particular situation and the environment of the aircraft as being contributing factors.

c. In *R v Lawrence Charles Oliver* (1998),[27] the defendant appealed unsuccessfully against a custodial conviction for his unruly behavior while aboard an aircraft from Florida to Manchester. The defendant was ordered by cabin staff to remove his child from the cot and hold it due to the prevailing flight conditions. A dispute erupted and the defendant struck his wife. The crew then intervened

---

25 [2001] EWCA Crim 1230.
26 TLR 20/10/2000.
27 TLR 13/10/1998.

and the defendant became abusive to them. The court refused the appeal and held that due to the defendant's actions, they could have jeopardized the safety of the aircraft and safety of other passengers. Therefore, the actions of the defendant in that environment were serious enough to order an immediate custodial sentence.

d. In the case of *R v David James McCallum* (2001),[28] the defendant appealed against a sentence of 12 months imprisonment for an affray aboard an aircraft. The defendant was convicted for his unruly behavior aboard an aircraft from Newcastle to Tenerife[29] as a result of being drunk.

The cases above indicate the type of behavior that a flight crew may have to endure and the possible legal consequences not only for the passengers, but also the related operational (and in some cases, mental health) effects experienced by the flight crew. In addition, issues of legal liability and the right to sue passengers for common assault and personal injury also need to be considered for any liability analysis.

Hijacking and Terrorism

The last few years and in particular, the events of September 11 on the World Trade Center (also know as the "WTC") have made airlines more conscious of flight crew and passenger health and behavior. Airlines have now imposed new regulations, equipped crew with repellent materials (and in some jurisdictions firearms and devices) and reinforced the flight crew doors from passengers. In addition, many airlines (as a result of governmental pressure) have employed air or sky marshals to act as armed guards aboard the aircraft during flight. These measures in turn, have caused flight crew and passengers more anxiety, stress and an increased consciousness of the lethal environment in which they travel.

Notwithstanding the above discussion, persons have used aircraft as a method to obtain publicity and in some cases, recognition for their cause against foreign governments, including those seeking asylum. This is demonstrated in the case below.

In *R v Mustafa Shakir Abdul-Hussain et al* (1998),[30] the defendants appealed against convictions for the hijacking of an aircraft from the Middle East and diverted to the United Kingdom. The defendants raised the argument that they were under duress in that they were Iraqi nationals trying to flee persecution in Iraq. The court held that when such a defence was supported by appropriate evidence, the defence of duress was available for hijackers, although it had to be considered in light of the circumstances prevailing at the time. Issues of proportionality also had to be considered.

---

28 [2001] EWCA Crim 2352.

29 See also *R v Glen Ronald Hunter* (1998) TLR 26/02/1998.

30 TLR 26/1/1999. Note *R v Moussa Membar & Ors* (1983) TLR 1/06/1983 regarding the jury's consideration of the possibility of the captain being involved in the conspiracy.

In addition, there have been a number of instances where flight crews have had to deal with illegal immigrants or detainees that were disruptive prior to departure and are often escorted off aircraft as a result. Additionally, there have been cases of illegal immigrants causing major disruption aboard commercial aircraft while in flight in the hope that they will not be deported back to their land of origin.

It should be borne in mind that the new security regulations for airlines and crew (including reinforced steel cockpit doors and air/sky marshals) have added to the stresses that a flight crew experiences. No longer do the flight crew have free access to the cabin and the sense of continued isolation from the rest of the crew and passengers during long flights must also be taken into account for any crew.

All the above issues are important to airlines and flight crew as they have an effect upon airline liability in the event of an accident or incident.

Flight Crew and Passenger Health: DVT and Other Medical Conditions

An important issue has arisen in the last few years that has plagued airlines, flight crew (especially pilots) and passengers alike with great concern, namely DVT.[31] This issue has been highlighted recently after a number of high profile deaths from passengers who had flown long-haul flights from Australia and SE Asia to Europe and the United Kingdom. There are currently class action proceedings being undertaken in the United Kingdom and recently concluded in Australia with respect to this issue. It should also be noted that although not publicized, there are also a number of flight crew who have also suffered DVT or related symptoms. Therefore not only is DVT a passenger problem, but also one that potentially affects the flight crew.

We shall now turn to some recent cases on the issue of aircraft conditions.

A recent case that considered the issue in the United Kingdom was *Horan v JMC Holidays* (2002).[32] In this case, Mr. Horan, who travelled from Manchester to Calgary in Canada, argued that he suffered DVT after an 8 ½ hour flight in *cramped conditions*. The court held that as Mr. Horan did not provide any medical evidence that he suffered DVT, he could not succeed. The court did however concede that he did suffer discomfort and awarded him GBP500 in compensation.

Legal Liability of the Carrier

As outlined above, for an airline to be liable for a passenger's personal injury, it must be shown that the passenger suffered bodily injury during the course of flight operations or during embarkation or disembarkation. The main hurdle that the passenger must overcome (in legal terms) is that of causation – i.e. was the injury due to an "accident" in the normal operations of an aircraft? According to the

---

31 For an exposition on DVT, see BUPA "Deep Vein Thrombosis" May 2001 Newsletter. See also "Deep Vein Thrombosis" BBC News Online 11 May, 2002.

32 *Horan v JMC Holidays Limited* – Macclesfield Crown Court, England (Unreported: January 2002).

current legal position, if the plaintiff/claimant passenger can show that there was some "accident" during the flight and that they suffered personal injury, then they will most probably succeed. Of course, the issue of quantum is a more combative exercise between the parties.

At present, there is legal precedent in Europe for DVT claims. Generally, such claims are rejected on the basis that there was no "accident." It must be considered that the lack of clear medical evidence to show the direct links between long-haul flight and DVT may also be a factor in mounting a successful claim at present. Such an approach has been taken by a number of courts in the European Union, the United States and Australia.

In addition, there is a greater awareness of DVT to the extent that airlines and third party agents provide information on how to avoid suffering such conditions. This is often provided by the agents who sell the tickets or as set out in the in-flight magazines. Clearly providing such information to passengers is not an admission of liability by airlines, but rather as assistance to those passengers who have a predisposition to suffering such conditions.

With respect to the current case law on the subject, there have been decisions in different jurisdictions, but the leading example emanates from Australia. In *Povey v Civil Aviation Safety Authority, Qantas Airways & British Airways*[33] the High Court of Australia considered whether DVT was an "accident" pursuant to Article 17 of the Warsaw/Hague Convention. The High Court ruled that the claimant had failed to demonstrate that DVT was an "accident" caused while in flight. Similar decisions appear to have been made in other jurisdictions and the House of Lords in the United Kingdom DVT Group Litigation case.

From an industry point of view, there had been some debate as to whether there are any links between DVT and the consumption of alcohol. Birmingham University recently conducted a survey[34] on this matter and concluded that reducing the level of alcohol available on board flights may have a positive impact on the incidences of DVT on flights.

As this matter appears to have fallen on the aspect of *legal causation*, it is unclear how cases (if they are brought by crew members themselves against their employer) for actions based upon *increased risk of cancer or cosmic radiation, anxiety or work stress* would succeed. It is arguable the extent to which an airline has a real exposure to its crew, where the Civil Aviation Authority, the Health & Safety Executive and interested third parties have had considerable input into the working conditions aboard and have undertaken all the necessary precautions. In addition, airlines generally have stringent legal contracts of employment and it is certainly arguable that these *conditions* are inherent risks of flight and are therefore part of the accepted working environment.

---

33 High Court of Australia (Unreported 23 June 2005).
34 See S Keenan, "DVT study calls for an end to in-flight alcohol" Times Online 16 May, 2002.

Notwithstanding the above, airlines have also faced litigation from a number of passengers, and in some circumstances, crew as a result of injuries they sustained during the course of a flight or directly thereafter. These fall into a different category to those set out above and are relevant to the current discussion as they deal with specific instances of events that are unique to the aviation experience and should be considered further.

As noted above, the general proposition is that an airline will not be held liable for a passenger that suffers injury that is not as a result of an "accident" as stated in Article 17 of the Warsaw Convention (as amended). It must be remembered that when considering the *accident*, a court will look to the causal nexus of the occurrence, rather than the effect. This particular aspect or interpretation was considered in the leading United States case of *Air France v Saks*.[35] This approach has been considered both in Australia and the United Kingdom and was approved in *Chaudhari v British Airways Plc* (1997).

In *Fischer v Northwest Airlines Inc* (1985),[36] a passenger suffered a heart attack while on a flight. The court held that the passenger's predisposition needed to be considered in holding that the carrier was not liable, as there was no accident.

In *Kotsambasis v Singapore Airlines* (1997)[37] and *American Airlines v Georgopoulos (No.2)* [1998],[38] these cases involved the question of whether nervous shock could be claimed under the Warsaw Convention without attribution to physical injury. The court considered all the prevailing cases around the world and held in each case that *bodily injury* as stated in the Convention does not cover purely psychological injury. In *Kotsambasis*, the question arose as to whether there was liability on the carrier for a back injury sustained whilst the passenger was carrying luggage whilst a passenger with the respondent.

In comparison, in *Magnus v South Pacific Air Motive Pty Ltd* (1997),[39] an aircraft crashed into Botany Bay (Sydney, Australia) and claims were brought against the operators and other parties for psychological injuries by those persons who viewed the accident. Claims were also made by parents of passengers. The claims of the passengers were dealt with under Part IV of the Civil Aviation (Carriers Liability) Act 1959 (as amended),[40] while those people not categorized as *passengers*, were not restricted to the types of recoveries as under the Warsaw Convention.

While courts have been reluctant to hold carriers liable where there has been no accident, on occasions there are situations where passengers have suffered injury as a result of inaction on the part of the crew. In the case of *Husein v Olympic*

---

35 (1985) 16 Avi 18, 538. For a discussion on Article 17 cases, please see P Martin et al., Shawcross & Beaumont on Air Law Issue 87 (2002) and T. Philipsson et al., *Carriage By Air* London: Butterworths, 2001.
36 623 F Supp 1064 (ND Ill, 1985).
37 (1997) 148 ALR 498.
38 [1998] NSWSC 463.
39 Federal Court of Australia (Unreported BC200101881).
40 See *South Pacific Air Motive Pty Ltd v Magnus* [1998] 156 ALR 443.

Airways,⁴¹ a passenger suffered breathing difficulties when exposed to secondary smoke emanating from passengers nearby during a long-haul flight with the carrier. The passenger asked the crew to be moved, but was refused, despite there being availability elsewhere aboard the aircraft. As a result of the crew members' disregard, the carrier was held liable.⁴² This case is important to consider where passengers and/or flight crew may not wish to undertake certain flights for a number of reasons, whether it be for security, health or other such matters.

In the case of *Moran v Leisure International Airways Ltd* (2001),⁴³ the plaintiff's claim (together with a number of other passengers) resulted from an aircraft accident in May 1998, after a heavy landing at Ibiza following a flight from Manchester. The plaintiff/claimant suffered physical and psychological injuries as a result of the accident. The defendants submitted an offer that was accepted, thereby ending proceedings before a decision was made by the court.⁴⁴

In *Milner v British Midland Airways Limited* (2001),⁴⁵ the claimant received £136,000.00 by way of an out of court settlement for posttraumatic stress, as a result of being at the scene of the 1989 Kegworth aircraft accident. The claimant was one of the first people on the scene of the accident and assisted the survivors, as well as removing the bodies from the aircraft for five and a half hours. The claimant later suffered from a psychological condition as a result of attending the scene of the accident.

These cases demonstrate that psychological injuries are claimable under the present legal regime in certain circumstances, but not all. The question then becomes to what extent can an operator exclude any of its liability to its crew? The answer is not clear. There have been examples of crew suffering posttraumatic stress disorder or similar disorders, which have led to crew members being *grounded*. It is certainly not clear what direction a court would take with respect to issues affecting crew health based upon passenger disruption or similar – is it an accepted part of the contract of employment? What is important to consider in such a case is what action, if any and what procedures did the airline and/or operator take into order to consider that the crew were of sound mind and were psychologically approved for operations. These issues would no doubt play heavily on any airline that suffers an incident or accident and in turn any liability that may ensue. A review of the psychological profiling and crew pairing would no doubt be taken into consideration as well as any confidential reporting of that pilot or crew in the airline's records.

---

41 (2000) 21 Avi 18,024.

42 Note *Qantas Airways v Cameron* (1996) Federal Court of Australia (Unreported: BC9604068) on the issue of duty of care to passengers within cabins.

43 LTLPI 23/08/2001 Unreported.

44 See also *Moran v Leisure International Airways Limited et al.* (2001) LTLPI 10/92001 Unreported, *Moran v Leisure International Airways Limited et al* (2001) LTLPI 28/8/2001 Unreported.

45 LTLPI.

Conclusion

Aviation and flights on aircraft have long been a romantic ideal in human society, even though being a passenger on an aircraft has not always been possible for those with little income. With the advent of low cost carriers and more frequent flights to popular destinations, people who once were unable to use air transportation, have now been able to travel to destinations across the globe. The additional demand for flights has in turn, added pressure to carry more economy passengers. This has also caused a number of associated problems, as more people have been confined to ever smaller spaces, and often for many hours at a time. As more and more people were made to sit in expansive economy cabins, arguably the quality of air has decreased due to inadequate or inefficient airflow systems. At the same time that space and comfort issues plagued the flying public, problems of alcohol misuse coupled with disruptive behavior, produced dozens of cases of air rage and associated behavioral problems, which have led to some disastrous consequences.

The resultant problems have merely increased the anxiety and stress levels of flight crew as more and more of these issues become a daily occurrence. Flight crews have had to deal with an array of new circumstances post September 11 and the advent of low cost carriers. Air carrier risk managers have also noticed a steady increase in liability exposure, due largely to the uncontrolled increase of psychological issues (caused by third parties or external factors) on today's commercial aircraft. It is not clear to them how much they will continue to be an operational issue, much as in the same way that cockpit doors have been reinforced since the September 11 attacks.

Flight crews have also come under extensive pressure with respect to operational requirements (more flying time and tighter schedules) and also face the added stresses of aviation security and safety issues. In conjunction with the current liability regimes, both pilots and airlines need to be proactive in ensuring that their employees are of the highest quality both operationally and most critically, psychologically to ensure the safe operation of aircraft.

In the event of accidents or incidents, investigators, technicians and lawyers will be turning to the operation of the aircraft, its maintenance and flight crew and most importantly, the actions of the flight crew. In the case of breaches of safety and operational protocols, airlines may face extensive litigation and increased damages claims. Whatever the outcome, the pilot's attitude to safety and security is paramount, since these flight crewmembers have the "last clear chance" to avoid disaster.

Note

This chapter is written taken into consideration the law in 2005.

Chapter 18

Aviation Psychology in a Developing Country: South Africa

Johann Coetzee

Developing Mental Health Awareness: Before 1988

As one looks backward from the beginning of the twentieth century, it is inescapable how pre-emptive human conflict has been on fascinating and necessary research in all conceivable disciplines. The two world wars of the twentieth century were perhaps the most intrusive to those engaged in non-military research. World War I particularly influenced the research and development into the field of applied psychology within the military realm and formal academia alike. The development of psychometrics for the selection of specific military applications, assessment, counseling and therapy of soldiers, were all stimulated by a world at war. The initiatives within the Republic of South Africa (RSA) military sphere are profound in this regard and both formative and pioneering in the field of the applied behavioral sciences. The formation of the Military Psychological Institute and its applied derivatives within the South African Air Force, are not only historically noteworthy, but contemporarily functioning.

With the development of psychology, particularly during the last century, the then medical council of South Africa, recognized the various psychological disciplines (clinical, industrial, educational, research), which are by the publishing of this chapter controlled by the Professional Board for Psychology. Significantly, there is no formal aviation psychology in South Africa, despite the predominance of the work done in this area by the military over the years. Consequently, very little has been published on aviation psychology and in rare instances only theses and papers have been produced to suggest the emergence of a new dimension of psychology, namely aviation. In this regard the initiatives and work of Professor Johan Scheepers of the Department of Industrial Psychology of the University of Johannesburg (formally the Rand Afrikaans University) must be noted. Also, the establishment of an undergraduate program in aviation management and administration by Professor Leo Vermeulen at the University of Pretoria has served as the basis for the accelerated recognition of formal academic training in aviation management, leadership and administration.

The exclusivity with which historical South Africa, under a different regime, managed itself, resulted in preferential opportunities provided for White candidates only. This applied to the awarding of bursaries, specialized education, training and development and even access to apprenticeships. Within this segregated ideology and practice, *non-White* South African citizens were also denied the opportunity to serve in the national defence force while virtually all young White male South Africans were conscripted into national service. This obviously excluded all Black people from the South African Air Force who may have had any aspiration or intention for a career in aviation. Blacks were employed in the South African Air Force, but merely as laborers in support functions.

Exactly the same applied to the South African Airways and other smaller operators alike. In isolated cases only talented and aspiring candidates found their way overseas in order to become trained aviators. Some of these exceptional pilots and engineers have returned to South Africa since 1994. Since the change of government and the advent of democracy in South Africa, the philosophy and conditions have not only changed dramatically, but have in fact normalized. The South African National Defence Force, South African Airways, the Police Air Wing and other private operators have all aggressively engaged in recruitment campaigns to identify and select candidates to be trained for professional careers in aviation. Moreover, funds have been made available for bursaries via the South African Department of Transport TETA (Transport Education and Training Authority) and much success can already be recorded. But, it is obvious that those deprived of opportunities for literally centuries, will continue to suffer from vocational and career deprivation to the extent that aspirations and achievement motives will have been neutralized to the extent that they actually suffers from oblivion. This all has to be rekindled at school level, so as to indicate to those natural latent aviators, that such potential actually exists and that aspirations will be genuinely accommodated.

The cadet scheme at 43 Air School in Port Alfred is a singularly outstanding example of how much can be achieved within a decade. This success story is not only tangible evidence of a proudly South African commitment and achievement, but also a superb model to be emulated by other developing countries. The extent to which this has been successful even outstrips the tempo of many other transformation endeavours. It is almost revolutionary considering the zero base from which it all originated around ten years ago.

Historically, aviation-oriented psychology has been a practice where industrial, clinical and vocational psychology found its application within the South African National Defence Force. Such specialties within the respective categories of psychology rendered significant services, not only in the selection practices of potential aviators, but also in counseling and therapy, notably during the Angolan conflict in the 1980s. The Military Psychological Institute (MPI), is not only a pioneer in this regard, but also engaged in much research and development and the practice of psychology within the military sphere of the South African National Defence Force in general, and in the South African Air Force specifically. Exactly the same model and practice has existed in the civilian sphere as well. Psychologists from the various disciplines have been retained by air lines to provide psychological services, specializing in selection and assessment.

Developing Mental Health Awareness: 1988–Present

Whilst the term *developing*, as in *developing country* implies just that, it can also be construed as somewhat belittling if not judgmental. But, it is undeniably true that aviation in Africa can best be defined as emergent and indeed developing. Africa is an aviation growth region. The most aviation-untraveled population per 100,000 in the world resides in Africa.

The growth and need for specific aviation psychological services, specifically in counseling and therapy, suggest not only an escalation in the need for such services, but also its potential recognition as a specialized field of Applied Psychology. Aviation psychology is, after all, an abstraction of a number of sub-categories of psychology, each finding its niche within the broader world of aviation research. Therefore, we see aviation psychology as a conceptualization of the very necessary inner workings of psychology among those who desperately require such assistance.

The initiative taken by a number of practicing psychologists involved in counseling, therapy, as well as education and training within aviation culminated in the inevitable establishment of the Institute for Aviation Psychology. The very existence of this Institute suggests an *address*, status and service facility for the aviation industry to which individuals can be referred for help. Within the developmental context therefore, it is recommended that where the demand for such services increase as dramatically as has been the case in South Africa as a developing community, it is wise to establish an entity (address) which would signal to the aviation industry, the professional recognition of the need but also the existence of a facility rendering a universal aviation psychological service.

In 1988, the South African Department of Civil Aviation was approached to consider the recognition of holistic pilot assessment as a statutory prerequisite for flying competence. It was also suggested at the time that aviation medical examiners should receive compulsory training in basic behavioral psychological assessment as part of a more comprehensive and holistic assessment of pilot health during mandatory periodic medical evaluations. Whilst the submission and motivation were most sympathetically received and even endorsed, it was similarly complex if not impossible to consider it any further at the time.

International consultations subsequently followed as did participation at conferences such as the Aviation Psychology Conference at Manly, Sydney, Australia in 2000. Again, a similar reaction was received from the audience, highlighting specifically the complexity of unilaterally prescribing non-technical criteria for flight competence, i.e. psychological wellness.

An educational strategy and tactic was followed. This resulted in education and training programs, presented for South African Airways and sanctioned by the parent organization Transnet. Also, volunteer pilots, as members of the Air Line Pilots Association (ALPA) were trained as lay counsellors to assist in their Critical Incident Response Program (CIRP). Remarkable breakthroughs were established at both institutional and enterprise levels during this time, resulting in the President of the then Health Professions Council of South Africa indicating the recognition of aviation psychology, should it obtain sufficient support, membership and substance. This momentum was recognized and embraced by the South African Society for Aerospace & Environmental Medicine (SASAEM), resulting in the official recognition of aviation psychology as a wing (chapter) during their 2004 international congress, held at Sun City, South Africa.

Throughout this slow, yet evolutionary process, the Institute for Aviation Psychology continued its work as a professional counseling and therapeutic facility for aviators. Major airlines, smaller aviation organizations as well as individuals recognized the significance of this professional service provided and even established formal associations employing the staff of the Institute as consulting psychologists. In many instances this work still remains covert and unannounced for reasons of stigma and potential career disqualification.

The National Institute for Personnel Research (NIPR) and the test commission (psychometrics) have played a very significant role in the development of applied psychology in general and the development, standardization and controlling of psychometric tests in South Africa over the past 40 years. The work of Professor Gordon Nelson in this respect is particularly noteworthy. The standardization of such tests made way for the application of imported tests and test batteries of both a general but also aviation specific nature.

However, all the tests developed in South Africa are standardized on ethnic and cultural populations. As ludicrous as it may sound, the same tests were standardized on White, Black, Colored, Indian, and Oriental norm groups. Consequently, the validity and relevance of all these tests have not only been rendered morally suspect, but also statutorily banned for obvious reasons of discrimination and racism. In post-Apartheid South Africa, a new integrative and holistic approach for all dimensions of human assessment prevails, differentiating on biographical and demographical imperatives though specifically not by color or creed.

With the development of cockpit resource management programs (CRM) (a statutory prescription), and the recognition of human factors in aviation, aviation psychology has become a recognized discipline in the core syllabus of CRM programs presented by most aviation education and training agencies.

Despite the relentless commitment to the relevance of aviation psychology by those involved in its promotion and implementation as well as the informal recognition thereof by large organizations and (strategic) individuals alike, aviation psychology is at best still in its relative infancy.

Aviation Psychology: Practice in South Africa

Any professional psychologist engaged in diagnostic and prognostic work regarding the flying proficiency of a pilot remains in a most precarious and invidious situation. On the one hand the livelihood of the pilot is fundamentally determined by productive flying hours in an aircraft. On the other hand, the life of the pilot, crew and passengers are at stake. Duty of care and accountability are sometimes complex and confusing and cannot be easily unraveled, but should always be argued with safety uppermost in mind.

Medical aviation examiners are not trained behavioral scientists, nor psychologists. They are mostly general practitioners in private practice, registered with the Civil Aviation Authority as doctors, licensed to conduct medical examinations of pilots and professionals in other ancillary disciplines. Consequently, such medical examinations are essentially of a mechanistic and *physical* nature, conducted within a prescriptive statutory format. Apart from the comprehensive general medical and health assessments, the concentrated focus remains on cardiographic and audiological assessments. Psychological assessment is mostly carried out by observing or noting the pilot's behavior, utterances and overall conduct in the physical assessment. Scrutiny of the prescribed documentation confirms the absence of a comprehensive and incisive behavioral and psychological assessment.

The aviation medical report of the Civil Aviation Authority of South Africa (CA67/02(a)) of June 2003 serves as the proforma or checklist that has to be completed by all aviation medical examiners. Under the section Physical Examination 23 aspects are scrutinized. Item 23 reads: *Psychological evaluation* and the response is a tick in one of two columns namely Normal or Abnormal. In the category Medical History Assessment 48 aspects are assessed in this protocol with items 18 and 19 reading: Any mental psychological disorder (18); Suicide attempt (19). In both instances a tick is placed in a column either Yes or No. This constitutes the sum total of behavioral, mental and conduct assessments.

This should serve to emphasize the necessity for a far more comprehensive and holistic aviation psychological and behavioral assessment to truly assess whether an aviator is conclusively *well* enough to fly. It can be concluded that aviator health assessment is most definitely not comprehensive, holistic or inclusive. It must also be stated that it is at best selective, exclusive and therefore inconclusive.

During presentations in this regard at numerous national and international conventions, the case for comprehensive and all-inclusive health assessment has been motivated. In every instance the issues of holistic aviation medical assessment and the education, training and development of aviation medical examiners were presented and motivated. Except for isolated cases of concern, both the philosophy of the relevance of the motivation and the imperative nature of such practice, were supported by every audience. Introduction and application, but also moral and statutory defensibility, remain the key militating factors.

The hesitancy by statutory bodies, airlines and individual aviators to insist on comprehensive and holistic health assessment, requires a far more resolute and fearless approach by all bodies concerned. Also, excessive concern for issues such as validity and reliability need to be *rationalized* in the interest of aviation safety in general and aviator protection specifically. Indecision puts lives at risk.

It is often contended that honesty (at enterprise level) can be career limiting. A spirit and culture of *volunteering* is advanced here, which would be beneficial to the volunteer and in no way personally or professionally disqualifying, whatsoever. In the absence of such a culture and practice, individuals are often materially compelled to disguise both emotional and psychological unwellness as a fear for blemishing themselves and consequently tarnishing careers. What is also prevalent, is the establishment of strong personalized relationships with medical practitioners over many years in such a way that certain concessions are made or even disqualifying pathologies *overlooked* as a result of the enduring relationship and the *apparent* universal health condition of the client-patient.

The only way around this dilemma is one of collaboration. Employers, the Civil Aviation Authority, the Health Professions Council, representative bodies (e.g. unions) of aviators and individuals need to consult and collaborate in a way that would advance the philosophy and practice of applied aviation psychology as an intrinsic component of aviator wellness and its assessment.

Moreover, the universality of such an introduction and application is vital in the interest of consistency, as well as comprehensibility. If certain medical practitioners were to practice these aspects of aviation psychology as part of the normal health assessment, and others not, then obviously this would continue to fragment the process and also continue to establish particular preferred alliances between certain practitioners and clients. Again, the fundamental issue here is one of initiative. It is (probably) incumbent on the presiding Civil Aviation Authority to commence a process of education as it registers its intent to expand wellness assessment to incorporate psychological, emotional and behavioral wellness as an integral part of pilot/aviation proficiency. But, this will require both circumspect and expansive education and collaboration so as not to be guilty of unilateral, or authoritative intervention and legislation, consequently creating aloofness between the statutory body and the institutions and individuals which it controls and wishes to serve.

Case Profiles

As a consequence of the establishment of the Institute for Aviation Psychology, there has been increasing and continuing utilization of such psychological counseling facilities by large and smaller aviation enterprises. During 2004, significant growth in applied aviation psychology counseling was experienced. A selection of a few pertinent case studies is incorporated here to illustrate the nature and extent of some of the clinical work carried out within the Institute.

Case 1 An aerobatics display pilot who is also a Captain at a major airline, witnessed an accident of one of his team members during a major air show. Later that day he finds himself on a scheduled domestic flight. Just before take-off from a coastal city, he phones to record his emotional anguish and self admitted psychological state and receives counseling over the phone before taking off and returning to Johannesburg International Airport.

Case 2 A young commercial pilot works on a freelance basis for a number of aviation companies. At the insistence of his passengers (must-get-home syndrome), he departs with his aircraft with erratic fuel gauges, believing that his visual check of fuel levels was adequate. He was involved in an accident, in which two of his passengers died, but he survives and receives comprehensive counseling (also family therapy). After full recuperation and with much difficulty he procures another freelance appointment and he is involved in a further accident during his first flight. The loss of self confidence, esteem, a tarnished record and the realization of unemployability ensues.

Case 3 A self confessed drug addict not only admits his addiction and its dangerous consequences, but also the fact that he essentially has become an in-flight *pusher*, even internationally, due to the fact that he transports his own substances with him. He cannot report to a rehabilitation centre for fear of jeopardizing his status, and therefore, income generating potential.

Case 4 A captain with a major airline engages in an altercation with senior management and summarily resigns. During counseling he admits that the acute nature of his marital instability and domestic *violence* has precipitated this extreme degree of irrationality and lack of self control resulting in this disastrous incident. Through various interventions, counseling and facilitating representations, he ultimately procures an adequate appointment after comprehensive marital therapy.

Case 5 An airline captain serves as spokesperson on behalf of a group of colleagues who believe that discriminatory practices have left them in jeopardy. His subsequent suspension and victimization resulted in the spontaneous request for professional help.

Case 6 A captain with a smaller airline reports for counseling voluntarily, admitting that he lacks competence in two critical disciplines, mostly related to flying in weather. He states that he always ensures that his co-pilot is of particular competence in these disciplines so as to underpin his non-declared inefficiencies in the event of a weather encounter.

Case 7 A senior captain experiences the death of a child, resulting in excruciating grief and emotional instability. He engages in protracted professional help, but insistently requests absolute confidentiality (psychologist-client integrity) so as not to jeopardize his career.

Case 8 Numerous senior pilots, in a variety of airlines, report for professional help, regarding infidelity (extra-marital affairs), notably on overseas trips. In two instances blackmail has resulted, resembling the pathology of *fatal attraction* and *dangerous liaisons*.

Case 9 Two gay flight attendants request counseling and career guidance, given the fact that they have heard of the expansive research presently conducted on the impact of HIV/AIDS on human behavior and the potential destabilizing effect thereof, notably for people associated with aviation.

Case 10 A PPL pilot reports for intensive counseling due to his self-admitted exhibitionistic and consequent reckless nature, resulting from his recently acquired PPL rating. He admits two near misses, yet cannot release himself from the lure of any audience witnessing his (amateur) flying.

Case 11 A senior officer associated with the 1987 South African Airways *Helderberg* (747 classic) disaster in which all 159 on board perished, requests counseling at the time when it is intimated that the inquest dossier will be re-opened.

Case 12 Four wives of aerobatics display pilots, who have been involved in accidents, privately and confidentially request counseling (individually) in order to secure a strategy to influence their partners to terminate aerobatics display involvement.

Case 13 A senior captain involved in an extra-marital affair, subtly requests an assessment *certification* of his wife as abnormal in order to use such a professional report during divorce litigation.

Case 14 The decision by South African Airways to move from Boeing to an entirely new Airbus fleet has had profuse (and acute) psychological impact on many affected pilots and flight engineers alike. The early career terminations of 120 flight engineers as a consequence have resulted in a number of the *victims* seeking both emotional and adult career counseling. Four of these affected persons are currently in counseling/therapy.

Case Study In the case below, the pilot was grounded while psychological therapy was undertaken.

Kim is 41 years old, divorced and presently living with her parents. She originally trained as a nurse but has been in aviation for the past 10 years, first as a trainee and presently as a first officer on a 737-200 with a major airline. She was referred because of "a recent serious change in her personality."

After thorough investigation, it was established that Kim suffers from an acute family relationship manifesting as a personality problem, emanating from a devouring jealousy of her sister who refuses to grant Kim the so-called status and stature associated with being a female pilot in South Africa. The resultant conflict has a serious impact on the functionality of the family and even a pathological possessiveness by the parents who believed that this conflict amongst the siblings assumes such a destructive and dangerous level that Kim should not be allowed to live on her own. This precarious family situation has caused uncontrollable distress within Kim due to the fact that she wishes to obey her parents on the one hand, whilst in another sense she has an insatiable yearning for freedom.

After deeper analysis and as a consequence of creating a conducive interface of trust, Kim reveals her intense hatred for men and her stated decision never to get involved with a man again.

She shared with her chief pilot some of the family saga. He subsequently alerted a number of captains who flew some sectors with her and all reported that she seemed emotionally and mentally preoccupied, notably during the final stage of approach for landing. During the subsequent sessions the focus moved from the family dimension to issues such as cognitive overload in acute conditions and optimal and rational judgement and decision making in stressful situations.

The need and the requirement of the employer was purely to obtain an objective and professional assessment on Kim's capability to re-engage her flying (at all) or whether an adjustment in her career path should be engineered.

Whilst the assessment seemed appropriate, prognosis was uncertain. The issue here is not only a matter of professionalism but indeed consequentionalism. Whilst the future of the client seems uppermost on the therapeutic agenda, there is also the matter of approving further flying, but also the impact of this decision on passengers. Moreover, there are also ancillary issues such as impact on peers, disclosure to the insurer, fair labour practice and the whole issue of psychological stigmatization which exists.

The protracted counseling and therapy process guided Kim towards re-establishing her confidence and faith in both genders. Collectively, her prejudicing and pre-disposed mentality had to be normalized within the cockpit where she perceived the male captain (in command) synonymous with the conduct and impact of her first abusive husband and also domineering father figure at home. After 14 months of intensive psychological treatment, Kim was re-admitted to flying.

Education, Training and Development

For Aviation Psychology to become a recognized discipline, it is vital to establish such specialized training at tertiary level. It is true that some aspects of human factors and certain aspects of behavioral wellness are incorporated in existing programs at Master's level (e.g. MBA or MSc) throughout the world. Also, most crew resource management (CRM) programs will include some aspects in this regard in the syllabus, not necessarily to provide comprehensive training or teaching on the subject, but sooner to sensitize delegates as to its relevance and significance. A Masters Degree (MSc) in Aviation Management and Leadership has been designed and will be presented by the Da Vinci Institute for Technology Management as from April, 2005 in collaboration with the Institute for Aviation Psychology of Henley Air.

Far more specific and incisive would be the establishment of a specialized field of study at post graduate level in aviation psychology. This is currently being considered in South Africa as aviation psychology has now been established as a recognized discipline. As this body establishes itself both constitutionally and also operationally, so too it will advance its credibility and relevance in pursuit of such recognition, ultimately finding its way into academia where such practitioners can obtain specialized training. Such negotiations will commence during 2005 in South Africa with interested academic institutions in order to sensitize such prospective collaborators about the aviation psychology momentum being obtained and the fact that specialist training will have to be considered in South Africa. Again, this will all have to become an integrative and comprehensive process of consultation with all interested parties in order to ensure that inclusivity prevails and therefore universal support mandates every future move. With the sanction of the Professional Board for Psychology in South Africa, academic institutions may be encouraged to pursue the potential development of a masters program in aviation psychology, equipping such candidates for ultimate professional registration as recognized aviation psychologists, as is the case with clinical, educational, industrial psychologists, etc. Only then, will aviation psychology truly be recognized as a viable and generally accepted discipline and practice and all the areas of concern listed above effectively addressed.

Conclusion

As a developing country, South Africa is assuming much initiative and gaining momentum in the field of the applied behavioral sciences and aviation psychology. Momentum gained over the past three years specifically has brought South Africa to a dynamic threshold of a significant breakthrough in the evolving field of aviation psychology which will positively shape such practice and alert authorities and agencies alike. What South Africa has achieved over the past 75 years, in many respects serves as a benchmark. The formidable impact of the South African Air Force, Civil Aviation and all the ancillary disciplines, have all converted into a model to be emulated by those emergent aviation industries in countries where *the wing need not to be reinvented.*

Countries established and experienced in modern aviation should assist the emergent aviation fraternities as they strive to establish themselves. It is logistically and operationally relatively simple to deploy first world practitioners and technologists into such developing environments and to call that a domestic aviation industry. Self sufficiency, independence and the development of a local aviation culture is imperative for the establishment and maintenance of credible and impeccably functional procedures, standards and more specifically appropriate mindsets.

It is in this latter regard that much work has to be done. Whilst aviation is fundamentally a utility, it is essentially *flown* by an appropriate mentality. It is in this respect that continuing education, training and development will have to be provided and where South Africa particularly can render an invaluable service. The mechanistic and systemic provisions are relatively easy to formulate and implement. It is the normative/behavioral dimensions which constitute the real challenge as we strive to shape the requisite and conducive aviation mind, all in the interest of operational efficiency and impeccable safety.

The preparatory interventions are very often more important than ultimate establishment. Therefore, a process and practice of localized research, development, simulation, experimentation and then implementation, will serve as a realistic precursor and template ushering ultimate aviation implementation.

South Africa already provides a vast array of aviation services into Africa. Not only is South African expertise exported into neighboring territories and regions further afield, but pilots and technologists alike come to South Africa for orientation, education, training and development, both at national, as well as private institutions. It is this spirit of reciprocity and sharing, which is conducive to the orderly expansion and establishment of aviation frontiers in the emergent world. But, such action must morally strive to reach a terminal point for self sufficiency and indeed independence.

Ultimately, the philosophical context, and even aspirational status of any emergent aviation fraternity, will only remain dreams and hopes if not underpinned by financial resources. This requires, more than anything, a mentality and practice of affordability, and therefore an infrastructure which will secure the requisite resources. In this respect, South African Airways is a classic example, as are various other European and of late, American Airlines, who have fallen victim to the "feast and famine" nature of aviation as an industry so vulnerable to incidents and the emotions that surround them. Establishing an airline, procuring aircraft and employing personnel externally, does not constitute sustainable aviation. It is false and terminal! From an African perspective, South Africa qualifies to educate authentically within this particular case profile as it not only seeks to assist others, but also currently strives to rehabilitate its national airway provider.

Boldness and the spirit of collaboration are advocated for those academic institutions in Africa courageous enough to consider the establishment of formal training in aviation at tertiary level. The establishment of such facilities in South Africa, at both undergraduate and masters levels can serve as a resource base for such potential collaborators. It is both impractical and unwise to globally centralize a few specialist aviation institutions only. The decentralization and localization of such services will not only serve as an excellent delivery channel but also assist in the establishment of a proper local aviation culture and mentality at both academic and industrial levels. Naturally, state support will lend credence to such developments.

Chapter 19

How We Explain Misfortune: Psychological Implications for Mental Health in Aviation Professionals

Todd Hubbard

As I was contemplating what I might write on the subject of how we explain misfortune, the flow of those thoughts was halted by a CNN news broadcast that described, with incredible detail, the extent of death and destruction caused by hurricane Katrina, along the southern coastal regions of Louisiana and Mississippi in the United States. Eight months before, the world had witnessed the destructive power of a tsunami that decimated parts of Indonesia, Thailand, and India, killing thousands and leaving tens of thousands homeless. In our recent memory, we can recall the events of 9/11, assassinations of political leaders by terrorists, hijackings, deaths by famine, deaths by protracted wars between countries in Africa, or the Near East, or Southwest Asia. The list of horrific events in human history is long and the emotional memories that we have stored as a race continue to haunt us even now (Pillemer, 2004).

How we explain misfortune directly influences how we carry on in the wake of disaster. If the narrative of misfortune is a sociological issue, then coping with the ravages of disaster is a psychological issue. Rather than wait for the next global catastrophe, crisis management agencies should be preparing their citizenry for the rebuilding process, both physically and emotionally (Deahl, 2000). Critical incident stress debriefings, which have been widely used in the days and weeks after a traumatic event, are not always effective and in some cases appear to do more harm than good (Lewis, 2002 July/August). However, this does not suggest that we do not provide these debriefings, but rather that we prepare the debriefers more thoroughly before the event (Richards, 2001).

We know that traumatic events have far reaching affects, beyond those directly involved (see Chapter 7 by Chung in this text). Therefore, as countries study how to prepare for the inevitability of misfortune, they should be prepared for the psychological after shock, when people present with posttraumatic stress weeks, months, or years after the event. I would also encourage countries to look at the insurance rules that govern who can or cannot receive psychiatric care after a disaster and the duration of that care. As Chung pointed out in his chapter, there are at least six types of victims following traumatic events. We would want all of these victims to be eligible to receive government aid, and not restrict aid to only the primary victims.

If we explain misfortune as the inevitable result of life on this planet, we need to take care of its citizens when they present with psychological problems that do not neatly fit in an Axis 1 diagnosis (Diagnostic and Statistical Manual of Mental Disorders. [*DSM-IV-TR*] 2000). As Morse pointed out in her chapter (in this text), clinicians often review alternative diagnosis options, when presenting behavior appears to be relationally oriented. The downside of Axis 4 (psychosocial and environmental) diagnoses for psychosocial problems is that insurance companies are less likely to pay for relational problems, but are willing to cover treatment of Axis 1 disorders. This presents an ethical dilemma for clinicians who wish to care for their patients, but cannot provide care unless they falsely present the diagnosis.

There is a more profound problem associated with Axis 1 diagnoses, especially if these diagnoses falsely represent the true nature of the psychological problem (Regier, 2003). Cultural taboos and regulatory sanctions have governed what the public will tolerate in mental illness, especially among airline pilots. If, for example, you are seeing a patient who is a commercial airline pilot, and he or she is presenting with spousal relational problems, typical of V-code 61-10 (DSM-IV-TR, 2000), you might follow a diagnosis and treatment plan based on the V-code and Axis 4 (see Morse chapter in this text for a more expansive view of this subject). However, when you file the insurance for your patient, the insurance company refuses to cover the expenses (O'Day, Killeen, Sutton, & Iezzoni, 2005). In the unlikely event that after further sessions you decide that the primary focus is bipolar disorder and that the relational problems were secondary, you can pursue an Axis 1 diagnosis; however, you must bear in mind that the change in diagnosis will have a drastic affect on your patient's ability to pass a flight physical.

Morse and Bor (in this text) tells us that the United States Federal Aviation Regulations stipulate that pilots with bipolar disorder cannot pass their flight physical, which means your patient loses his job, if you or your patient discloses this information. The dilemma is clear. Since there is no incentive for a pilot to disclose his or her symptoms to his or her employer, a pilot might choose to avoid the subject altogether. However, nondisclosure can jeopardize safety. If the pilot does share his or her symptoms in a clinical session with a psychologist or psychiatric, there is a chance that the pilot will not disclose his or her diagnosis to the company, due, in part, to suggestions by the clinician. If you, as the clinician, choose to hide the diagnosis on the insurance form, but retain clinical notes that associate bipolar disorder with your patient and you medicate your patient accordingly, you might put yourself and your patient in jeopardy. As Appelbaum (2002) pointed out, leaks of confidential medical information are occurring more often as filing systems become digitized and are electronically stored. Appelbaum said that what patients reveal to the clinician is based on their trust that what was confessed behind closed doors will not come back to haunt them in the future. Bipolar disorder is not the only means to exit flying, fear of flying as related to a traumatic event might also jeopardize one's career, as pointed out in Iljon Foreman, Bor, and van Gerwen (Chapter 6 in this text). In terms of pilots, if they believe that their diagnoses will be made public, one might reason that they will not seek your help, and this is regrettable.

In this chapter, I will show how social explanations for misfortune have had psychological implications on the mental health of pilots. Our first stop is to the distant past; where over thousands of years individual actions have been subsumed within the overarching social structure, suggesting that we respond more freely within groups than as individuals. Our next stop is in the present, where psychological problems become manifest in pilots; determined, to a great extent, by organizational beliefs, taboos, and rituals. Finally, I shall end with a suggestion. It is my hope, that as a result of this chapter, pilots will seek to resist automatic urges to respond to social pressure by squelching their individual needs. Further, it is my hope that pilots will have the courage to resist the taboo of psychological counseling and pursue help when it is necessary. I also wish to present my case before clinicians, who have the authority and duty to terminate a career with the stroke of the pen, or support a pilot while he or she is in the throes of psychological stress. Pilots are caught in the middle, between organizations that will terminate them for having psychological problems and clinicians who follow the letter of the law without seeing the gray.

When Misfortune Happens

To understand how we behave today, Goleman (1995) and others tell us to look backwards, at the past thousands of years of human trial and error. Goleman contended that the 21st-century human brain is only the most recent version of what he called the designer brain. He explained that as humans developed mentally, emotionally, and physically, we attached emotional memories to each life event. Over thousands of years, the shared emotional memories of societies of individuals formed dominant behavior patterns, which we now recognize as the underlying basis for our modern society (Goleman, 1995). We are at the same time the product of all human emotional development (Adolphs, 2003) and the product of our own emotional development (Dolan, 2002). What part of our behavior is governed by memories from our ancestral past and what part is governed by more recent episodic memory is still being debated (Roediger & McDermott, 2000; Tulving, 2002). However, most scientists agree that one's understanding of personal value in relation to an organization is partly discerned by emotional memory and partly by a sense of self in relation to others (Adolphs, 2003). These emotional memories are intensified when accompanied by strong feelings of pleasure and pain (Dolan, 2002), which suggests that personal value is a product of emotional messages that confirm either a positive or negative standing within an organization.

Lynn Struve (see Pillemer, 2004) has evidenced strong historical roots in emotional memory in her work as it regards historical China. Displays of posttraumatic stress disorder are mentioned to describe the mental state of Zhang in 1651 after a particularly horrific battle where men were slaughtered without mercy. Pillemer pointed out that Struve's observations from the historical record could not be verified as one would in a clinical setting today. However, the writings of history do present psychological issues of prominent leaders of our distant past. As we read history it is difficult to remain aloof and detached from the events. As Goleman (1995) pointed out, those having emotional intelligence possess the ability to empathize and this feeling is often the gateway for our emotional connection to the events lived by others, past or present. It is not uncommon for persons to so thoroughly engage in empathy that they present with psychological problems unrelated to their present environment (Pillemer, 2004).

Dolan (2002) believed that the stronger the emotional event, the more likely that an individual would attend to that event. This emotion-attention link, according to Dolan, had been established over all the years of human development, not just in the near term with one individual. This was exactly the point made by Goleman (1995), that over thousands of years, humans have been making decisions about how they relate to their environment and to other individuals by a complex interaction between emotional memories and declarative facts. These fused memories have also helped us determine our place within an organization and have helped us form alliances with other group members (Beaty, 1995; Frith & Frith, 2001).

Investigations performed by anthropologist Mary Wood (in Dekker, 2005), suggested that people form organizations along the lines of how they explain misfortune. Since misfortune is often accompanied by strong emotions such as pain and anger, it is likely that individuals would align themselves with those organizations that adequately soothe their emotional upset and effectively resolve the event. In other words, organizations that know how to take care of their people, in the face of misfortune, have a very loyal following.

How We Explain Air Disasters

Deadly air disasters pique our interest as we attend to the news or read the headlines. We are often more detached from these events, unless we know someone involved. Even though air travel is touted to be safer than driving one's automobile, transporting people by air is not completely safe. When things go badly in flight, they go badly for hundreds of men, women, and children. Air disasters, unlike traffic accidents, capture worldwide attention, as the gruesome details are televised around the world. It is the number of people involved that stirs our emotions. Lockerbie still reminds us that air travel can be hazardous to innocent people. We are putting more and more people on single aircraft, as is the case for the new Airbus 380. Disasters involving wide body aircraft, such as the Boeing 747 and Airbus 380 have the potential of killing 500 or more passengers and crew. Such catastrophes assault our conscience. We wonder if we could have done something as a race to prevent these disasters.

For every event, we are taught, there is a cause, and if there is a cause then there is one who caused the event. When there is an aircraft accident, I doubt you naturally care for the pilot or pilots on the flight, but rather concentrate your interest on the passengers. After all, there are more of them. When we count the bodies and view the wreckage, we often cry for justice, or at minimum, an explanation for the misfortune. So we save our hate and contempt, in these events, for those we hold accountable, which is all too often the pilots of the flight. They are easier to hate than airline companies, because we can personalize our hate for them, and we find pilots better targets than entire organizations.

Contempt and hate are difficult to ignore. Where there is no shared, corporate accountability for an accident or incident, pilots are left to fend for themselves. Pilots who survive an accident are not only overcome with grief for the lost, but they suffer physically and emotionally for the rest of their lives, regardless of the results of the investigation. It is difficult to keep from personalizing the event. Tortured by day-time and night-time rehearsals of the events leading to the accident, many pilots cannot find a place in their minds or in their environment where they can have peace. Taken to extremes, this behavior will eventually destroy a pilot's career and their lives. It has also resulted in failed marriages or failed relationships with those who could have provided support. If these debilitating events are not arrested by psychiatry or clinical psychology, the person's brain chemistry may horribly change, preventing any short-term solution and prolonging any long-term relief.

Origins of the Blame Game

The blame game is as old as human society (Dekker, 2005). You could say that it is in our genes (Dolan, 2002). If an airplane crashes, we are apt to blame the pilot, the airline company, the manufacturer, the maintenance crew, the air traffic controllers, or anyone else who had "last clear chance"[1] to alter the course of events that led to the accident (for legal perspectives see chapter by Anthony Frances in this text). When those we count on to save us do not save us, we not only hold them responsible, we give ourselves permission to detest them, not just their actions. They are no longer worthy of our concern and care, because we believe that they have caused our suffering and should be punished. What is worse is that after the courts have exonerated those we have reviled, we refuse to make amends or show remorse over our vile treatment of these aviation professionals.

As a society, we have the tendency to maintain a course of thought or action, unless made to change direction. Adolphs (2003) said it is the automatic mode of brain-based thinking that keeps us moving on without altering course. However, we also have a self-regulatory feature in our brains, due to thousands of years of development. Self-regulation keeps us from reacting only by instinct. Blaming people for pain in our lives is part of our automatic mode and we find that blame comes easy to us all. Dekker (2005) captured the central issues of the societal blame

---

1 The US District Court used the last clear chance doctrine to assess that the Captain and First Officer of TWA Flight 514 had the last clear chance to avert a disaster, but did not. See *Marlene J. Brock, Donna M. Krescheck v. United States of America* and NTSB Report NTSB-AAR-75-16 for more details.

game in *Ten Questions about Human Error*. He warned the reader about the dangers of organizational indictments of errant individuals when organizations adopt a moralistic code to assess blame.

> External enemies of the system are to blame for misfortune, a response that can be observed even today in the demotion or exile of failed operators: pilots or controllers or technicians. These people are ex post facto relegated to a kind of underclass that no longer represents the professional corps. (p. 197)

Dekker's main interest has been aviation organizations so his comments are largely directed at airline management and the pilots that make up an airline company. Based on more than a decade of observation, Dekker has found a troubling occurrence within large aviation companies. When an accident happens, and we know that they will despite our best efforts, airline management often seeks to isolate the erring pilot or pilots, suggesting that it was the pilot's error, and not the organization's error that caused the damage to life and property. Once a pilot is reduced to *the error maker*, he or she will continue to be dehumanized by the company and the pilots, which eventually leads to criminalization of the victim. Widely publicized events, such as the accident in the Florida Everglades in the 1970s, where a perfectly good airplane was driven into the ground, often find their way to the courtroom, and the flying public cries for blood and money. Airline companies that wish to remain in business find it easier to criminalize the pilot than to accept corporate responsibility for the accident (Dekker, 2005).

Airline companies are made up of lots of people; some of them are pilots, some of them are cabin crew, some of them are mechanics, and some of them serve the flying public as baggage handlers, or ticket-takers. However, they all have one thing in common; they share a common social history, and it is that social history that disturbs us when we are made to explain misfortune. We have a difficult time including ourselves as error makers, because we cannot bear to imagine the consequences of our error and the punishments in store for those who cause misfortunes.

Will Durant's (1950) dated, yet relevant thoughts on the effects of moral codes on the ethics of decision making, might help us understand why we see such extremes in human response to misfortune. His comments are important to this discussion because he draws his examples from his prodigious research on how all civilizations behave, not just on how the modern Western civilizations have behaved. Durant noticed that at the birth of Christianity, more than at any other time in human history, humankind had to make a choice between two schools of thought. One school of thought supported individual choice, the extremes of which focused on instinctual behavior without any curbing standard. The emerging school, supported by the new religious philosophy of Christianity, held that human behavior had to be regulated by a moral law that was based on ancient texts. Both sides had different opinions on how instinct should be regulated, but they both agreed that for the betterment of a community, instinct had to be regulated. Those who supported individual choice as tempered by instinct found that moral codes choked the life out of people. As Durant pointed out, instinct alone would disintegrate community and moral codes would extinguish life. To overcome the extremes, successful communities have put the two views in balance, allowing room for instinct-based behavior while allowing moral codes to set the limits for that behavior.

Within the aviation community in the United States, the federal government regulates all pilots, aircraft mechanics, and air traffic controllers. And although Title 14 of the Code of Federal Regulations is not a moral code, persons often respond to these regulations in much the same manner as persons have reacted to moral codes from within every civilization in human history (Dekker, 2005; Durant, 1950). For example, the federal regulations establish acceptable and unacceptable behavior, which is also present in moral codes. We know that the Federal Aviation Administration acts as a policing agency when federal regulations are violated. This too is similar to moral codes, because moral codes must be enforced. I believe that messages of right and wrong, created by the federal aviation regulations, have been translated by individuals and groups within aviation to mean good (worthy) and bad (unworthy), thus establishing an assessment standard of behavior that seamlessly connects federal law with an invisible moral code.

Unconscious Notions of Misfortune

We have all had moments when we knew the right thing to do, but we could not explain how we knew. This is what Goleman (1995) and Gladwell (2005) called *gut feel*. Goleman said it was an actual sensation in the gut that was stimulated by the emotional brain. Malcolm Gladwell's explanation is more mysterious. He claimed that what starts in our adaptive unconscious as an unassociated array of thousands of pieces of information becomes associated in the presence of a unifying cue, such as an object of art.

Gladwell told the story about the Getty Museum's purchase of a statue that was allegedly from ancient Greece. After 14 months of intense investigation, which included core sampling and other techniques to judge authenticity, the museum bought the statue from an individual. When the statue was displayed at its first showing, a well-known expert in ancient Greek sculpture, who was not part of the former investigation, examined the sculpture and had the immediate impression that it was a fake. The word *fresh* came to his mind. Not one, but many experts had similar first impressions, although the sight of the sculpture evoked different manifestations of doubt. It is interesting that the Getty staff overlooked these other experts.

The Getty research team had taken suitable precautions and approached the task with utmost professionalism. However, what they did not fully appreciate was the influence of behavior interdependence (Cranno & Messé, 1982), where the actions of one member are in response to another member, and behavior conformity, where members of a group comply with norms "irrespective of one's personal beliefs" (Marshall, 1998, p. 36). Without any outside intervention, individuals within groups can operate solely within a factually confined space, devoid of new thoughts or objections to previously held positions. The Getty staff wanted the statue to be authentic and that desire was subtly communicated to the team. Based on Gladwell's explanation of the events, the team reached consensus perhaps a bit too quickly. Minor findings were not rigorously attacked. Instead, these minor findings became the principal evidence for authenticity. It is my contention that they unconsciously forfeited their individual right to object. Instead of doubting the evidence, they welcomed it. When individuals abdicate their personal right to disagree, they often become the unwitting member of a larger conspiracy of ignorance.

Organizational Explanations of Misfortune

In 2001, Dekker gave a talk in Columbus, Ohio, at the 11th International Symposium on Aviation Psychology. He made the point that human error is inevitable. More importantly, all humans participate in error-making, which he suggested has become a social norm. Rarely, do we find that a single person has made an error in isolation of other events or activities. He said that the findings of Fitts and Jones in 1947 were based on the notion that errors had a solitary source and could be eradicated by removing the errant individual. He spent the next hour telling the assembled group that errors have a social quality. He said that many persons have an opportunity to stop errors, but do not.

In the modern cockpit, pilots find themselves sitting within a system: the human-machine system of which they are merely a part. Smith (2000) suggested that the consequences of human fallibility within the system would be enough to upset the outcome of any flight, if the automated systems did not overcome these tendencies. However, Smith was quick to point out that technology alone cannot always be depended upon to save the day, even when the pilots are on the top of their game. Readers may remember the outcome of Flight 232, the episode some years ago in which a DC-10 was partially controlled to a crash landing at Sioux City, Iowa. Smith argued that pilot's have the ability to adapt to their environment, but that they also can make errors while doing it. These errors, whether by omission or commission, do occur and are the shared characteristic among all humans (Reason, 1990). In the case of Flight 232, the fault had been in the redundant hydraulic system that was never to fail completely. Instead of being a story of error making, Flight 232 has become the story of skill and determination in the face of impending disaster. Due to some clever flying techniques that are not found in normal training programs, Captain Haynes and his crew and many of the passengers survived the crash. This episode, however, does not diminish the warnings made by Smith that human error is to be expected. Perhaps acceptance of error-making is one of the first steps toward an explanation of misfortune among aviators.

Reason (2000) has been widely quoted in error management circles, because he points to the shared nature of error. He said that organizations have error traps that, unless noticed, will cause persons within the organization to make errors. These errors need not be catastrophic to be significant. Reason demanded that management create a reporting culture so that these error traps could be exposed and avoided. It is no surprise that the aviation community has taken notice of Reason's views on error management and has incorporated his models into their own risk management systems. Although Reason did not invent the idea of system defenses to foil human error, his work and that of Helmreich (2000) have been more widely accepted and used by aviation professionals and the medical profession. It is doubtful that work by Lofaro and Smith (2001) on operational decision-making would have been as widely accepted had it not been for the studies conducted by Reason (2000) and Helmreich (2000).

Dekker and Hollnagel (2004) also suggested that we rethink our views on human error. They posited that human error is an abstraction that is often cited as the cause of an accident, but fails to precisely describe what went wrong. Human factors scientists have grown used to using abstractions, such as complacency, loss of situational awareness, and ineffective communication as descriptions of human behavior, but none of these abstractions give us a list of behaviors that we can observe and measure. Complacency might be better described as monitoring failure, or "reduced awareness of danger" (p.81), "a state of confidence plus contentment" (p. 81). Certainly complacency, as described more precisely here, is part of human error, but it does not completely describe it. As Dekker and Hollnagel (2004) pointed out, when we as a community of scientists fail to insist on specificity and empirical evidence, we create folk models of non-observable constructs which then get us into trouble when we insist that these models help us make sense of behavior.

As I have suggested, I believe pilots are predisposed by historical social psychology to act by the conventions most acceptable by the culture in which they live. Further, I believe that pilots also act out in relation to the social understandings of the organizations they serve. If these social conventions are not questioned from time to time, they can lead to psychological problems for all the employees. As Dekker (2005) indicated, airline managers often use social models to help them explain misfortune. When these organizations use a moral model, errant pilots are often dehumanized and then criminalized.

While human factors specialists rally to alter society's explanations for misfortune and the handling of those who make errors, we are faced with a prevalence of aviation organizations that refuse to embrace the systems explanation for human error and would rather continue making an example out of the guilty party. We are now aware, by new evidence from studies performed by moral psychologists and neuroscientists, that organizations have been imprinting employ's emotional memories at the moral level. Dekker (2005) suggested that moral models of handling error seem to dominate the minds of many aviation organizations. When an organization embraces the moral explanation of error that condones punishment and dehumanizing activities, with a view toward criminalizing the victim, it is likely that each employ's emotional brain will be scripted to react in a morally punishing manner when errant persons are revealed. Moll et al. (2002) said that moral emotions "are intrinsically linked to the interests or welfare either of society as a whole or of persons other than the agent" (p. 2730). They went on to say that "moral emotions are readily evoked by the perception of moral violations" (p. 2730), which enable persons within an organization to rapidly, automatically, and unconsciously appraise an event.

I have suggested that the link between right and wrong behavior has already been interpreted by the aviation community to mean good (conforming to federal regulations) and bad (disobeying regulations) behavior, which enables pilots, aircraft mechanics, and air traffic controllers to automatically determine if a person is good or bad, guilty or innocent. I believe that the individual response to right and wrong behavior happens so quickly and so automatically that an individual might not even recognize why he or she reacts in affirmation or condemnation.

Persons who identify strongly with an organization tend to fall under its spell. If an organization reinforces moral determination of professional worthiness by rewarding error-free behavior and punishing error-ridden behavior, we can expect individuals within the organization to assess professional worth to those who are error-free and assess professional unworthiness to errant members of the group. This was exactly what Shultz (2000) claimed would happen if rewards became goals and goals controlled motivation.

Psychological Implications of the Misfortune Narrative

Workplace violence need not be overt to be dangerous (Coombs & Holladay, 2004). There are surely overt episodes of violence in the workplace today. Brennan (2005) reported that 35,000 workers are violently attacked each year in the United Kingdom, and the problem is getting worse there. Evolution and natural selection are nice ideas until we see them in action. It is a messy process to have people competing for the same scarce resources. Jobs are not easy to get in some places, and those with little education are destined to work in environments that are only slightly better than prison.

In recent years we have become more aware of the damaging effect of individual and corporate rejection of errant members. When individuals and organizations act out in moral retribution toward errant members, there can be serious consequences. Davenport, Schwartz, and Elliott (1999) cited a claim by Hornstein that as many as 20 million workers in the United States workforce suffer from a phenomenon called mobbing. Mobbing has a particularly nasty effect, in that the force of a crowd of people who have it in for an individual can actually cause the victim to commit suicide. The pressure is that great. One figure from Sweden that is very alarming is that 15 percent of the suicides in that country are attributed to the effects of mobbing (Davenport, Schwartz, & Elliott, 1999).

Perhaps I am not stating the psychological issue clearly enough. In regards to the literature on bullying behavior and workplace violence, we see a rise in stress related poor health, which has been found to be related to the conditions of the workplace (Kary, 2003). Steve Jex (as cited in Kary, 2003) attributes much of the psychological violence to "the stagnant economic climate" (p. 12). In the airline industry, we cannot help but notice that most airlines are operating in the red. Pilots and mechanics, who have been with their companies for years, are being forced to take pay cuts in order to keep the company operating. Many pilots are being furloughed and are not being asked back. For those left in the company, there has to be an incredible pressure to perform, and lack of performance might mean early termination. To understand what might be happening within the airlines, one can examine a study by the Scripps Howard News Service. In their report they said that 88 percent of those polled said they were more prone to emote now than in years past (Loafmann, 2001). This finding suggests that people are giving themselves permission to act out, and not curb their emotions.

Dekker (2005) claimed that many within the aviation community have been caught up in a particularly cruel pattern of scapegoating. Instead of explaining misfortune as a natural outcome of a society that is error-prone, some organizations that employ pilots, technicians, and air traffic controllers have elected to protect their corporate narcissism by ousting errant members or pressuring the weak ones to seek other occupations.

On the Nature of Pilots

During the symposium on aviation psychology in the spring of 2001 in Columbus, Ohio, I was standing in the company of an airline captain and a human factors scientist when the Boeing 737 pilot expressed his reaction to the studies he had listened to over the previous two days. He said he was deeply honored to find so many researchers spending so much time studying what he did for a living; "but (he added) you still do not know who I am as a pilot." It appears, at least in this one pilot's mind, that psychology only addresses part of what it is to be a pilot. As social beings, we act out our psychology while satisfying our association needs in the workplace. Perhaps more should be done in the psychosocial area of pilot studies.

Attempts have been made in decades past to understand pilots. For example, John Lauber, Charlie Billings, and George Cooper conducted interviews of airline pilots and flight engineers in the 1970s to get their perspectives on human factors issues (see remarks by Lauber in the foreword of Wiener, Kanki, & Helmreich, 1993). Surveys and interviews have exposed pilot behavior in studies of personality (Helmreich & Foushee, 1993), learning (Kanske, Brewster, & Fanjoy (2003), communication (Uhlig et al., 2001), leadership (Kutz & Brown, 2001), and decision-making (Holbrook, Orasanu, & McCoy, 2003); but in all of these studies, pilots were not assessed in their natural environment. They were made to behave in an artificial environment, which eliminated the cues and references normally available to pilots to stimulate appropriate behavior.

Airline pilots are different from their business counterparts (Helmreich & Foushee, 1993). Firstly, their office is most often found above the Earth's surface, where Boyle's Law of gases has a more dramatic effect. Secondly, a pilot's office is sustained by the laws of aerodynamics, which means that any disruption of those laws can put a pilot's life is jeopardy. Thirdly, a pilot must always, and in all

Table 19.1 Comparison of Industries

Aviation Industry	Other Industries
1. Structured & bounded	1. Uses many different organizational structures
2. Problems are similar between airlines	
3. Flight crew breakdowns are visible	2. Problems don't often relate between companies
4. Range of decisions are constrained and can be incorporated in a simple model	3. Reasons for breakdowns are diffuse
	4. Range of decisions are not as constrained

circumstances, show up for work with a clear mind, a motivation to do well, and the knowledge, skills, and abilities necessary to complete every task perfectly. When pilots are employed by an organization, say airline companies, the rules established by airline management become the standard operating procedure for each pilot, in addition to all regular flight duties.

The differences between airline pilots and executives from other industries are more clear-cut when we examine them along the lines of resource management. As you begin to appreciate the information in Table 19.1, notice the key words in the left column. A pilot's world is *structured* and *bounded*, *problems* encountered by pilots *are similar* to problems encountered by other pilots, if there is a *breakdown in communication* it *is visible*, and *decisions are largely scripted* by lessons learned from previous events if the pilots are reading the same accident and incident reports (Helmreich & Foushee, 1993).

Beyond the technical skills to fly an airplane, a pilot should be physically fit and should be psychologically fit as well. Airline pilots must pass physical examinations by an aeromedical examiner every six months. Any and all circumstances that put a pilot's physical or psychological fitness in jeopardy are regarded seriously and if any indication appears to question the fitness of a pilot, he or she can be grounded, never to resume flying duties (Edwards, 1997). And yet, notwithstanding advances in psychological assessment of pilots and other crew members, psychological fitness is still often assessed in highly rudimentary, subjective ways.

By the time pilots are finally employed by an airline, they have spent at least 500 hours in high performance aircraft, performing complex tasks in instrument meteorological conditions and showing technical ability to handle an array of emergencies. After receiving a provisional acceptance to an airline, pilots are made to complete another rigorous training program, which involves numerous knowledge tests, simulator sessions, and final evaluations in the type of aircraft flown by the company. If all went well, the pilot will begin flying duties with various line captains.

When pilots are paired together to complete a scheduled flight, they often share their duties with strangers, albeit similarly skilled strangers. There is no guarantee that the mix of personalities will add to or detract from the flight environment. There is also no guarantee that pilots will communicate easily with each other or that decisions will be accurate and timely.

Beaty (1995) indicated that communication between captains and first officers was similar in nature to a troop of gorillas. The silver back, or dominant male, exerted his power and authority over the others based on his size and on his ability to intimidate the others. If you ask first officers if captains are like silver backs, you will find that most first officers will say they have flown with *silver back* captains from time to time. Although airline management determines power and authority by rank and not size, we cannot be too sure that size has no importance at all.

I believe that some of the psychological problems faced by pilots stem from fractured relationships with other pilots and the flying organization. As Beaty pointed out, we are not far from the jungle in our behavior, nor are we far from the jungle in our psychology. What can occupy my mind will have an influence on my social standing as well. Dominant behavior by alpha male pilots, and perhaps alpha female pilots among other female pilots, is firstly social, but is always psychological. Being driven from the group, as is the case during accident investigations, not only separates the pilot from his or her affiliation with the group, but also puts in question his or her ability to make a living. As Dekker (2005) pointed out, licenses are taken away from the allegedly problem pilot as the first step toward termination. Once stripped of the license, a pilot has little to fall back on. Most pilots do not have a fall back option for employment.

Pilots use communication as a means of control, particularly in the captain-first officer setting. Kanki and Palmer (1993) cited work by Billings and Cheaney that indicated that effective communication has been forestalled in past by either not sharing information that was available or not accurately transmitting the message. Goleman (1995) tells the story of the air crew over Portland, Oregon where the first officer and flight engineer weakly indicated to the captain a shortage of fuel to land at their destination. Apparently, the captain was a *silver back*, dominant male, who intimidated his crew. By the time the captain acknowledged the lack of fuel, it was too late. The aircraft crashed short of the airport, and some lives were lost.

When crew members are intimidated, they may hesitate to share vital information, even when their lives are being risked by the captain. This can only confirm the notion that pilots have a sense of themselves within an organization and that that sense is more powerful than self-preserving messages.

Kern (1998) added that breakdowns in communication can be caused by hero worship of individuals who have beaten the odds. In his appraisal of the 1979 Downeast Airlines Flight 46 accident, no one on the doomed aircraft questioned the decision to press the minimums due to obscured visibility in thick fog. What had developed within Downeast Airlines was a cult of ignoring weather minima, ignoring mechanical problems, and ignoring overweight aircraft. Unchecked, this behavior persisted within the group, indicating again that pilots can succumb to organization pressures to conform, while ignoring their instinctual need to survive.

Lofaro and Smith (2001) pointed to communication as the necessary ingredient for proper decision making. In their presentation of the Operational Decision

Making (ODM) model, they said that the pathway toward error is represented by a series of smaller errors in judgment that could have been interrupted if someone had intervened. In 1978, an Air Force crew on board a KC-135 had just finished the refueling leg of the flight and was beginning the *nav leg* to fulfill the requirements for the navigator's checkride. The evaluator navigator was riding in the jump seat so that he could monitor the co-pilot's TACAN fixes (plots of points that indicate the aircraft's position in reference to the line depicting the filed course). The navigator was navigating via radar. The entry to the nav leg began shortly after a 135 degree left turn. Well before the actual turn point, the navigator commanded the left turn. The co-pilot, in good crew resource management style, notified the navigator that he disagreed with the turn command. The aircraft command (similar to an airline captain) borrowed the co-pilot's chart to see where the co-pilot thought the aircraft was. He told the navigator to check his position, to which the navigator countered with, "I have a good fix on radar and we need to turn now." The co-pilot continued to object, but the aircraft commander followed the navigator's guidance and ordered the turn. The co-pilot, knowing that the crew would be violated for flying outside the planned track, quickly notified the air traffic control center of their need to change the plan and fly direct to the entry point of the navigation leg. By the time the center controller approved the new track, the aircraft had already flown outside the protected corridor. The navigator's checkride went further downhill and the evaluator failed him. He also failed the aircraft commander for not ensuring that the aircraft remain on the approved track.

What went wrong here? I believe that the aircraft commander allowed his judgment to be influenced by his decision to give the navigator control over the aircraft and thus abdicating his own responsibility for the safe operation of the aircraft. Moreover, he held his co-pilot's technical skill in so little regard that he was willing to refuse any input from this other pilot.

When a pilot's technical skill is in question, it can mean the end of flying altogether. What muddies the water here is that technical skill might not be the problem, but rather sociability. If a pilot does not seem to get along with other pilots, he or she is left alone, very alone. There are those pilots who are unfazed by the rejection, and so do not feel it. However, there are pilots who must be members of the group to feel whole. What we cannot see in these pilots' minds is the reason why some have a longing to belong, while others are content to remain alone.

The strength of inter-pilot identity is largely caused by the fusion of emotional memories with actual flight events. Since flight training is very demanding, emotional responses to the environment are etched into the memory of flight events while flying with instructors, often remaining very vivid for decades. The emotion-packed environment also allows pilots to develop strong feelings toward instructors and fellow student pilots. Once these bonds develop between pilots, each pilot tries to maintain that bond by behaving in ways that are acceptable to the group. For example, when pilots die in aircraft accidents, pilots mourn, because there is a tear in the emotional and professional fabric of the community of pilots.

When pilots survive, those pilots that read the reports pay attention, because they learn from others' mistakes and try very hard not to repeat the errors of other pilots.

When professional integrity has been breached by a pilot or pilots involved in an accident or incident, pilots personalize the event. Someone has paid dearly for his or her mistakes and we know that accidents caused by pilot error are as high as 70 percent (Helmreich & Foushee, 1993). In the aftermath of an accident, each pilot within the larger community of pilots has to resolve that accident emotionally and professionally. Pilots will have to conclude that either it would not have happened to them, or they would have to conclude that had the same events occurred, they would have died as well. Those who believe they would have escaped move on without any self-doubt. However, those who believe they would have perished in the accident, are duty bound to learn from the accident and to change their behavior. It is only after dozens of successful flights that the self-doubting pilots once again feel emotionally and professionally ready to fly.

Self-doubting behavior can linger for months, even causing regression in technical skills. Self-doubting pilots often confess that they have *lost the picture*. Since flying is predominantly visual, losing the picture is very literal. Some pilots lose the landing picture, and for months they bang the aircraft on to the runway, jarring the passengers and upsetting service. During this time, pilots doubt not only those skills, but others as well, and their upset can spill over into their personal life. Until a pilot's technical performance is settled, life with that pilot will be difficult. Although these episodes are only temporary, lasting up to two or three months, they last long enough to work on the patience of a spouse or partner. It is in these cases that the primary focus of treatment might not be relational, but psychomotor. Once the coordination is regained, relationships have a chance to be renewed, and you can expect pilots to be very apologetic. Sadly, repeated episodes of emotional distancing while working through problems, can destroy a marriage, especially if the pilot expects all to be forgiven after an apology.

Pilots, by training and by social convention when in groups, police their own ranks. Policing behavior in pilots is as formal as standardization/evaluation divisions, check airmen, and FAA examiners and as informal as gossip. Formal modes of culling the rank and file of pilots are based on federal regulations, standard operating procedures, pilot operating manuals, and technical orders, with standards of performance and minimum passing scores on written and practical tests. Decidedly more objective, formal modes of preserving the reputation of pilots are preferred over the less formal modes.

Operating within the formal system of policing pilots is the informal system. The informal system is largely controlled through story telling. As information becomes available, pilots will process what they hear and read and form opinions, which are then shared with other pilots – except, of course the pilot being talked about. Truth is not always prized above conjecture, and the object of contempt is often pictured as the villain, even if unsupported by the official report. If you have ever watched schools of fish swimming above the reef, you've observed how that sometimes they all suddenly divert in the same direction. You cannot see what startled them, but when one fish turns, they all turn. This reef behavior is similar to what happens in the informal policing network of pilots. When a respected member of the group forms an opinion of another pilot, the other pilots tend to form the same opinion. We see here more echoes of Beaty's (1995) insistence that pilots are not much different in behavior to groups of gorillas.

Conclusion

Compassion and ridicule are neighbors in the community of humankind. We are driven to compassion in the wake of tsunamis, hurricanes, and air disasters, but we are also as likely to be driven to ridicule after the same event. Some one or some thing must be blamed for the pain and suffering. I doubt that the *supreme other*, or God, is frightened by shaking fists after a natural disaster. Divine entities are not compelled to show up in court if accused of causing harm. However, ordinary people are not as fortunate. As a society, we have a blind spot. It seems as though we cannot see all the needy victims in a crisis. This is particularly true when it concerns pilots. When we read of an air disaster that claimed everyone's lives, we quickly search for the cause. If the pilots are in any way to blame, we do not pity them their lives, but rather feel justified to relish in their deaths as payment for the suffering they caused. If the pilot escapes, we, as a society seek to exact revenge, either directly in court or indirectly by mobbing behavior in the workplace.

In this chapter, Dekker (2005) reminded us that the blame game is as old as humankind is. Dolan (2002) said this behavior was in our genes. If we allow ourselves to be driven by our instincts, we will act out from our emotional memory (Adolphs, 2003; Goleman, 1995). Durant (1950) suggested that since the advent of Christianity and its influence on world culture, moral codes have become intertwined with our emotional memory to form a hybrid moral emotion. Dekker said that these moral models dehumanize pilots so that it becomes easier later to criminalize them.

Given the backdrop of human society, I discussed the nature of pilots, with the hope that the reader would understand how the internal pilot culture polices its own ranks. We know that following an air disaster pilots mourn the loss of life and the loss of another aircraft, because each accident points to weaknesses in humans and the aircraft they build. To diminish the debilitating affect of human error among pilots, the pilot culture seeks to purge from its ranks those who are less worthy to fly. There are overt means to affect this purge, which include removal of licenses and other certifications; however, there are also covert means to purge the rotten apples. These methods include emotional abuse, malicious gossip, and unsubstantiated claims of wrongdoing.

An insidious problem among pilots is their use of gossip and conjecture to pressure weaker pilots to resign. Where this behavior is allowed to continue, employees are harassed and emotionally abused. Clinicians who have worked with pilots are aware that this internally inspired abuse leads to alcohol misuse or drug misuse in victims of harassment. It can also lead to suicide (Davenport et al., 1999). So what are we to do?

I believe Dekker (2005) had the right idea. Airline management should investigate and reflect on how they explain misfortune. To paraphrase Dekker, managers of pilots should stop pretending that only weak pilots commit errors. Errors are within the domain of all humans, despite our best efforts. A pilot error is a system error and the company should be ready to accept responsibility (Helmreich, 2000; Reason, 2000).

Doubtless, there are voices from within airline management that are insisting that the corporation should be more accountable for errors made by company pilots, relieving pilots of the total emotional burden of misfortunate events. It is my hope that captains and check airmen come to the aid of those within their ranks who are suffering from abuse. I also hope that they are discouraging informal harassment and that they will promote due process through formal actions taken by company human resources personnel.

As I have suggested, societies of people often run along in a direction until made to change course. Reef fish suddenly turn direction when confronted with danger; otherwise they continue to swim in the same direction. People are not very different. Besides our social preoccupation with *running with the pack*, we often *run with the group* in our emotional perceptions of people within the organization. Pilots will have perceptions of other pilots, with or without cause. When pilots group together to pressure another pilot out of the organization, because of an error-centered explanation of misfortune, they might be behaving in the automatic mode without much conscious thought. However, these mentally unhealthy behaviors will remain unless someone intervenes. I suggest that we encourage each other toward self-regulation from within the group. As Reason (2000) suggested, we are all part of a system; and like it or not, any error made within that system is as much your problem as mine. More than any other time in recent memory, we need to hear this message.

References

Adolphs, R. (2003, March). Cognitive neuroscience of human social behavior. *Neuroscience, 4,* 165–178.

American Psychiatric Association (2000). *Diagnostic and statistical manual of mental disorders* (4th edn., text revision). Washington, DC: Author.

Appelbaum, P. S. (2002). Privacy in psychiatric treatment: Threats and responses. *The American Journal of Psychiatry,* 159, 1, 1809–1818.

Beaty, D. (1995). *The naked pilot: The human factor in aircraft accidents.* United Kingdom: Airlife Publishing Ltd.

Brennan, W. (2005). When colleagues turn. *Occupational Health, 57,* 4, 18–19.

Coombs, W. T. & Holladay, S. J. (2004). Understanding the aggressive workplace: Development of the workplace aggression tolerance questionnaire. *Communication Studies, 55,* 3, 481–497.

Crano, W. D. & Messe, L. A. (1982). *Social psychology: Principles and themes of interpersonal behavior.* Homewood, IL: Dorsey Press.

Davenport, N., Schwartz, R. D., & Elliott, G. P. (1999). *Mobbing: Emotional abuse in the American workplace*. Ames, IA: Civil Society Publishing.

Deahl, M. (2000). Psychological debriefing: Controversy and challenge. *Australian and New Zealand Journal of Psychiatry, 34*, 929–939.

Dekker, S. W. A. (2005). *Ten questions about human error: A new view of human factors and system safety*. Mahwah, NJ: Lawrence Erlbaum Associates, Inc., Publishers.

Dekker, S. & Hollnagel, E. (2004). Human factors and folk models. *Cogn Tech Work, 6*, 79–86.

Dolan, R. J. (2002, November 8). Emotion, cognition, and behavior. *Science, 298*, 1191–1194.

Durant, W. (1950). *The story of civilization: The age of faith*. New York: MJF Books.

Edwards, D. (1997). *Fit to fly: Cognitive training for pilots*. Brisbane, Australia: Copywright Publishing Company Pty Ltd.

Frith, U. & Frith, C. (2001). The biological basis for social interaction. *Current Directions in Psychological Science, 10*, 5, 151–155.

Gladwell, M. (2005). *Blink: The power of thinking without thinking*. New York: Little, Brown and Company.

Goleman, D. (1995). *Emotional intelligence*. New York: Bantam.

Helmreich, R. L. (2000). On error management: Lessons from aviation. *British Medical Journal, 320*, 781–785.

Helmreich, R. L. & Foushee, H. C. (1993). Why crew resource management? Empirical and theoretical bases of human factors training in aviation. In E. L. Wiener, B. G. Kanki, & R. L. Helmreich (eds.), *Cockpit resource management* (pp. 3–45). San Diego, CA: Academic Press.

Holbrook, J. B., Orasanu, J., & McCoy, E. (2003). Weather-related decision making by aviators in Alaska. In R. Jensen (ed.), 12th International Symposium on Aviation Psychology, *100 years of Flight*. Columbus, Ohio.

Kanki, B. G. & Palmer, M. T. (1993). Communication and crew resource management. In E. L. Wiener, B. G. Kanki, & R. L. Helmreich (eds.), *Cockpit resource management* (pp. 99–136). San Diego, CA: Academic Press.

Kanske, C. A., Brewster, L. T., & Fanjoy, R. O. (2003). A longitudinal study of the learning styles of college aviation students. *International Journal of Applied Aviation Studies, 3*, 1, 79–89.

Kary, T. (2003). Down and out on the job. *Psychology Today, 36*,3, 12.

Kern, T. (1998). *Flight discipline*. New York: McGraw-Hill.

Kutz, M. N. & Brown, D. M. (2001). Technical myopia in aviation leadership: Does research support the notion that technical expertise is imperative to effective aviation leadership? *International Journal of Aviation Research and Development, 1*,2, 149–166.

Lewis, G. (2002, July/August). Post-crisis stress debriefings: More harm than good? *Behavioral Health Management, July/August*, 23–25.

Loafmann, B. (2001). Taking out violence. *Occupational Health & Safety*, *70*,1, 32–57.

Lofaro, R. J. & Smith, K. M. (2001). Operational decision-making: Integrating new concepts into the paradigm. In R. Jensen (ed.), 11th International Symposium on Aviation Psychology. *Focusing attention on aviation safety*. Columbus, Ohio.

Marshall, G. (ed.) (1998). *Dictionary of sociology*. Oxford: Oxford University Press.

Moll, J., de Oliveira-Souza, R., Eslinger, P. J., Bramati, I. E., Mourão-Miranda, J., Andreiuolo, P. A., & Pessoa, L. (2002). The neural correlates of moral sensitivity: A functional magnetic resonance imaging investigation of basic and moral emotions. *The Journal of Neuroscience*, *22*, 7, 2730–2736.

O'Day, B., Killeen, M. B., Sutton, J., & Iezzoni, L. I. (2005). Primary care experiences of people with psychiatric disabilities: Barriers to care and potential solutions. *Psychiatric Rehabilitation Journal*, *28*, 4, 339–345.

Pillemer, D. B. (2004). Can the psychology of memory enrich historical analyses of trauma? *History and Memory*, *16*, 2, 140–154.

Reason, J. (1990). *Human error*. Cambridge: Cambridge University Press.

Reason, J. (2000). Human error: Models and management. *British Medical Journal*, *320*, 768–770.

Regier, D. A. (2003). Mental disorder diagnostic theory and practical reality: An evolutionary perspective. *Health Affairs*, *22*, 5, 21.

Richards, D. (2001). A field study of critical incident stress debriefing versus critical incident stress management. *Journal of Mental Health*, *10*, 3, 351–362.

Roediger, H. L. & McDermott, K. B. (2000). Distortions in memory. In E. Tulving & F. I. M. (eds.), *The Oxford handbook of memory* (pp. 149–162). Oxford: Oxford University Press.

Shultz, W. (2000). Multiple reward signals in the brain. *Neuroscience*, *1*, 199–207.

Smith, D. (2000). On a wing and prayer? Exploring the human components of technological failure. *Systems Research and Behavioral Science*, 17, 6, 543–559.

Telfer, R. & Biggs, J. (1988). *The psychology of flight training*. Ames, IA: Iowa State University Press.

Telfer, R. A. & Moore, P. J. (1997). The roles of learning, instruction and the organization in aviation training. In R. A. Telfer & P. J. Moore (eds.) *Aviation training: Learners, instruction and organization* (pp. 1–13). Aldershot: Avebury Aviation.

Tulving, E. (2002). Episodic memory: From mind to brain. *Annual Reviews Psychology*, *53*, 1–25.

Uhlig, P. N., Haan, C. K., Nason, A. K., Niemann, P. L., Camelio, A., & Brown, J. (2001). Improving patient care by the application of theory and practice from the aviation safety community. In R. Jensen (ed.), 11th International Symposium on Aviation Psychology. *Focusing attention on aviation safety*. Columbus, Ohio.

Wiener, E. L., Kanki, B. G., & Helmreich, R. L. (1993). *Cockpit resource management*. San Diego: Academic Press.

Wood, J. N. & Grafman, J. (2003, February). Human prefrontal cortex: Processing and representational perspectives. *Neuroscience, 4,* 139–147.

Epilogue

Robert Bor and Todd Hubbard

We have addressed many of the psychological issues in air transportation, but this text does not address all of the participants, nor does it fully address the issues as they relate to each aviation occupation. We have concentrated on flight deck crew, cabin crew, and passengers. These groups represented a significant number of individuals and capture the larger picture of psychological implications. That said, we would have liked to have presented more about aircraft mechanics, air traffic controllers, aviation managers, aerospace engineers, and logisticians, but were constrained by the lack of published research on these occupational groups. We also hoped to include work from notable persons from Australia and New Zealand. It happened that these authors were not available at this time. We hope that further research on the important topic of aviation mental health will emerge from this book and that future editions will redress any omissions. One especially important topic for future research is what happens to pilots over time when they are made to shut out their domestic life in order to attend to their work. This investigation should also include the other groups as well, since it appears that some aviation occupations require one to make extraordinary allowances for occupational disruptions.

As a reminder to the reader, we welcome your comments on this text. It is our hope that your comments and suggestions will help us focus on the direction for future editions. If you are interested in contributing to these future editions, please let us know the direction of your contribution. We can be contacted at robertbor@hotmail.com. We have enjoyed this project and hope that you will find our thoughts to be relevant to your view of aviation mental health.

Index

ab initio selection 162, 164, 165, 166, 171, 184
    complex criteria 175
    job samples 178
    mechanical methods 176
    MICROPAT 179
    personality tests 177
    predictive validities 169
Ability Screening On-Line 176
ability tests 176–7, 204–5
abnormal perceptions 133, 134
absenteeism xxi, 116
Access to Health Records Act (1990) 150
accidents 195, 261, 285, 318–19
    blame 319–20, 330
    Captain overruling of flight crew 290–1
    cognitive impairments 28, 29
    legal liability 299, 300, 301
    military pilots 60
    pilot error 329
    posttraumatic stress disorder 114–15
    prevention 262
    sleep loss relationship 264–5
    see also critical incidents; disasters
achievement motivation 249, 250, 251
acute stress disorder 85, 86, 115
acute stress reaction 49, 134, 140
adaptive tasks 215, 216, 217, 218, 219
ADD *see* attention deficit disorder
addictive behavior 234
ADHD *see* attention deficit hyperactivity disorder
adjustment disorders 109–10, 121
adjustment reactions 159
Adolphs, R. 319
adrenaline 138–9, 140, 142–3
advice 22, 23
aeroneurosis 4, 60
Aerospace Medical Certification Division (AMCD) 111
affective psychosis 133, 134, 148

AFOQT *see* Air Force Officer Qualification Test
aggression 19, 20, 41, 47
    see also air rage; anger
agoraphobia 55, 56, 57, 70, 74, 114
Air 2000 286
Air Botswana crash (1998) 120
Air Florida B737 crash (1982) 291
Air Force Officer Qualification Test (AFOQT) 176
Air France 185n3
Air France A340-300 crash (2005) 285
Air Line Pilots Association (ALPA) 305
Air Navigation Order (2000) 295
air quality 16, 289–90, 301
air rage 19, 40, 44–6, 292, 301
    alcohol consumption 20
    cabin crew management 197
    legal cases 295–6
    media coverage 13
    personality traits 21–2
    prevention 47
    stressors 23
    see also disruptive behavior; violence
air traffic controllers 6
air-conditioning 33
airport language 59
Aitken, R.C.B. 60
ALAPS *see* Armstrong Laboratory Aviator Personality Survey
Albert, Ferenc 7, 195–208
alcohol xx, 45–6, 47, 285
    addictive types 234
    air rage 19, 20, 22, 23, 44, 45, 301
    anxious passengers 20, 22, 49, 53
    cabin crew 211, 235
    court cases 295, 296
    deep vein thrombosis link 298
    dependence 148
    depression 42
    disaster victims 87, 91, 92

economy class 292, 293
negative impact on performance 276
pilots 4–5, 116–17, 119, 330
posttraumatic stress disorder 141
reduction of intake 23
sleep problems 217
stress 110
suicide 119
alcohol altitude syndrome 45
alertness 211, 265–7, 272, 274, 275, 276
Allen, G. 75
Alm, T. 76–7
ALPA see Air Line Pilots Association
alternative treatments 122–3
AMCD see Aerospace Medical Certification Division
American Airlines 313
American Psychiatric Association (APA) 2, 70, 149
amygdala 139
analogue environments 243, 250–1
Anderson, H. Graeme 4, 60
anger 17, 18, 19, 21, 39, 47
    acute stress reaction 49
    disaster victims 91–2
    posttraumatic stress disorder 84, 115
    see also aggression
Annas, P. 54
antibiotics 34
antidepressants 112, 120, 121–2, 123, 148
antihistamines 31, 33–4
antimalarials 32–3
Antony, M.M. 76
anxiety
    acute stress reaction 49
    adjustment disorders 110
    air rage 45
    alcohol use 45
    anticipatory 16, 40, 70
    cabin crew 40, 213, 217, 234, 236
    coping strategies 22
    disaster victims 84, 85, 86, 87–8, 89, 92, 94
    elderly passengers 43
    empathy with passengers 202
    gender differences 20–1
    health problems 20
    hostilities from passengers 288
    legal issues 298
    medication side-effects 32
    neuroses 135
    panic tolerance 200–1
    passengers 13–14, 16–18, 20, 22, 40, 41–2, 50
    psychiatric emergencies 48
    psychological assessment 149
    psychological debriefing 95, 97
    treatment 3
    work stress 5, 209
    see also fear of flying; stress
anxiety disorders 78, 110, 113–15, 116
    disaster victims 87, 88
    medication 121, 122
    psychotherapy 121
    recovery 135
anxiolytics 14, 42, 122
APA see American Psychiatric Association
Appelbaum, P.S. 316
applied psychology 305, 306, 308
APSS see Automated Pilot Selection System
Armstrong Laboratory Aviator Personality Survey (ALAPS) 177, 187
assessment xx, 5, 6, 7, 123, 127–44
    astronaut selection 243, 244–5, 248
    cabin crew 204–6, 224
    case study 311
    disqualifying conditions 147–50
    fear of flying 70–1
    major depressive disorder 112
    medication 31–2
    mental state examination 133–6
    passengers 42, 43, 47
    pilot selection 161, 162–3, 165, 168, 176–9
    psychotic disorders 111
    reporting 7, 145–60
    South Africa 305, 306–8
    see also diagnosis; medical examinations
assessment centers 168, 206
ASTB see Aviation Selection Test Battery
Astemizole 33
astronauts 8, 243–54
Atkinson, Greg 8, 255–84
atovaquone 32–3
attention deficit disorder (ADD) 117
attention deficit hyperactivity disorder (ADHD) 117

Automated Pilot Selection System (APSS) 178
aviation industry 324–5, 326
aviation psychology 8, 303, 304, 305–8, 311–12
Aviation Security Act (1982) 294–5
Aviation Selection Test Battery (ASTB) 179
avoidance
  anxiety disorders 78
  bereavement 232
  diagnosis of PTSD 115, 140
  disaster victims 84, 85, 86, 88, 91, 94
  fear of flying 55, 59
  Impact of Event Scale-Revised 142
Axis I psychiatric disorders 118, 316

baggage reclaim 16, 17
  see also luggage
Bakal, P.A. 59
Bartholomew, C.J. 29, 30
Bartram, D. 178, 181
Bartsch, P. 30
Basic Attributes Test (BAT) 178
BDI *see* Beck Depression Inventory
Beaty, D. 327, 330
Beck, A.T. 56
Beck Depression Inventory (BDI) 141
behavioral therapy xxi, 72, 121
benzodiazepines 115, 122
Berne, Eric 130, 131
Bijlmermeer air crash (1992) 87, 97–8
Billings, Charlie 325
biological stressors 137
biopsychosocial factors 136–7, 142, 143
bipolar disorder 111, 112–13, 134, 148, 316
Bloom, W.A. 57
BMIBaby 286
body clock 256–60, 268, 269, 271, 276, 278–80
body temperature 256, 257, 258, 259–60, 278, 280
  alertness 266, 267
  jet lag 271
  mental performance and mood 263, 274
  'shift workers' malaise' 271, 272
Bonnon, M. 29
Bor, Robert 1–9, 335
  air rage 45
  cabin crew 231
  disruption of relationships 234
  fear of flying 7, 42, 53–68, 69–82
  pilots 7, 107–25
  psychological assessment and reporting 145–60
borderline personality disorder 111, 118, 119
Borkovec, T.D. 75
Bornas, X. 73, 75
Borrill, J. 56, 58, 77
Boston air disaster 89
Brabant, M. 28
brain 139, 143
Brandberg, M. 76–7
Branson, Richard 53
breastfeeding 43
Brennan, W. 324
brief solution-focused therapy 42
Briner, R. 237–8
Britannia Airways 286
British Airways 227–8, 229, 230, 235, 236–9, 286
British Midland B737 crash (1989) 195
Bucove, A.D. 61
Burke, E.F. 169, 170
Burrough, Brian 247
business class 292–3
Butcher, J.N. 64

CAA *see* Civil Aviation Authority
cabin crew
  company culture 230–1
  Crewcare counseling service 227, 231, 235, 236–9
  fear of flying 62, 63–4
  female xx
  group dynamics 229–30
  hostility from passengers 287–8
  jet lag and fatigue 216–18
  legal issues 287–91, 299–300, 301
  nursing background 70
  onboard behavior 286–7
  personality types 231–4
  posttraumatic stress disorder 49–50
  psychological problems 5, 227–39
  relationship problems 220–1
  security regulations 297
  selection xix, 195–208
  stereotypic images 196–7
  stress 39, 40, 209–26

terrorism 288–9
therapeutic alliance with 47
training 41, 50, 64, 196
well-being 7–8, 209, 212, 213–14, 220, 221, 224
work-life balance 218–20
cabin managers 197, 204
CAE *see* Computer Assisted Exposure
caffeine 42, 46, 48, 275
Caldwell, J. 215
California Psychological Inventory 177
Canadian Automated Pilot Selection System (CAPSS) 178
cancer 289, 298
Cantopher, T. 42
Capafons, J.I. 54
CAPS *see* Clinician-Administered Posttraumatic stress disorder Scale
CAPSS *see* Canadian Automated Pilot Selection System
cardiac problems 13
cardiovascular illness 19, 21, 40
Carretta, T.R. 179, 182
Carriage by Air Act (1963) 294
'caseness' 141, 149
CBGT *see* Cognitive Behavioral Group Treatment
CBT *see* Cognitive Behavioral Therapy
Cetirazine 33
Challenger space shuttle disaster (1986) 93
Chapman, D.S. 180
charter operators 286
Chidester, Tom 249, 250
children 21
Chung, Man Cheung 7, 83–104, 315–16
cigarette smoking 22, 42, 44, 45, 46, 141
ciprofloxacin 34
circadian rhythms 31, 137, 138, 142, 211
alertness 265–7
body clock 256–60
desynchronization 273, 280
mental performance and mood 263, 273, 274
naps 261
'shift workers' malaise' 271
CIRP *see* Critical Incident Response Program
CISD *see* Critical Incident Stress Debriefing

Civil Aviation Authority (CAA)
South Africa 307, 308
United Kingdom 54, 214, 289, 298
Clarke, J.C. 57
claustrophobia
air rage 45
Computer Assisted Exposure 74
fear of flying 56, 57, 58, 70, 77
flight delays 17
Cleland, Jennifer 7, 27–38
clinical psychology xx, xxii
fear of flying treatments 77–8
psychological assessment and reporting 145, 146, 152
clinical selection 162, 165, 171, 176
Clinician-Administered Posttraumatic stress disorder Scale (CAPS) 142
co-pilots 195, 196, 203
Coetzee, Johann 8, 303–13
cognitive appraisal 215
Cognitive Behavioral Group Treatment (CBGT) 75
Cognitive Behavioral Therapy (CBT)
fear of flying xxi, 42, 59, 63, 72–3, 76, 78
pilots 121
trauma-focused 50
cognitive impairments 28, 29–30, 43, 46, 117
cognitive psychosis 133–4
cognitive restructuring 72, 73, 76
cognitive skills 27, 162, 167, 177
CogScreen 168, 178
communication 131, 132, 327
communication skills 197, 199, 204
community disaster victims 87, 91, 93–4
company culture 210, 230–1, 238
competences 163–5, 166, 167
complacency 323
Computer Assisted Exposure (CAE) 73–4
computer-based assessment 168, 178–80
Computerized Test Battery (NATO) 179
concentration 28, 110, 211
confidentiality xx, 129
confusion 40, 43, 48
conscientiousness 248, 251
consultation teams 94–5
Cooper, C. 109

Cooper, George 325
cooperative ability 203
coping
    cabin crew 209, 211, 215–16, 222, 224, 234
    fear of flying 71
    negative strategies 17
    pilots 61, 108–9
    posttraumatic stress disorder 94
    sleep disturbance xxii
    stress 2, 15, 16, 20, 22, 23
cosmic radiation 285, 289, 298
counseling xx, 3, 121, 159, 232
    adjustment disorders 110
    case studies 308, 309, 310, 311
    Crewcare service 227, 231, 235, 236–9
    South Africa 305, 306, 308
counseling psychology 145, 152
court cases
    deep vein thrombosis 297, 298
    hijacking 296
    liability 286
    passenger injury 299–300
    pilot psychological reports 145–6, 152
    unruly passengers 295–6
Craig, K.D. 61, 63
crew augmentation 277–8
Crew Resource Management (CRM) Training 50, 196, 274, 290–1, 306, 311
Crewcare counseling service 227, 231, 235, 236–9
crisis management 315
Critical Incident Response Program (CIRP) 305
Critical Incident Stress Debriefing (CISD) 95–7, 315
critical incidents 237, 261–2, 295
CRM *see* Crew Resource Management Training
Croft, A.M. 32
cultural diversity 203
Curtis, K. 48

Daly, R.J. 60
DAT *see* Differential Aptitude Test
Data Protection Act (1998) 150
Davenport, N. 324

Davidson, J. 140
DCM *see* Design Cycle Model
De Silva, P. 61, 62
Dean, R.D. 56
death anxiety 94
debriefing 49, 95–8
    Critical Incident Stress Debriefing 95–7, 315
    disaster personnel 91
deep vein thrombosis (DVT) 13, 14, 19, 21, 23
    court cases 297, 298
    media scares 40, 285
Defense Mechanism Tests (DMT) 177
defense mechanisms 135, 137, 200
defensiveness 118
DeHart, R. 16
dehydration 43, 45, 229
Dekker, S.W.A. 319–20, 322, 323, 325, 327, 330, 331
Delta Airlines disaster (1985) 89
delusions 111, 133, 134
dementia 43
Dempster, C. 55
depression
    adjustment disorders 110
    alternative treatments 122–3
    astronauts 248
    bipolar disorder 112–13
    cabin crew 213, 217
    circadian rhythm disruption 137
    depressive disorder NOS 113
    disaster personnel 89, 90
    disaster victims 84, 86, 87–8, 91, 94, 114
    flight crew 40
    major depressive disorder 111–12
    medication 32, 121–2
    passengers 18, 21, 42–3
    psychological debriefing 95, 97
    psychological tests 141
    recovery 135
    rumination 140
    substance misuse 116
    suicidal pilots 120
    treatment 3
    work stress 209
derealization 134

Design Cycle Model (DCM) 173–4
Desloratidine 33–4
diagnosis 3, 78, 316
    attention deficit disorder 117
    major depressive disorder 112
    passengers 47, 50
    personality disorders 118, 119
    pilots 107, 108, 111, 142
    posttraumatic stress disorder 85, 140–1
    see also assessment
Diagnostic and Statistical Manual of Mental Disorders (DSM-IV) 2–3, 77, 78, 108
    'caseness' 149
    fear of flying 55, 70
    posttraumatic stress disorder 84, 98
diazepam 42, 47, 49
Diekstra, R.W.F. 18, 76, 77
Differential Aptitude Test (DAT) 176
Diphenhydramine 33
disasters 62–3, 315, 318–19
    crisis interventions 94–5
    posttraumatic stress disorder 7, 83–104, 114
    psychological debriefing 95–8
    see also accidents; victims
disclosure 316
dispositions 163, 164, 165
disputes 45
disruptive behavior 19, 46, 47, 48, 292, 293, 294–6
    see also air rage
dissociation 85, 86, 134–5
dizziness 32
DMT see Defense Mechanism Tests
Dolan, R.J. 318, 330
Downeast Airlines Flight 46 crash (1979) 327
doxycycline 33, 34
dreams 140
drug use xx, 84, 115–16, 148, 149, 285
    addictive types 234
    alertness-enhancing 276, 278
    case study 309
    passengers 45, 46, 49
    pilots 330
    see also medication; substance misuse
DSM-IV see Diagnostic and Statistical Manual of Mental Disorders

Dunn, L.A. 64
Durant, Will 320–1, 330
duty of care 285–6, 306–7
DVT see deep vein thrombosis
dysthymic disorder 84, 113

EAA see European Airline Association
East Tennessee State University 86
eastward flights 271, 273, 279–80
EasyJet 210, 286
economy class 292, 293, 301
education 50, 198, 203, 308, 311–12
    see also training
Edwards, Ben 8, 255–84
Edward's Personal Preference Scale 177
ego 131
Egypt Air Flight 990 crash (1999) 119, 288
ejection 61, 86, 114
elderly passengers 21, 40, 43
Elliott, G.P. 324
Elsey, Adrian 8, 255–84
EMDR see eye movement desensitization and reprocessing
emergency personnel 83, 89–91, 96, 97–8
Emery, G. 56
emotional memory 317, 318, 330
emotional stability 200, 246, 248, 251
emotional support 6, 49, 62–3, 109
emotions
    acting out 325, 330
    cabin crew 234
    contagion 238
    moral 323–4
    posttraumatic stress disorder 86, 87
    suppression of 129
    see also mood
empathy 202, 246
epilepsy 148
Eriksen, Carina 7–8, 209–26
errors 8, 195, 320, 322–4, 327–9, 331
ESA see European Space Agency
EU see European Union
European Airline Association (EAA) 54
European Space Agency (ESA) 246
European Union (EU) 289, 294
'evidence-based psychological treatment' 73
exhaustion 138
experience 198
expressivity 249

eye movement desensitization and reprocessing (EMDR) 50
Eysenck Personality Inventory 177, 245

FAA *see* Federal Aviation Administration
FAM *see* Flight Anxiety Modality questionnaire
families
  cabin crew 233–4
  case study 310–11
  disaster victims 89
  pilots 108, 109
  see also relationship problems
FAS *see* Flight Anxiety Situations questionnaire
fatigue xx, 2, 48
  adjustment disorders 110
  cabin crew 210–11, 213–14, 216–18, 219, 221, 222–3, 224
  layovers 273
  major depressive disorder 112
  mental performance 263, 265
  see also sleep deprivation; sleep disturbance
fear of flying xxi, 5, 7, 53–68, 316–17
  air crew 60–4, 201
  air rage 45
  alcohol use 45
  cabin crew reassurance 40, 41, 202
  causes 59
  implications 53–4
  nature and characteristics 55–7
  passenger stress 14, 16, 17, 18
  prevalence 54–5
  subject of fear 57–8
  treatment 57, 63–4, 69–82
  see also anxiety
Federal Aviation Administration (FAA)
  alternative treatments 122
  anxiety disorders 115
  assessment of pilots 148
  attention deficit disorder 117
  Guide for Aviation Medical Examiners 107–8
  medical standards 107, 147
  medication 31–2, 33
  mood disorders 112–13
  personality disorders 119
  psychotic disorders 111

regulations 316, 321
substance abuse 116
suicidal pilots 121
Figley, C.R. 83n2
financial resources 313
Fine, E.W. 45
FIRO-B *see* Fundamental Interpersonal Relations Orientation-Behavior Scale
first officers 327
Fisch, H.U. 30
Fischer, H. 54
flashbacks 140
Flight Anxiety Modality (FAM) questionnaire 71
Flight Anxiety Situations (FAS) questionnaire 71
flight attendants *see* cabin crew
flight delays 17, 39, 45
flight simulators 72, 76, 168, 178, 262
Florida Everglades crash 320
Flybe 286
Flynn, Chris 246
Foushee, Clay 249
Fowler, B. 28
Fowlie, D.G. 61, 62
Frances, D Anthony 8, 285–301
FRASCA 178
Frazer, A.G. 83n2
Frederickson, M. 54
Freedom to Fly course 77
frequent flyers 292–3
Freud, Sigmund 59, 131
frustration 17, 18, 19
Fullana, M.A. 73
Fullerton, C.S. 90
Fundamental Interpersonal Relations Orientation-Behavior (FIRO-B) Scale 251

Gagarin, Yuri 245
Gander air crash (1985) 90–1
GAT-B *see* General Aptitude Test battery
gender
  aggressive male behavior 41
  Challenger disaster responses 93
  fear of flying 58
  stress differences 20–1
  see also women

gender reassignment surgery 149–50
General Aptitude Test battery (GAT-B) 176
General Health Questionnaire 141
generalized anxiety disorder 78, 87, 114
Gladwell, Malcolm 321, 322
Goebert, D. 44, 50
Goleman, D. 317, 318, 321, 327
Goodman, Tracy 8, 227–39
Goodwin, Tony 7, 39–52
gossip 329, 330
Gotch, O.H. 60
Greco, T.S. 55, 58
Green, R. 273
Greenberg, R. 56
Greist, G.L. 53
Greist, J.H. 53
ground staff 46
group dynamics 229–30
Guide for Aviation Medical Examiners 107–8

hallucinations 111, 134
hand luggage 40, 45
Harris, Thomas 133
Haugli, L. 211
hayfever 33, 34
health
    definition of 41
    legal issues 286, 287, 297–300
    physical 19–20, 27–38, 40
    stress impact on 19–20
    see also deep vein thrombosis
health promotion 6, 287
health risks
    cabin air quality 289–90
    cabin crew 210, 211, 213
    cosmic radiation 289
    passengers 13, 14, 16, 23
    see also risk
Helderberg disaster (1987) 310
Helios Air B737-300 crash (2005) 285, 289
Hellesoy, O.H. 211
Helmreich, Robert L. 249, 250, 323
Henderson, D. 54
herbal medicines 31
Hermans, Pieter H. 7, 161–93
Herxheimer, A. 31
high-fidelity (HF) job samples 168, 178

hijacking 288, 296
    see also terrorism
hippocampus 139–40, 143
Hoffman, S.G. 76
Hofstee, W.K.B. 177
Hogan Personality Inventory (HPI) 251
holistic assessment 305, 307
Holland, Al 246
Hollnagel, E. 323
honesty 307
hormones 256
Howard, W.A. 57, 73
HPI *see* Hogan Personality Inventory
Hubbard, Todd P. 1–9, 315–34, 335
Hughes, R. 234
human error 8, 195, 320, 322–4, 327–9, 331
Hunter, D.R. 169, 170
hyperalertness 86, 87, 91
hyperarousal 84–5, 98, 115, 140
hyperventilation 42, 64
hypochondria 40, 41
hypoglycaemia 39
hypothalamus 257
hypoxia 27–30, 40, 43, 45

id 131
IES-R *see* Impact of Event Scale-Revised
Iljon Foreman, Elaine 7, 53–68, 69–82
immune system 19
Impact of Event Scale-Revised (IES-R) 141–2
in vivo exposure 73, 74, 75
independence 202, 205
Indianapolis air crash (1987) 87
industrial psychology 145
infection risks 13, 21
Institute for Aviation Psychology 305, 306, 308
instrumentality 249
insurance 148–9, 315, 316
International Space Station (ISS) 248–9, 252
internet testing 179–80, 188
interviews 169, 179, 184, 187, 188, 205–6
intrusive thoughts
    diagnosis of PTSD 115, 140
    disaster victims 84–5, 86, 87, 89, 91, 94, 97
    Impact of Event Scale-Revised 141–2
ISS *see* International Space Station

JAR *see* Joint Aviation Regulation
JAROPS *see* Joint Aviation Requirements Operations
jet lag 2, 31, 41, 271, 274
    body clock changes 259
    cabin crew 209, 210, 211–12, 216–18, 219, 222–3, 224, 229
    long-haul flights 214, 268, 269
    see also time-zone changes
Jex, Steve 324
job insecurity 3, 5, 210
job samples 168, 177–8, 187
Johnson Space Center (JSC) 243
Joint Aviation Regulation (JAR) 147, 148
Joint Aviation Requirements Operations (JAROPS) 228
Jones, D.R. 2, 61–2, 64, 76, 108
JSC *see* Johnson Space Center
judgment 199

Kahneman, D. 223
Kanas, Nick 248
Kanki, B.G. 28, 327
Kantor, J.E. 179
Karlins, M. 109
Katchen, M. 108
Kelleher, Herb 195
Kern, T. 327
Kirsch, I. 75
Kleber, R.J. 83n2
KLM Royal Dutch Airlines 178, 182–4, 269
KLM/Pan Am B747 crash (1977) 86, 291
Korean Airlines B747 crash (1999) 291
Krijn, M. 74
Kutek, A. 237

language skills 197, 199
Lauber, John 325
Lauria, L. 215
layovers 33, 268, 273, 274, 278
leadership 204
legal issues 8, 285–301
    cabin air quality 289–90
    cosmic radiation 289
    crew selection and resource management 290–1
    employee hostilities 288
    legislation 294–5
    passenger hostilities 287–8
    passenger injury 297–300
    terrorism 288–9, 296–7
    unruly passengers 293, 294, 295–6
    see also court cases
liability 286, 287, 288, 293, 294–5, 301
    cabin air quality 290
    cosmic radiation 289
    Crew Resource Management Training 291
    infectious diseases 290
    passenger injury 297–300
    terrorist incidents 289
    unruly passengers 295, 296
licensing 2, 3, 149
Lieberman, P. 28
lifestyle
    body clock 256, 257, 259
    cabin crew 212, 219, 221, 222–3, 227–8, 229
light 257–8, 264, 278–80, 281
Link Trainer 178
Lister, J.A. 60
Llabres, J. 73
Lockerbie disaster (1988) 88, 114, 286, 289, 318
Lofaro, R.J. 323, 327–8
long-haul flights 39–40, 41, 207
    passenger upset 292
    sleep and mental performance 267–70, 272, 274
Loratidine 33
Louro, C.E. 57
low-cost airlines 206, 207, 210, 286, 295, 301
low-fidelity (LF) job samples 168, 177–8
Lucas, Graham 7, 39–52
luggage 40
    see also baggage reclaim

MAB *see* Multidimensional Aptitude battery
McCarroll, J.E. 90
McCarthy, G.W. 61, 63
McIntosh, Iain B. 7, 13–25, 55
McNally, R.J. 57
major depressive disorder (MDD) 111–12
Maltby, N. 75
Manchester Airport disaster (1985) 85
Marioriello, R.P. 61
marital counseling 63, 158, 159, 309

Marks, M. 61, 62
Martinussen, M. 169, 170, 181
Matsumoto, K. 44, 50
maturity 200
maximum performance principle 167, 168
Mayers, M. 75
MDD *see* major depressive disorder
media 13, 40, 50, 285
medical certification 3, 113
    disqualifying conditions 147–50
    medication 121–2, 123
medical examinations 1, 5, 112, 148–9
    FAA Guide 107–8
    physical fitness 326
    South Africa 305, 307
    see also assessment; diagnosis
medical records 150, 316
medication xxii, 121–2
    adjustment disorders 110
    anxiety disorders 115
    depressive symptoms 112
    fear of flying 76
    mood disorders 113
    passenger anxiety 14, 22, 42
    personality disorders 119
    pilots 108
    pre-flight passenger assessment 47
    psychological consequences 27, 31–4
    psychosis 43
    sleeping pills 217
    see also drug use
mefloquine 32, 33, 43
melatonin 31, 256, 257, 258, 275–6, 278
memory 28–9, 30
    dissociative amnesia 134
    emotional 317, 318, 330
    hippocampus 139, 140
    jet lag impact on 211, 216
mental health xx, xxi, xxii, 1–2, 4
    cabin crew 213–14, 215, 221–2
    Diagnostic and Statistical Manual IV 2–3
    industry suspicion towards 6
    personal relationships 212
    pilot selection 184–7
    South Africa 305–6
mental health consultation teams 94–5

mental performance 255, 261–81
    advice to air crew 274–81
    body clock 259
    circadian rhythms 263
    environmental factors 264
    jet lag 271
    modeling 265–7
    'shift workers' malaise' 271
    sleep loss 264–5
    time on task 264
mental state examination (MSA) 133–6, 145
Mercury Program 244, 245
meta-analysis 169, 170, 181
micro-sleeps 273
    see also naps
Microcomputerized Personnel Aptitude Tester (MICROPAT) 178–9
military aviation 60–2, 111
    astronauts 244
    selection 169
    South Africa 303, 304
    stress 272
Military Psychological Institute (MPI) 304
Miller, M.R. 169
Minnesota Multiphasic Personality Inventory (MMPI) 177, 184, 187, 205, 244, 245
Mir space station 247–8
MMPI *see* Minnesota Multiphasic Personality Inventory
mobbing 324, 330
Moll, J. 323
monotony 264, 272, 274
Montreal Convention (1999) 294
mood
    body clock 259
    circadian rhythms 263
    disaster victims 86
    emotional contagion 238
    environmental factors 264
    hippocampal damage 139
    jet lag 271
    medication side-effects 32
    mental performance relationship 262
    'shift workers' malaise' 271
    sleep loss 264–5, 273
    time on task 264
    see also emotions

mood disorders 5, 110, 111–13
    bipolar disorder 111, 112–13, 134
    medication 121, 122
    pilot suicide 120
    psychotherapy 121
    substance misuse 116
moral codes 320–1, 330
moral emotions 323–4
Morgenstern, A.L. 59
Morse, Jennifer S. 7, 107–25, 316
motion sickness 2
motivation 203–4, 207, 274–5
Mount Erebus disaster (1979) 89
MPI see Military Psychological Institute
MSA see mental state examination
Muhlberger, A. 74
Mulder, H.W. 177
multi-modal treatment 76
multicultural environment 203
Multidimensional Aptitude battery (MAB) 176
Murphy, S.M. 57
Musson, David M. 8, 243–54

naps 260–1, 267, 269–70, 273, 275, 276–7
National Aeronautics and Space Administration (NASA) 243, 244–5, 246, 247–50, 251–2
National Institute for Personnel Research (NIPR) 306
Nelson, Gordon 306
NEO Five Factor Inventory (NEO-FFI) 551
nervous system 148
neurobiology 138–40, 142–3
neurosis 60, 133, 135–6, 148
Newman, M.G. 76
Nicholson, Anthony N. xix–xxii
night work 268, 271–2
NIPR see National Institute for Personnel Research
Noel-Jorand, M. 29

obsessional behavior 40, 41, 232
obsessive-compulsive disorder (OCD) 84, 114, 118
Occupational Personality Questionnaire 177, 183
occupational psychology 152

OCD see obsessive-compulsive disorder
O'Connor, P.J. 60
ODM see Operational Decision Making model
Olympic Airways 287–8
Operational Decision Making (ODM) model 327–8
organizational culture 210, 230–1, 238, 244
Ost, L.G. 76–7, 78

PAI see Personality Assessment Inventory
Palmer, M.T. 327
Pan Am/KLM B747 crash (1977) 86, 291
panic attacks 42, 48, 113, 114
    cabin crew 234, 236
    fear of flying 55, 57
panic disorder 42, 57, 74, 77, 78
    disaster victims 84, 88
    pilots 113–14
    substance misuse 116
panic tolerance 200–1, 204
paranoia 118–19
Partridge, Chris 8, 227–39
passengers
    angry 213
    cabin crew as maternal substitutes 231
    disruptive behavior 46, 47, 48, 292, 293, 294–6
    empathy with 202
    fear of flying 53–9, 64, 69–79
    hostilities 287–8
    legal issues 292–3, 294–6, 297–300
    long-haul flights 267–8
    onboard behavior 286–7
    psychological problems 4, 7, 39–52
    stress 13–25
    see also air rage
Patterson, J. 108, 109
Penfield, Wilder 131
performance see mental performance
Perry, Carlos 4
personal appearance 199, 228
personal maturity 200
personality
    assessment and reporting xx, 157, 158
    astronauts 249–50, 251
    cabin crew 197, 231–4
    disaster victims 94

passengers 40
pathology 56, 57
pilot selection 163, 165, 167–8, 171, 177
psychiatric morbidity 46
stress 15, 21–2
Transactional Analysis 131, 136
see also personality disorders; personality tests
Personality Assessment Inventory (PAI) 157
personality disorders 84, 111, 118–19, 148, 310
Personality Research Form 177
Personality Scale (Taylor) 245
personality tests 157, 165, 167–8, 177, 184–5, 188
  astronaut selection 250
  clinical selection 171
  flight attendants 205
  KLM selection system 183, 184
  low predictive validity 186
  pre-selection stage 186–7
  see also personality
pharmacotherapy 63
phobias xxi, 14, 18, 21, 55, 70
  air crew 63
  classification of 56–7
  Cognitive Behavioral Therapy 73
  coping strategies 22
  disaster victims 86
  see also fear of flying
physical abuse 40, 49
  see also air rage; violence
physical health 19–20, 27–38, 40
PILAPT *see* Pilot Aptitude Tester
Pillemer, D.B. 317–18
Pilot Aptitude Tester (PILAPT) 168, 178
pilots 5–6, 325–30
  adjustment disorders 109–10
  alcohol use 4–5, 45
  anxiety disorders 113–15
  attention deficit disorder 117
  blame for accidents 319–20, 330
  Captain overruling of flight crew 290–1
  case studies 308–11
  crew augmentation 277–8
  disqualifying conditions 147–50
  errors 8, 195, 320, 323, 327–9, 331
  fear of flying 60–3

formal policing 329
industry pressures 324–5
informal policing 329–30
inter-pilot identity 328–9
legal issues 301
medication 31, 33
mental performance 255, 261–81
mood disorders 111–13
non-disclosure 316–17
organic mental disorders 117
performance and competence 163–5, 166, 167
personal relationships 212
personality disorders 118–19
psychiatric disorders 5, 107–25
psychiatric evaluation 127–44
psychological assessment and reporting 5, 7, 145–60, 306–8
psychotic disorders 110–11
selection xix, 7, 161–93
social pressures 317, 323, 330, 331
stress model 136–8
substance-related disorders 115–17
suicide 119–21
PND *see* post-natal depression
POR *see* Program of Requirements
post-natal depression (PND) 42–3
posttraumatic stress disorder (PTSD) 7, 49–50, 62, 83–104, 315
  behavioral therapy xxi
  cabin crew 40
  court cases 300
  delayed-onset 135
  derealization 134
  diagnosis 140–1
  historical China 317
  pilots 114–15
  psychological debriefing 95–8
  recovery 135
  sleep problems 41
  substance misuse 116
  tests 141–2
  see also victims
Power, K. 55
pre-travel health advice 22, 23
prevention 6, 46–8
  accident 262
  Critical Incident Stress Debriefing 96
  work stress 209–10

Prlic, H. 28
Program of Requirements (POR) 174, 176, 188–9
proquanil 32–3
Protopapas, A. 28
psychiatric co-morbidity 21, 84
psychiatric emergencies 40, 44, 48–50
psychiatrists
    astronaut aversion to 244
    biopsychosocial formulation 136
    evaluation of air crew 127, 128–9, 130, 140, 142, 143
    passengers 47
    pilot suspicion of 1, 128
    Transactional Analysis 131, 132–3
    see also psychologists
psychoanalysis 72
psychodynamic approaches xxi, 72
psychologists
    astronaut aversion to 244
    pilot suspicion of 1, 147
    qualifications 146
    reports 146, 147, 150–1, 152
    South Africa 305, 306–7
    see also psychiatrists
psychometric testing 5–6, 146, 147, 306
    see also tests
psychomotor skills 27, 162, 167, 168, 177, 182
psychopathology xx, xxi, 145, 149
    clinical selection 162, 165
    personality testing 157, 171, 177, 184, 186, 205
psychosis 40, 43, 148
    in-flight emergencies 44, 48
    medication side-effects 32
    mental state examination 133–4
    psychotic disorders 110–11
psychotherapy
    fear of flying 63, 72–3
    mistrust of 135–6
    mood disorders 112
    personality disorders 119
    types of 121
    see also counseling; therapy
PTSD *see* posttraumatic stress disorder
pursers 197, 204

Qantas Airways 286
questionnaires
    flight anxiety 71–2
    General Health Questionnaire 141
    Occupational Personality 177, 183
    sign instruments 167

racial discrimination 303–4, 306
radiation 285, 289, 298
Raeside, F. 55
RAF *see* Royal Air Force
rapid eye movement (REM) sleep 260
Raschmann, J. 109
Rayman, R. 48
re-experiencing 84, 115, 140
Rea, M. 108
Reason, James 2, 322–3, 331
Ree, M.J. 169, 182
Reilly, Thomas 8, 255–84
relationship problems 2, 3, 5
    cabin crew 213, 215, 216, 220–1, 222, 224, 229
    diagnosis 316
    pilots 108–9, 212, 329
    psychological report sample 154–5, 156, 159
    suicide 120
    see also families
relaxation techniques 73, 76, 121
    air crew 64
    passenger stress 22
    sleep problems 211, 217
REM sleep *see* rapid eye movement sleep
reporting xx, 7, 145–60
    purpose of reports 150–2
    sample report 153–60
    shortcomings 147
responsibility 202–3
reviews 130
Richards, Paul 7, 27–38
Right Stuff 249, 250
risk
    perception of 14, 21, 213, 285
    psychological assessment of pilots 147
    see also health risks
risk management 291, 293, 323
Roe, Robert A. 7, 161–93
Rorschach Inkblot Test 177, 244, 245

Rose, Robert 250
Rosekind, M. 276
Roth, W.T. 56, 57, 59, 76
Rowe, P.M. 180
Royal Air Force (RAF) 60, 61, 86, 114, 127
Royal Air Maroc crash (1994) 120
Ryanair 210, 286

S-Adenosylmethionine (SAMe) 122
safety xix, 2, 3, 39
    accident prevention 262
    cabin crew 207, 214
    confidential reporting systems xx
    pilot psychiatric disorders 107
    pilot selection 185
    psychological assessment and reporting 145
St. John's Wort 122–3
SAMe *see* S-Adenosylmethionine
sample instruments 165–6, 168, 187, 188
Santy, Patricia A. 244, 245, 246, 250
SARS epidemic 290
Scandinavian Airlines (SAS) 200, 204
scapegoating 325
scheduled operators 286
Scheepers, Johan 303
schizophrenia 43, 47, 84, 133, 134, 137
Schlaepfer, T.E. 30
Schofield, G. 109
Schultz, W. 324
Schwartz, R.D. 324
SCN *see* suprachiasmatic nuclei
screening 46, 291
Scrignar, C.B. 57
security regulations 293, 296, 297
sedatives 33, 53, 148
selection
    ability tests 176–7
    assessment methods 176–9
    astronauts 243–54
    cabin crew 195–208, 227–8
    cabin managers 204
    clinical 162, 165, 171, 176
    critique of procedures xix–ix
    employment 162, 164–5, 166, 171, 176, 177, 178
    forms and functions 162–5
    initial assessment 145
    interviews 169

KLM example 182–4
legal issues 290–1
pilots 7, 161–93
predictors 165–72, 174–5, 180, 182
recommendations 187–9
robust methods 3
sample instruments 165–6, 168, 187, 188
sign instruments 165–6, 167–8
statistical issues 180–2
systems design 172–6, 187–8
test delivery methods 179–80
see also ab initio selection
Selective Serotonin Reuptake Inhibitors (SSRIs) 115, 121–2, 148
self-concept 71, 158
self-doubting behavior 329
self-esteem 138, 197, 198, 201
self-knowledge 202
self-regulation
    blame 319
    cabin crew 218, 219, 220, 221, 222, 223
    encouragement of 331
September 11 2001 terrorist attacks 13, 14, 41, 54, 285
    fear of flying 69
    impact on cabin crew 235–6
    legal issues 286
    passenger anxiety 50
    security regulations since 296, 301
    see also terrorism
Serotonin Norepinephrine Reuptake Inhibitors (SNRIs) 115, 121–2
sexual dysfunction 5
shift work 209, 211, 212, 216, 217, 224
'shift workers' malaise' 259, 268, 269, 271–2
SHL test batteries 176, 183
short-haul flights 267
Shuttle-Mir 247–8
sick leave 155
Sides, J.K. 75
sign instruments 165–6, 167–8
Silk Air B737 crash (1997) 119–20
simulators 72, 76, 168, 178, 262
    see also Virtual Reality Exposure Therapy
Sixteen Personality Factor (16PF) test 205

skills xix, 163, 164
　　astronauts 248, 252
　　cabin crew 197, 198, 199, 204–5, 227, 228
　　cognitive 27, 162, 167, 177
　　psychomotor 27, 162, 167, 168, 177, 182
Skogstad, A. 211
sleep deprivation
　　cabin crew 211, 213, 214, 216–18, 222, 223, 229
　　eastward flights 271, 273
　　long-haul flights 268, 269
　　mental performance 264–5, 272–3
　　psychiatric morbidity 41
　　'shift workers' malaise' 271
　　short-haul flights 267
　　see also fatigue
sleep disturbance xx, xxi–xxii, 5, 31
　　body clock 256–60
　　cabin crew 210, 217, 224
　　disaster victims 86, 90
　　major depressive disorder 112
　　medication side-effects 32
　　pilots 255–61
　　see also fatigue; jet lag
sleep inertia 261, 276, 277
sleeping pills 217
Sloan, S. 109
slow wave sleep (SWS) 260, 261, 266
Smith, D. 322
Smith, K.M. 323, 327–8
smoking 22, 42, 44, 45, 46, 141
SNRIs see Serotonin Norepinephrine Reuptake Inhibitors
social anxiety/phobia 55, 56, 57, 77, 78, 84
social flexibility 203, 248
social stressors 136–7
social support 109
somatization 136
Sosa, C.D. 54
South Africa 303–13
South African Airways 304, 305, 310, 313
Southwest Airlines 210, 286, 292
Soviet cosmonauts 245–6, 247
'space analogue' environments 243, 250–1
Space Ship One 53
space shuttles 93, 246–7

Spence, Janet 249, 250
Spencer, Mick 8, 255–84
SSRIs see Selective Serotonin Reuptake Inhibitors
Standard Operating Procedures 274
sterile cockpit concept 195
stigmatization 3, 40, 311
stress xx, 2, 7, 136–8, 272
　　acute stress disorder 85, 86, 115
　　acute stress reaction 49, 134, 140
　　antisocial behavior 44
　　anxiety 5, 113
　　cabin crew 40, 201, 204, 209–26, 234
　　coping strategies 22
　　counseling 3
　　demographic differences 20–1
　　fatigue 48
　　legal issues 298
　　loading 136, 137
　　mediators 20–2
　　neurobiology 138–40, 142–3
　　neuroses 135
　　passengers 13–25
　　personality traits 21–2
　　physical health 19–20
　　pilots 108–9, 142, 212
　　psychological outcomes 18–19
　　psychological report sample 157, 158, 159
　　reactions 136, 137, 138–40
　　sources of 14–15, 16–17
　　transactional theory 15–16, 23
　　vulnerability-stress model 59
　　workplace conditions 324
　　see also anxiety; posttraumatic stress disorder
stressors 15, 16, 23, 40, 136–7
　　adjustment disorders 109, 110
　　air rage 19
　　cabin crew 209, 223, 224
　　secondary 137, 139
Strongin, T.S. 64
Struve, Lynn 317–18
substance misuse 84, 115–17, 120, 140, 148, 149
　　see also drug use
suicidal thoughts 5, 43, 112
suicide 119–21, 288, 324, 330

superego 131
suprachiasmatic nuclei (SCN) 257, 258
survivor guilt 85, 91, 92
Swanson, Vivien 7, 13–25, 55
Swanson, W.C. 57
SWS *see* slow wave sleep

TA *see* Transactional Analysis
taboos 316, 317
TAT *see* Thematic Apperception Test
Taylor, A.J.W. 83n2
TCAs *see* tricyclic antidepressants
technical competencies xix
Terfenadine 33
terrorism 13, 288–9, 296–7
    air rage 45
    fear of flying 69
    Lockerbie disaster 88
    media scares 40
    posttraumatic stress disorder 49–50, 114
    see also September 11 2001 terrorist attacks
tests 118, 141–2, 151
    ability 176–7, 204–5
    astronauts/cosmonauts 244, 245
    delivery methods 179–80
    integrated test batteries 178–9
    job samples 177–8
    maximum performance 167, 168
    psychometric 5–6, 146, 147, 306
    see also personality tests
TFCBT *see* trauma-focused cognitive behavioral therapy
Thematic Apperception Test (TAT) 177, 244, 245
therapy
    cabin crew 224
    case study 310, 311
    fear of flying 63, 72–9
    South Africa 305, 306
    see also counseling; psychotherapy
Therme, P. 29
time on task 264
time-zone changes 41, 211, 216, 224, 267–71, 278–81
    see also jet lag
Timm, S.A. 64
Tortella-Feliu, M. 73
toxins 117, 289–90

training 1, 23, 196, 326
    British Airways cabin crew 228
    emotional environment 328
    mental health consultation teams 95
    pre-accident 64
    selection 161, 162, 163, 164
    South Africa 305, 311–12, 313
    variation in 41
    work stress 213–14
    see also Crew Resource Management Training; education
Transactional Analysis (TA) 130–3, 136, 142
transactional theory of stress 15–16, 23
trauma 139, 140, 141
trauma-focused cognitive behavioral therapy (TFCBT) 50
tricyclic antidepressants (TCAs) 122
Tuckman, B. 229
turbulence 40, 41
Turnbull, Gordon J. 7, 127–44
Type A personality 19, 21

ultra long-haul operations 39–40
Unified Personality Indicator (UPI) 177, 183, 186–7
United Airlines Flight 232 crash (1989) 90, 94–5, 322
United States Air Force 91–2, 176, 179
United States space programs 243, 244–5, 246–8
UPI *see* Unified Personality Indicator
Ursano, R.J. 90, 98

van Gerwen, Lucas J. 7, 18, 53–68, 69–82
verbal abuse 40, 44, 49
Vermeulen, Leo 303
victims 83–98
    access to care 315–16
    crisis interventions 94–5
    primary 83, 85–8
    psychological debriefing 95–8
    quaternary 83, 91–3
    quinternary 83
    secondary 83, 89
    sesternary 83, 93–4
    tertiary 83, 89–91
Vienna Test System 178
Vietnam War 60–1

Vina, C.M. 54
violence
    passengers 44, 213, 292, 293, 295–6
    workplace 324
    see also air rage
Virgin Blue 286
Virtual Reality Exposure Therapy (VRET) 73, 74–5
volunteering culture 307
VRET *see* Virtual Reality Exposure Therapy
vulnerability-stress model 59

WAIS *see* Wechsler Adult Intelligence Scale
Walters, L.C. 169
Warsaw Convention (1929) 294, 298, 299
Waterhouse, Jim 8, 255–84
Weathers, F.W. 142
Wechsler Adult Intelligence Scale (WAIS) 176, 244
Whitaker, K.M. 56
Wik, G. 54
Wilhelm, F.H. 56, 57, 59, 76
Willenhall disaster (1994) 93–4

Williams, M.H. 59
Wolfe, Tom 249
WOMBAT 178
women
    anxiety 200–1
    breastfeeding 43
    cabin crew xx
    fear of flying 54, 58
    post-natal depression 42–3
    stress 20–1
    see also gender
Wood, Mary 318
work-life balance 218–20, 222, 224
workload 2, 29, 210, 264, 272
World Trade Center attacks *see* September 11 2001 terrorist attacks
Wright, K.M. 90

Yeager, Chuck 249
Yule, W. 61, 62

zeitgebers 137, 257, 272
Zuckerman, Jane 7, 27–38